AF805143

International Perspectives on Aging

Volume 14

Series editors

Jason L. Powell
University of Lancashire, Preston, Lancashire, United Kingdom

Sheying Chen
18 th Floor, Pace University, New York, New York, USA

The study of aging is continuing to increase rapidly across multiple disciplines. This wide-ranging series on International Perspectives on Aging provides readers with much-needed comprehensive texts and critical perspectives on the latest research, policy, and practical developments. Both aging and globalization have become a reality of our times, yet a systematic effort of a global magnitude to address aging is yet to be seen. The series bridges the gaps in the literature and provides cutting-edge debate on new and traditional areas of comparative aging, all from an international perspective. More specifically, this book series on International Perspectives on Aging puts the spotlight on international and comparative studies of aging.

More information about this series at http://www.springer.com/series/8818

Thibauld Moulaert · Suzanne Garon
Editors

Age-Friendly Cities and Communities in International Comparison

Political Lessons, Scientific Avenues, and Democratic Issues

Springer

Editors
Thibauld Moulaert
School of Public Health
Université Catholique de Louvain
Louvain-la-Neuve
Belgium

Suzanne Garon
School of Social Work and Research Centre on Aging
University of Sherbrooke
Sherbrooke, QC
Canada

ISSN 2197-5841 ISSN 2197-585X (electronic)
International Perspectives on Aging
ISBN 978-3-319-24029-9 ISBN 978-3-319-24031-2 (eBook)
DOI 10.1007/978-3-319-24031-2

Library of Congress Control Number: 2015950453

Springer Cham Heidelberg New York Dordrecht London
© Springer International Publishing Switzerland 2016
This work is subject to copyright. All rights are reserved by the Publisher, whether the whole or part of the material is concerned, specifically the rights of translation, reprinting, reuse of illustrations, recitation, broadcasting, reproduction on microfilms or in any other physical way, and transmission or information storage and retrieval, electronic adaptation, computer software, or by similar or dissimilar methodology now known or hereafter developed.
The use of general descriptive names, registered names, trademarks, service marks, etc. in this publication does not imply, even in the absence of a specific statement, that such names are exempt from the relevant protective laws and regulations and therefore free for general use.
The publisher, the authors and the editors are safe to assume that the advice and information in this book are believed to be true and accurate at the date of publication. Neither the publisher nor the authors or the editors give a warranty, express or implied, with respect to the material contained herein or for any errors or omissions that may have been made.

Printed on acid-free paper

Springer International Publishing AG Switzerland is part of Springer Science+Business Media (www.springer.com)

To Our Red Lovers

Foreword

I am delighted to provide a foreword to this book. It represents an important contribution to the debate on age-friendly environments, drawing upon a wide range of theoretical and methodological perspectives linked with case studies from a range of countries. Given the rapid adoption of the age-friendly perspective, following its development by the World Health Organization, the critical assessment offered in this volume is especially welcome. There can be little doubt as to the importance of the issues which underpin many of the chapters in the book. The twin forces of population ageing on the one side and urbanization on the other are influencing a wide range of discussions within the broad field of public policy. Added to this are pressures facing many rural areas, faced with the loss of population as people migrate to cities in search of employment and better prospects for themselves and their families.

By 2030, two-thirds of the world's population will be residing in cities; at that point many of the major urban areas in the developed world will have 25 % or more of their populations aged 60 and above. Cities are now regarded as central to economic development, attracting waves of migrants and supporting new knowledge-based industries. However, the extent to which the new 'urban age' will produce 'age-friendly' communities, creating opportunities for older people as well as strengthening ties across different age and social groups, remains uncertain. Such concerns may also be linked to the emergence of conflicts within urban space arising from inequalities in access and influence amongst different social and generational groups.

A significant contribution made by this book is both to highlight the range of achievements of the age-friendly movement and some of the obstacles and problems which it is beginning to encounter. Certainly, there can be no doubt that it does represent a significant 'movement' given that it now stretches to over 250 communities in 28 countries worldwide. And one of the merits of this book is to provide examples of this diversity, with examples from the Asia-Pacific, Latin America, Europe and North America. The book also provides for consideration theoretical frameworks within which the age-friendly approach can be considered and developed, notably those associated with environmental gerontology and the perspective of 'active ageing' as adopted by the European Union. The various chapters further emphasize a variety of principles around which age-friendly work needs to be built, not least of these being the importance, first, of *political leadership*, i.e. gaining the support of key actors who are able to influence decision-making within the urban system; second, *co-production*, i.e. the empowerment of older people themselves in driving forward the policies and agendas associated with the age-friendly movement.

Equally, the book does bring out some of the challenges facing those working to develop age-friendly environments in urban and rural communities. Three in particular might be mentioned: first, work with older people continues to be hampered by an overly 'medicalized' approach, reflecting the dominance of healthcare systems within industrial/post-industrial societies. Of course, good quality health care is essential to achieving good quality of life of all age groups—not least older people. But the issue here is the way in which the need for health (and social) care can come to redefine identities in later life—often in highly negative ways. In this respect, the age-friendly perspective offers a complementary approach, emphasizing the wide range of issues around which work with older people needs to be based.

Second, as a number of the chapters in the book indicate, age-friendly perspectives need closer integration with issues relating to urban development and regeneration. The processes associated with these—new housing developments, gentrification, changes to city centres—will be crucial in determining the way in which cities are experienced. Yet older people (and indeed most other age groups) are often excluded from crucial areas of decision-making, with feelings of alienation and lack of power often the result. So age-friendly cities have to be about ensuring new sources of empowerment for older people, especially in respect of control over key areas of urban planning and policy.

Finally, all of the various chapters face difficult questions posed by the impact of globalization on the one side, and economic austerity on the other side. The debate around the creation of age-friendly environments that have to be embedded in the political and economic realities which arise from these powerful forces. To what extent are age-friendly communities possible given the economic crisis affecting many societies? How is the debate affected by the rolling back of the welfare state in many European countries? Is there a danger that age-friendly policies will be used as a substitute for well-resourced social services delivered at a local level? These are important questions for the reader to consider alongside exploring the undoubted potential which the developments reviewed in the various chapters represent.

In conclusion, I very much welcome this book as an important addition to the literature on age-friendly communities. The various chapters represent a major survey of the field, providing a valuable summary both of current work and likely developments over the next decade.

The University of Manchester Chris Phillipson
June 2015

Preface

This book is a story of meetings. A meeting of a young postdoc researcher discovering a new application of the 'active ageing' concept in Age-Friendly Cities and Communities (AFCC) and of an experienced researcher in community social work exploring ageing as a new fieldwork. A meeting of European ways of thinking on "active ageing" and of Canadian and North American ways of promoting this concept through practices. A meeting of researchers, public policy experts and older people from Argentina, Australia, Belgium, Brazil, Canada, China, France, United Kingdom, United States and Switzerland. Based on a selection of chapters, this international meeting necessarily reveals some of the challenges for the global AFCC perspective, the absence of Africa and the focus on developed countries being a frequent result of such an exercise.

This book is written to give an in-depth and empirically based knowledge of AFCC, which is a global discussion of new avenues for research and practice joining ageing and environment in the same track. The insistence on the role of the World Health Organization supporting this perspective since 2005 recalls some facts; however, the discussions around it in Part I, the variety of international local practices in Part II and the common challenges beyond its scope in Part III are the cement of a deep research-oriented direction of this book. In so doing, AFCC is probably, at the moment, the most internationally discussed meeting point for anyone interested in innovative public policies on ageing trying to adapt to global demographic changes and global urbanization. Next to researchers, policy makers, experts and stakeholders in ageing, this book will also interest master's students searching the emerging field of knowledge and of practice. Similarly, anyone wishing to better know the AFCC "in practice" will be happy to read these pages and discover the variety of experiences.

Avoiding the isolated academic position, this book prefers to share academic expertise with the knowledge of practitioners and experts who are dealing with AFCC in their history (past, present or future). This is the reason that we select chapters both from established scholars (and give special attention to the voices of some emerging scholars too) and chapters written by or with practitioners confronted with AFCC in their daily routine. In each of the three parts, this focus has been maintained: in Part I, through a balance between scholar A. Walker and international expert on AFCC, Dr. A. Kalache; in Part II, through experiences from all over the world (from neighbour level to national and transnational levels) written by scholars and experts; and in Part III, by comprehensive open questions suggested by A. Scharlach.

This book is the result of teamwork. Beyond authors' generous contributions, it received the help of a series of colleagues who accepted to anonymously revise chapters: Catherine Bigonnesse, Tine Buffel, Kathleen Brasher, Nathalie Burnay, Linda Castagnola, John L. Lewis, Louise Plouffe, Samuèle Rémillard-Boilard, Thomas Scharf and Enrique Vega Garcia.

Writing an international book is also an opportunity to improve transnational mutual understanding. For some chapters, this was not possible without the inestimable help of our expert translator Claudia Delli-Colli from the University of Sherbrooke.

Contents

Contributors

Simon Biggs Brotherhood of St Laurence, School of Social and Political Sciences, Melbourne University, Parkville, VIC, Australia

Françoise Le Borgne-Uguen Laboratory for Research in Sociology (Labers), University of Western Brittany (UBO), Brest, France

Kim Boudiny FWO-Flanders, Faculty of Political and Social Sciences, CELLO (Research Center for Longitudinal and Life Course Studies), University of Antwerp, Antwerp, Belgium

Kathleen Brasher This day & age, Box Hill South, VIC, Australia

Ashley Carr School of Social and Political Sciences, Melbourne University, Melbourne, VIC, Australia; Brotherhood of St Laurence, Fitzroy, VIC, Australia

Grace M.Y. Chan The Hong Kong Council of Social Service, Wan Chai, Hong Kong, China

Suzanne Garon Research Centre on Aging, University of Sherbrooke, Sherbrooke, QC, Canada; School of Social Work, University of Sherbrooke, Sherbrooke, QC, Canada

Silvia Gascón Social Welfare and Social Services Management, Alcalá University, Alcalá, Spain; Iberoamericana Organization for Social Security (OISS), Management of Gerontology Services, ISALUD University, Ringuelet, La Plata Buenos Aires, Argentina

Lindsay Goldman The New York Academy of Medicine, New York, NY, USA

Arlene Groh Healing Approaches to Elder Abuse and Mistreatment, Waterloo, ON, Canada

Geneviève Houioux Health Observatory of Hainaut (Observatoire de la Santé du Hainaut), Mons, Belgium; School of Public Health, Université Libre de Bruxelles, Brussels, Belgium

Alexandre Kalache International Longevity Center-Brazil (ILC-BRAZ), International Longevity Center Global Alliance (ILC-GA), Rio de Janeiro, RJ, Brazil

Lisanne S.F. Ko Sau Po Center on Aging, The University of Hong Kong, Pokfulam, Hong Kong, China

John L. Lewis School of Planning, Faculty of Environment, Research Institute for Aging, University of Waterloo, Waterloo, ON, Canada

Vivian W.Q. Lou Department of Social Work and Social Administration, Sau Po Center on Aging, The University of Hong Kong, Pokfulam, Hong Kong, China

Meghan Lynch The New York Academy of Medicine, New York, NY, USA

Dana Martens The New York Academy of Medicine, New York, NY, USA

Thibauld Moulaert REIACTIS, Dijon, France; School of Public Health, Université Catholique de Louvain, Louvain-la-Neuve, Belgium; Human and Social Sciences Institute, Université de Liège, Liège, Belgium; School of Social Work, University of Sherbrooke, Sherbrooke, QC, Canada

Shauneequa Owusu The New York Academy of Medicine, New York, NY, USA

Anne-Sophie Parent AGE Platform Europe, Brussels, Belgium

Mario Paris Research Centre on Aging, University of Sherbrooke, Sherbrooke, QC, Canada

Simone Pennec Laboratory for Research in Sociology (Labers), University of Western Brittany (UBO), Brest, France

Louise Plouffe International Longevity Center Canada, University of Ottawa, Ottawa, Canada

Nélida Redondo University of Buenos Aires, Buenos Aires, Argentina; Sociology of the Aging, ISALUD University, Ciudad de Buenos Aires, Argentina

Samuèle Rémillard-Boilard School of Social Work, University of Sherbrooke, Sherbrooke, QC, Canada

Andrew E. Scharlach The Center for the Advanced Study of Aging Services, School of Social Welfare, University of California, Berkeley, Berkeley, USA

Caitlyn Smith The New York Academy of Medicine, New York, NY, USA

Anne Veil Research Centre on Aging, University of Sherbrooke, Sherbrooke, QC, Canada

Ina Voelcker International Longevity Center Brazil (ILC-Brazil), Rio de Janeiro, RJ, Brazil

Julia Wadoux AGE Platform Europe, Brussels, Belgium

Alan Walker Department of Sociological Studies, University of Sheffield, Sheffield, United Kingdom

Lisa Warth Department of Ageing and Life Course, World Health Organization, Geneva, Switzerland

Rachel Winterton John Richards Initiative, Australian Institute for Primary Care and Ageing, La Trobe University, Wodonga, VIC, Australia

Chapter 1
Introduction: Toward a Better Understanding of AFCC

Thibauld Moulaert and Suzanne Garon

This book presents one of the first international discussions of aging[1] *through* the World Health Organization's (WHO) "Age-Friendly City and Community" (AFCC) model. Even if the WHO's program is not the only one to support an "aging-friendly" perspective (Lui et al. 2009; Scharlach and Lehning 2013) and even if it is not the first one that has linked aging and environmental perspective and insisted on the need for policy support (Phillipson (2011) refers to UN-Habitat report and to the NORCS–"naturally occurring retirement communities"), it is however the first one that permits researchers to explore the various international translations of the age-friendly environments thinking, AFCC becoming a case study for such objective. In doing so, we ambition to offer a more comprehensive understanding of a phenomenon that is clearly making sense essentially in North America (United States and Canada) and more generally, in Anglo-Saxon areas, as illustrated by recent issues of *Generations* (Scharlach 2009), *Journal of Aging and Social Policy*

[1]Note from the editors: if we opt for English American according to the rules of the publisher, the international perspective of this book implies a compromise concerning the term aging/aging (and its association in "active aging/aging") because traditions may differ and also because "active aging" has been particularly discussed in Europe (Walker 2002) as an attempt to go beyond US concept of "productive aging" (Bass et al. 1993).

T. Moulaert (✉)
REIACTIS, Dijon, France
e-mail: tmoulaert@gmail.com

T. Moulaert
School of Public Health, Université Catholique de Louvain, Louvain-la-Neuve, Belgium

T. Moulaert
Human and Social Sciences Institute, Université de Liège, Liège, Belgium

T. Moulaert · S. Garon
School of Social Work, University of Sherbrooke, Sherbrooke, QC, Canada

S. Garon
Research Centre on Aging, University of Sherbrooke, Sherbrooke, QC, Canada
e-mail: Suzanne.Garon@USherbrooke.ca

© Springer International Publishing Switzerland 2016
T. Moulaert and S. Garon (eds.), *Age-Friendly Cities and Communities in International Comparison*, International Perspectives on Aging 14,
DOI 10.1007/978-3-319-24031-2_1

(Fitzgerald and Caro 2014), *Journal of Social Work Practice* (Biggs and Haapala 2015), *Public Policy Aging Report* (Hudson 2015), or the coming book *International Perspectives on Age-Friendly Cities* by Fitzgerald and Caro (2016).

This being said, we have the argument for the general agenda of the book: embedded in strong empirical experiences and data's from *all over the world* and fuelled by critical discussions on aging, active aging behind the WHO experience and on the links between knowledge, practices and public interventions, its objective is to explore the origins, the developments and the various measures and programs produced or inspired by AFCC worldwide, with the notable exception of Africa. As editors of the book, we consider the AFCC model only as a means, not as an end. They are the means of focusing on two key challenges when we consider the AFCC's environmental perspective.

The first challenge concerns **scientific aspects**. Precisely how to define what an AFCC is, to enable us to produce an **International Comparison**. While most of the literature on age-friendly environment contains an important critical dimension as the essence of scientific and human science discourse (Buffel et al. 2013; Garon et al. 2014; Lui et al. 2009; Menec et al. 2011; Phillipson 2011; Scharlach and Lehning 2013), little is proposed to go over the concentration of experiences (and of scientific discussion) within some developed countries, Canada, United States, and United Kingdom coming at the forefront of a national-based scene. Even if this book does not propose a totally satisfying answer to this dilemma and is also strongly alimented by the experiences and scientific knowledge of such countries (through various chapters and through a strong network whose members gently agree to review some of the chapters), it has nevertheless considered the international perspective as a basic condition to address the challenges of aging, environment, and intervention. Therefore, we didn't want to set these different experiences into a case study outlined with predetermined units of analysis. Even if doing so, it would have made comparisons easier and more usual. We preferred to let the different authors set their own framework to explain their specific experience into the process of becoming Age-friendly Communities. Consequently, we didn't even propose a single geographical unit as a city, a province or a country.

Clearly, such multiple case studies are a first step answer to the "cross-national and cross-cultural research efforts… needed to examine the potential influences of sociocultural and political-economic contexts" (Scharlach and Lehning 2013, p. 128) beyond Age-friendly Communities. Furthermore, the selection has been driven by a research conviction/question fuelled by 5 years of collaboration and comparative research between the editors (Moulaert and Garon 2015). We consider that *even if* the WHO Guide (2007) has been presented or described as promoting an "ideal city" (Buffel et al. 2012), we don't only need to change our view "by shifting the focus from questions such as 'What is an ideal city for older people?' to the question of 'How age-friendly are cities?'" (Buffel et al. 2012), but we should also consider addressing an intermediate question: "How are Age-friendly Cities (and Communities) developments experienced?" In doing so, we simultaneously avoid to fall into two trap positions: the **expert position** that would “defend” the Age-friendly Cities practices and discourses from a “helicopter view” and mitigate their limits and difficulties; on the opposite, the **lay position** would prefer the experience of “aging in place” to capture the person-environment fit. On the contrary, we promote a third position, linking these two positions, a **pragmatic practitioner position**. By practitioner, we include here a variety of actors playing a role into the AFCC, from researcher or consultant to policy officer, from older people’s representative to administration staff, all of them being active at one or different stage and level of these multiple applications of the AFCC model into a variety of programs. Therefore, AFCC are what such practitioners are locally making.

This challenge also refers to the kind of research that suits to AFCC experiments. If it’s known that the act of doing a research is shaping the object on which it works on, in case of AFCC, one can say that the research field is also largely influenced by the contacts with so many stakeholders, the importance of *political leadership* being recalled in the foreword. As we just mentioned, this book does not focus exclusively on the “cities” dimension, thus prioritizing a geographical or urban perspective as proposed by the coming book of Fitzgerald and Caro (2016). Although this dimension is essential, we regard it as a material aid to considering the “real” questions of AFCC: How do Age-friendly Cities create a space where people can remain citizens and social subjects as they age? How different practices of AFCC can be considered as an in vivo implementation of similar (but not identical) age-friendly environment principles?

While the plurality of case studies brings a series of divergences and similarities summarized in the next pages, a last milestone of the scientific challenge concerns the role of research in AFCC. Research is acknowledged as an essential condition to support the planning of AFCCs (Moulaert and Garon 2015). Researchers do not longer fall within the expert category, a scientific tradition that has been carried on for years and to which we have become accustomed; they are mobilized actors in the implementation of AFCC initiatives. While Moulaert and Garon (2015) have already identified this avenue within AFCCs, this book participates to the need of studies which aim at shedding empirical light on the different roles of researchers involved in AFCC initiatives, but also seeks to develop a common knowledge and sharing. In fact, the presence of researchers within AFCC initiatives is part of a **collaborative research** with partners of various sectors of the local community (Dagenais et al. 2009), including seniors themselves (Bindels et al. 2014). Collaborative research is the result of a joint work between people from local communities, various stakeholders, and municipal officials and researchers. These various groups must have a common interest—AFCC here—throughout the research process (Hills and Mullett 2000). As it has already been mentioned, the diversity of AFCCs on the international scene nevertheless requires avoiding roles and patterns of collaborative research that are too hermetic. As we can see, the issue is to create meeting and exchange places for research with local community partners.

For all these reasons, one of the main difficulties regarding the AFCC research field is and will be to provide a one-fit-all evaluation model. And because of the richness and the complexity of the various experiences it might not be accurate to do so. This particularity shouldn't stop the scientific community interested in AFCC to strive to develop research designs that take into account the outcomes of the living conditions of elderly of Age-friendly Communities. And so, not only because of the context of the New Public Management which asks to provide those kinds of data. The evaluation designs should also take into account the different perspectives of the divers stakeholders involved in the process.

The second challenge is about **Public and Political Intervention**. Even if our goal is not to deny the potential place and role of private sector and voluntary sector in public and political intervention to bridge aging and environment through a comprehensive vision and frame for actions, our second cornerstone structuring this book is the central role of politics when AFCC is transformed from an ideal or theoretical framework into real policies, when projects become realities; at the crossroads of research and policy, it calls into question the use of knowledge by policy-makers (Moulaert and Garon 2015). Such a centrality is particularly salient in a globalized and more interconnected society driven by neoliberal forces, as opportunely mentioned too in the foreword. In aging area, it has been illustrated by international organizations such as Organization for Economic Co-operation and Development (OECD) or World Bank promoting a productivist version of society and of "active aging" from the 1990s and European Commission until the mid-2000s (Moulaert and Biggs 2013), sometimes described as a neoliberal "rolling back of the state". Ironically, at the time of major challenges such as aging and climate changes, Welfare States, as a place of democratic debates, has not-yet-been replaced by a transnational democratic institution able to regulate the discussions and gather global solutions.

However, politics does not limit itself to Welfare States or to democratic and political institutions. It also refers to the general agenda of empowering older people and the call for a "society for all ages" promoted by United Nations (UN 2002) Madrid International Plan of Action on Aging (MIPAA). AFCCs are a particular response to the call of MIPAA which promotes the importance of participatory approach in the development and implementation of public policies. From the beginning of the millennium, the MIPAA has outlined the participation of seniors:

(a) "Take into account the needs and concerns of older persons in decision-making at all levels;
(b) Encourage, when they do not already exist, the establishment of organizations of older persons at all levels to, inter alia, represent older persons in decision-making;
(c) Take measures to enable the full and equal participation of older persons, in particular older women, in decision-making at all levels" (UN 2002, pp. 11–12).

At the crossroad with the first challenge, this call has become an issue of growing importance in research (Walker 2007; Bindels et al. 2014). The research on AFCC initiatives should not be an exception and a participatory research must be put in place to include seniors in the research design process, data collection, and analysis. AFCCs are driven by the enhancement and increased participation of the citizen and democratic engagement of seniors in the city. This avenue allows to create links with the "right to the city", to use the expression of Lefebvre (1968) repeatedly cited by Phillipson (2010, 2011). By contrast, the engagement of senior citizens within the democratic life of the city must not become a requirement for all AFCC initiatives. In fact, despite the honorable intentions to include seniors in city life, it remains a risk to normalize the old age experience. It is almost impossible to get out of the following imbroglio: to be a senior does not mean to be old. Specifically, the challenges of the third and fourth age show how the experience of advancing aging differs according to whether one is prior to old age or afterward. This difference has been widely documented in gerontology (Biggs et al. 2006) and sometimes in the study of AFCCs (Paris et al. 2013). It is then important, with seniors, to question the meaning of the democratic life within the city. By definition, it generally refers to the public sphere; however, is it located side by side with all citizens? And is it possible also to consider it within the private sphere of seniors? For our part, the role played by the neighborhood becomes increasingly recognized in the establishment of a citizenship which rests on local solidarity. It is important to distinguish the city as a democratic institution and as a community of values (Honneth 1995).

The general structure of the book reflects our purposes. The first part briefly summarizes the origins and contemporary developments of AFCC in the light of aging and of "active aging". The second part presents nine experimental AFCC programs worldwide from the local level of neighborhoods and cities to national case study and international developments such as those in progress at European level. While transversal themes appear in each of them—the place of research, policy, and older citizens—some chapters will specifically focus on particular domains. For example, the chapter about Quebec offers a final discussion on the role of evaluation while the chapter about Argentina insists on how implementation of programs has to assess different elements supporting quality of life of older people like the size of the cities, their level of human, economic and social developments, and their urban design. The third part presents three challenges for researchers, policy-makers, and older citizens through AFCC: Can AFCC be compared to similar program for Children and Dementia? Can "active and healthy aging" behind AFCC be critically discussed to propose future directions for AFCC? Can we make a radical evaluation of age-friendly cities issues? Beyond the strong international perspective, these challenges (part 3) represent another original contribution of the book, as they are based on and discuss how AFCC experiences are articulated and implemented (part 2) as practical and multiple answers to global aging (part 1).

1.1 Part 1: Age-Friendly Cities and Communities: From Ideas to Practices

One of the originality of the AFCC case study is that it might be considered as one of the first attempt to translate the comprehensive concept of "active aging" into practical experiences of local policies. It is not to deny the often reduced translation of "active aging" into public policies and local practices supporting longer working lives, particularly in European countries (for a resume of such debates see the chapter by Moulaert, Boudiny, and Paris), neither to deny the interesting focus on 'active aging' from an ethnographic and critical perspective in order to observe how older people live their own experience of "aging actively" (Lassen 2014; Marhankova 2011; Moulaert 2012a). All these efforts[2] are participating to the general discussions of the "production of aging and its relationship to contemporary norms" (Moulaert and Viriot Durandal 2013). Like other terms (successful aging, productive aging, healthy aging, aging well, etc.) "active aging" is simultaneously a scientific and political construction (Lassen and Moreira 2014; Moulaert and Paris 2013) a source of action and of knowledge. However, one of its originality is probably its international circulation through international organizations (Moulaert and Biggs 2013).

If international and cultural perspective on "aging-friendly communities" is needed (Scharlach and Lehning 2013), such a sociohistorical study is essential. And if "active aging" is linked since its origin to the "Age-friendly Cities" case study on which we focus in this book, it is then of the greatest importance to gather, into a dialog, the persons who have been central in those collective processes.

For such reason, **Louise Plouffe**, **Alexandre Kalache**, **and Ina Voelcker** present in Chap. 2 a critical review of the methodology promoted by the original WHO Guide in 2007. Defined as the "Vancouver Protocol", it was based on consultation process of a variety of older people in local community. It has also been associated with a "checklist" resuming the actions to be taken into consideration into the eight domains identified by WHO as pertinent for a complete description of environments supporting aging (outdoor spaces and buildings, transportation, housing, social participation, respect and social inclusion, civic participation and employment, communication and information, community support, and health services). Often criticized by practitioners and researchers as promoting an "ideal city" (Buffel et al. 2012), as a "top-down" perspective

[2]Concerning "active aging" we should also mention the recent *Active Aging Index* developed by a team led by professor Ashgar Zaïdi as a statistical tool to assess the level of European countries (data's from and comparison with countries outside Europe are also into progress) in four domains related to the concept (employment; participation in society; independent, healthy and secure living; capacity and enabling environment for active aging). For further details: (Online). Available from http://www1.unece.org/stat/platform/display/AAI/Active+Ageing+Index+Home. Accessed May 13, 2015.

(Barusch 2013; Lui et al. 2009) or as a tool unable to capture the diversity of aging experience, the authors answer to these critics through an evaluation of the implementation of such methodologies in an international perspective. In doing so, they conclude that "adaptations include refinement of the original dimensions [of the eight domains], with more emphasis on informal social support and on safety and security" and consider that "the checklist has been used in different ways, but not without a complementary consultation process"; finally, concerning the inclusion of a plurality of aging experiences, they note that "respecting diversity and ensuring engagement have been key considerations in many, although not all, initiatives."

While Chap. 2 presents the original methodology developed by WHO since the origin of the Age-friendly Cities project in the mid-2000s, the interview of **Lisa Warth** describes the actual developments of the WHO program since the WHO Global Network of Age-friendly Cities and Communities (GNAFCC) set up in 2010. Launched in 33 cities in the original standard methodology (WHO 2007), the WHO program has been adapted to its global ambition and it has grown to "over 250 cities and communities across 28 countries worldwide that are engaged in Age-friendly City initiatives. Together these cities and communities cover over 100 million people worldwide." In continuity with the WHO conceptual framework of "active aging" (WHO 2002) and the Global Age-friendly Cities Guide in 2007 (WHO 2007) and its participatory approach with older people, the GNAFCC connects all stakeholders from local areas who are interested in developing an AFCC approach but particularly insist of the importance of local government support. Converging with scientific critic (Keating et al. 2013) the network's name was changed from Age-friendly Cities to Age-friendly Cities and Communities (AFCC) in order to better accommodate smaller communities, including rural areas. Today, the network shares the local experiences of its members through a website with an Age-friendly World (www.agefriendlyworld.org). With the expansion of AFCC worldwide, the network has worked on some domains with more intensity. Beyond the language challenge, the networking dimension of GNAFCC is another of its achievement, Europe being an interesting case study to discover how different levels of actions must be articulated (this case study is further developed in Chap. 14 as the search for an "Age-Friendly Europe"). Nevertheless, some challenges remain important for GNAFCC. **Lisa Warth** explains how the sustainability of the network is one of the first challenges as WHO has very little resources and as the network is simultaneously expanding; similarly, the "question of what impact policy interventions have on the health and well-being of older people" is another recurrent challenge of AFCC.

These first two chapters present a continuous dialog between the persons at the origin of the WHO program and the person now in charge of organizing the continuity of the actions. A second dialog consists of the confrontation of Chaps. 4 and 5. The first one offers an academic description of "active aging" by **Alan Walker**, the second describes an empirical construction of "active aging" at the WHO level by **Alexander Kalache**. Both of them need to be read in parallel because they are the first ever simultaneous discussion of the same concept, the first ever work to present the closely intertwined relations between the academic world of knowledge and practices and the public policy world of practices and knowledge.

In Chap. 4, **Alan Walker** recalls how the realities of aging, even if diverse worldwide, are nevertheless shaped by a general trend of "a continuance of aging in the more developed global north coupled with very rapid aging in the less developed south." He then presents the main policy discourses, with a particular focus on "European region, which is the oldest in the world in terms of the proportion of its population in the oldest age groups." Presenting "active aging" in a comprehensive discussion, the author also illustrates the barriers met on its road, from political to ideological or cultural ones, and how to overcome these barriers. Most interestingly for our book and the dialog with the next author, the concluding remarks address key challenges for WHO program. Indeed, **Alan Walker** makes a provocative statement. If he agrees with the similarities between his comprehensive vision of "active aging" and the WHO perspective, he notes that

> too often, in practice, 'age friendly' means 'old age friendly' not 'aging friendly'. This is a fundamental fault line that must be eradicated if AFCs are to be anything more than just at the service of special interests rather than the centerpiece of responses to the challenge of aging.

In Chap. 5, we invite the reader to follow **Alexandre Kalache** in his personal travel between public health and public action, academic research and international organizations. Through these moves, we have an original longitudinal "inside view" of a key figure in the international development of Age-Friendly Cities who has been particularly active at the World Health Organization; this view permits to understand the relations between the "social" and the "health" inside the organization through the links between "healthy aging" and "active aging" (for a further discussion of these relations, see the chapter of Moulaert, Boudiny, and Paris). Clearly, what is presented here as "A personal account", is not a complaisant or uncritical biography. It is probably a key element to illustrate the hypothesis of "intermediaries of active aging" (Moulaert 2012b), that are people or processes organized to connect ideas and practices, to "interest" various stakeholders (the Latin origin of "interest" being *inter esse*, what the French clearly presents at "being in between"). Last but not least, by reconnecting this chapter with the forth one, we also have some arguments to push the idea of an "epistemic community" (Moulaert and Paris 2013) where the conceptions and the roles of ideas and practices beyond AFCC may differ but still progress in a similar direction.

A similar direction does not mean a unique direction. Following **Lisa Warth**, "there is no one-size-fits-all solution to meeting the needs of older people, there is no one-size-fits-all recipe for creating age-friendly environments." As a consequence, the Age-friendly World website promoted by WHO (www.agefriendlyworld.org) will probably present a variety of actions throughout general guidelines. If the transnational evaluation of the effective results and uses of such network might be welcome, this book suggests following existing experiences worldwide.

1.2 Part 2: Active Aging and Age-Friendly Cities: One Model, Many Programs

The second part of the book presents nine case studies ranging from a local case study of one city (the city of Waterloo, Canada in Chap. 6; the city of New York in Chap. 10), a case study of different cities (three cities in Western France in Chap. 11 ; three cities in Argentina in Chap. 9) to programs at a regional or national levels (the Quebec case in Chap. 7; the Hong Kong case in Chap. 8; the Walloon case in Belgium in Chap. 12; the Australian case in Chap. 13) and even at supranational level (with the European case study in Chap. 14). All of them have in common a link with the WHO program supporting AFCC, this link being strong or weak.

The first empirical experience of a local program inspired by AFCC from WHO (and by other sources from the *Livable communities* of AARP and the *AdvantAge Initiative* from *The Visiting Nurse Society of New York*) is the case study of Waterloo, Canada. **John Lewis and Arlene Grooh** present some of the results of the local consultation they conducted by interviewing 76 older inhabitants of the city about their vision of the "challenges associated with aging in the city of Waterloo." Referring to ecological theory and environmental gerontology, their qualitative analysis insists on various sources of "environmental press" (transportation; walkability, and housing) which confirm what the literature has demonstrated regarding barriers to age independently at home and in its neighborhood. The authors consider the "social press" with a greater interest as they note how older people consider "social exclusion" as a harder challenge to meet:

> the central message from the participants posits that while older people are capable of addressing environmental presses as their minds and bodies change, the 'social press' that is associated with the behavior and underlying attitudes of other community members can be the most difficult challenges to address.

In order to face this challenge **John Lewis and Arlene Grooh** go one step forward by exploring two avenues to act on "social press", ageism being the target to analyze and critically assess. One is "to call for greater public awareness through various forms of government sponsored social marketing campaigns, public outreach or professional training events." The second and more demanding path consists of discussing "our modern culture's tendency to simplify the aging experience and conceptualize it through a medical lens that emphasizes dependence, increasing ill-health, disability and decline." As a consequence, community planners and urban designers are encouraged to promote and think in terms of policies enabling "aging in community" because community "it's about the people."

In Chap. 7, **Suzanne Garon** and her AFCC team, **Anne Veil, Mario Paris, and Samuèle Rémillard-Boilard** propose a resume of the past 8 years of research of Age-Friendly Cities in Quebec and explore actual challenges in terms of evaluation. The authors first describe how the original participation of the city of Sherbrooke in the WHO initial study of 2007 has been an opportunity to propose a case study in seven pilot sites and how this program has rapidly generated an interest from cities locally and from central government until the AFCC governmental program launched in 2009. This contribution explores in details how a social science research can enhance a public policy by adopting the guidelines of community planning (older people involvement; cross-sectoral approach of aging; political and administrative staff implication through a steering committee; research-based practices). The last part of the argument illustrates what the challenges of such a research are, in particular through its capacity to produce an evaluation able to explain what happens into the "black box" of local actions supporting AFCC.

In Chap. 8, **Grace Chan, Vivian Lou, and Lisanne Ko** present the role of Hong Kong Council of Social Services (HKCSS) in leading an AFCC project in Hong Kong, partly because of its concerns for the development of social welfare. While the methodology is closely related to the one proposed by WHO (active aging framework; AFCC checklist; qualitative methods for consulting the various stakeholders including older people representatives), the chapter brings important messages from the Chinese context, a context where "the concept of Age-friendly City has not been concisely defined." For example, while the WHO seems comprehensive, there is little discussion "on how to provide older people with their basic social and income protection," about assessing the policy outcomes or about which groups of older people should be targeted by local authorities. Discussing their observations through international practices and strategies supporting AFCC, the authors consider that the importance of government

> does not lie on its dominant position in guiding and monitoring the implementation of all age-friendly related policies, but on its leading role in providing platforms and environments with which the public and NGOs can go hand in hand to advise the government on related policy making and cultivate age-friendly environments.

In Chap. 9, **Nelida Redondo and Sylvia Gascón** present the results of the implementation of an AFCC project in three cities from Argentine in the light of regional inequalities. While they consider that the size of cities or their level of human, economic and social developments are important determinants of the level of the quality of life of older people, interesting element also come through the reflexivity of researchers about the effects of "contexts where institutional uncertainty and financial restrictions prevail." Indeed, the three cases show little efforts made by local authorities to support an AFCC agenda. To manage this difficulty, the researchers insist on the importance of the "role of volunteer organizations run by elderly people… independent from the government sphere," on the vitality and originality of forums organized and their democratic distance with opinion surveys or market research and eventually they consider that private nonprofit organizations, like their own university, may play a central role in such context.

Chapters 8 and 9 have smoothly introduced a general concern of AFCC: what about the sustainability of local efforts to better take into account the various dimensions of a city or a community? And how can such efforts be supported? While the first part of the book and the Chap. 7 present some answers, the New York case study in Chap. 10 offers complementary elements. **Lindsay Goldman and her team (Shauneequa Owusu, Caitlyn Smith, Dana Marten, and Meghan Lynch)** from the New York Academy of Medicine (NYAM) describe the various stakeholders implied and the strategy of action: "Increase utilization of existing assets; develop local solutions to local problems; promote economic security and spending; leverage private sector resources". The many practical examples not only illustrate the search for "creative, low-cost solutions to reduce barriers to maximum social and economic participation" but are also the empirical evidence of locally and culturally-sensitive practices (while the bench initiative is found worldwide, the insistence on leverage private sector may be a specificity of a liberal tradition). Last but not least, we also find answers to the question of sustainability: coming with convinced local leaders (in administration, in elected officials, in local communities), the resources (funding and human) are dependent on the mobilization of knowledge (the role of research and advisory by NYAM) and outputs in communicable formats (showing the impact of walkability efforts in figures; selecting communicable actions like "Senior Splash", "CityBench program", or the renewal of bus shelters). But sustainability also means changes: through a continuous process of "top-down" *and* "bottom-up" approach, the second phase of implementation will be more sensitive to the places where older people are living (Brooklyn and Queens in place of Manhattan), to the variety of languages, to existing networks of care (to better include people with disabilities), and to the measure of health outcomes from AFCC practices.

At this point, the book ends a first series of case studies in which AFCC leaders such as the province of Quebec (Chap. 7) or the city of New York (Chap. 10) are regularly mentioned. Chapters 11, 12 and 13 present, on the contrary, case studies in which difficulties and challenges appear to be more common.[3] Chapter 14 has still another status: first, it presents a rarely discussed level of action (the European level per se); second, it is an interview of a central player of the European agenda, i.e., Age Platform representing the voice of people 50+ in Europe, presented by **Anne-Sophie Parent and Julia Wadoux**; third, its central message is one of translating the AFCC framework into the European context and its multiple levels of power. In doing so, it links the two series of empirical studies (the "good practices" ones vs the "challenging issues") from the book.

In Chap. 11, **Simone Pennec and Françoise le Borgne-Uguen** discuss the implementation of AFCC in three cities from Western of France. Recalling the original French connection between the AFCC agenda and the French "Aging Well plan" agenda (while the first one insists on "active aging" in a comprehensive perspective and on social participation, the second gives more attention to "successful aging" and highlights prevention), the authors present some tensions from their local empirical observations. First, are practices a source of real innovation or are they a labeling process of existing actions or partnerships? Is the diversity of expertise sufficient to build a cross-sectoral agenda or might it be the cause for a

[3]It might also be associated with the position of the authors of the chapters. In particular, the French chapter has been written by French sociologists who might be particularly critical about public policies and be more reserved about the role of research facing such policies.

"fragmentation of competences" contributing to the "blurring of priority objectives and methods for consultation?" In France, they conclude to the persistence of the dominance of a health and social sectors perspective on aging and of a medical gaze ("medicalization of aging"). They consider that the advantage of the "Aging Well plan" might conduct to a move from "individual prevention to the duty of social utility." Similar to other case studies, the challenges concern the inclusion of the diversity of older people (Where are the isolated persons?) and the long-term attention to their voice in participative approaches.

In Chap. 12, **Thibauld Moulaert and Geneviève Houioux** describe how Walloon government attempted to impulse an AFCC strategy in 2012 and how local actors reused the AFCC framework to meet their own understanding of senior's social participation. With a similar sociohistorical perspective as Chap. 11, the authors introduce the case by situating the existence of "municipality advisory councils of seniors" as a preexisting interest for social participation of older people. They then critically present the political Walloon appropriation of the WHO AFCC framework, insisting for example on the inadequate planned timing for consultations of older people. Based on empirical data's from a qualitative study in 12 cities that received public funding to locally adopt the framework, the chapter describes the conduct of assessments of needs and resources as "a relatively disparate manner from one municipality to another, due to a lack of time, experience and interest." Nevertheless, it also presents administrative staff as potential "active aging intermediaries", i.e., actors who are concretely operationalizing a theoretical model of "active aging" into local practices and experiences.

In Chap. 13, **Kathleen Brasher and Rachel Winterton** present Australia as another critical situation. Even if two cities were initially members of the WHO original project in 2007, the authors examine the factors which limited the implementation of AFCC in the State of Victoria, even with the initial provision of funds to support relevant policy and service reform. They demonstrate how a lack of vision and coherence within top-down approaches can inhibit the development of "bottom-up", localized age-friendly initiatives. While the discourse of 'active aging' has once been promoted in the nation through expert voices, intake of AFCC has failed due to a "lack of vision" at the different levels of government; terms such as "active aging" or "age-friendly" are adapted to existing programs with little place for their innovative dimension; the authors also insist on the "lack of multi-sectorial partnerships", would it be between the different departments of government, or through little attention to the consultation of older people processes. In Victoria, Council on the Aging (COTA), the advocacy organization for older people, has thus implemented only limited "bottom-up" peer-education model with its volunteers "reaching out to groups of older people across the state to inform them about the AFCC model."

1.3 Part 3: Challenges from and for Age-Friendly Cities and Communities

The last part of this book is dedicated to special and generic concerns produced by and around AFCC. Even if there are traces of these challenges to be found in previous case studies, we believe that their level of generality can generate discussion for a larger audience.

In Chap. 15, **Simon Biggs and Ashley Carr** discuss the age-friendly narrative, both in terms of a focus on active aging and of a claim to be "for all ages". This claim is compared to initiatives around child and dementia friendliness, both to help specify the focus of Age-friendly Cities and environments (AFE) and to examine the commonalities and differences between these groups when it comes to their access to public space. Empirical data's from Australia complete discussion of international sources. The authors then explore how existing assumptions about the negative consequences of public space on identity may be reversed for marginalized age groups, such as people living with dementia. Finally, some of the prerequisites for effective intergenerational interaction are explored through a discussion of propositions about identity and the meaning of generationally intelligent space.

In Chap. 15, **Thibauld Moulaert**, **Kim Boudiny and Mario Paris** discuss the links between active and healthy aging behind AFCC. After presenting some general trends and definitions within the fields of active and healthy aging and building on some overall limitations, the authors (re)introduce empowerment as a key element of active and healthy aging. By focusing on two central elements of empowerment—i.e., a multilevel perspective and an insider's view—a life course perspective on active and healthy aging is proposed in which the merits of both notions are integrated. In doing so, the frequent criticism of active and healthy aging discourses being so centered on individual responsibilities can be overcome. In conclusion, potentials of this perspective are presented for AFCC.

In concluding Chap. 17, **Andrew E. Scharlach** proposes a clear and strong resume of the main questions for the Age-friendly Cities agenda. The author himself recognizes that "many of these reflect unresolved issues concerning the very nature of Age-friendly Cities, including what they are, whom they are for, how they can be created, and indeed whether or not Age-friendly Cities are even desirable or possible." The chapter first examines overarching perspectives and underlying conceptualizations of Age-friendly Cities, then explores the characteristics that are indicative of age friendliness, follows with a critical examination of approaches to helping cities become more age-friendly, and closes with a number of issues that require further attention if we truly are to create livable communities for all ages.

1.4 Future Challenges

AFCC has a short and long history. While some challenges remain central throughout time, new ones emerge.

First, a **global North/South challenge**. One may consider different phases of AFCC. The first phase might be located into the Vancouver Protocol and the original WHO (2007) study on the 33 cities in the world. At that time, one of the main goals to develop an AFCC community was to respond to the concern of the rapid aging of the populations of the developing countries (i.e., countries of Africa) by creating a network of knowledge transfer with the northern countries that are already addressing this situation. Nairobi was the only city from Africa into the 33 original towns (WHO 2007). Yet it has to be recognized that this issue has not been solved adequately. It's not that the WHO has lost the focus but the fact that "aging fast" for the developing countries is one of the too many issues that they have to face in a context of globalization. Furthermore, the recent epidemic crisis with the Ebola fever in Africa just showed the enormous difficulties for countries to face a public health issue without the means of a structured State. That being said, the challenge of bringing along the developing countries in the AFCC is still complete. Further strategies should be put in place to make this network more inclusive by thinking of matching countries together for sharing experience or working with NGO that have some echoed on this continent.

The goal of **building bridges between the different cultures** must still be the main goal in the second phase, that have seen the development of the Global Network of Age-friendly Cities and Communities (GNAFCC) set up in 2010. Here, the consultative phase, central in first phase, is completed by a more complex framework to imply local officials and to join the consultative phase with an action plan. All the processes are more adapted to northern countries and may have to be adjusted to the developing countries to make sure that they will be able to participate. This brings us the question: How can a model be adapted into variable experiences and contexts without losing its specificity and without being considered as a pure "top-down" or "ideal" model? Also, this second phase has seen the extensions from the focus on urban context to the more general inclusion of various communities and to the many initiatives identified in the United States or in Canada (Scharlach and Lehning 2013) "reflecting societal values of independence, privacy, and personal control". In this book we tried to diversify the variety of contexts by including case studies from Argentina, Hong Kong, Brazil, or France as they are rarely discussed in international publications on Age-friendly Communities. These chapters are essential if one wants to avoid the North American bias.

From empirical work, a second challenge concerns **evaluation regarding AFCC**. How to take into account the diverse local realities related to cultural diversity, social, economic and knowledge resources inequalities, etc.? For instance, is it possible to do a mapping of the "good/best practices" without considering the sociological or political context in which those practices have emerged? If the evaluation is a centerpiece to any learning process (United Nations Evaluation

Group 2015), the conditions of leading an evaluation procedure are very different from country to country.

The last challenge brings us back to the first paragraph of this introduction. Furthermore, preparing this book revealed the **lack of a real international community of research on AFCC** where some basic questions could be discussed. These exchanges are essential to the development of communities of practices that can provide answers to the later challenges introduced. If this book can be a little piece of this issue, our efforts for a better understanding of aging through the lens of emerging public policies at the community and environmental levels were not in vain.

References

Barusch, A. S. (2013). Age-friendly cities: A social work perspective. *Journal of Gerontological Social Work, 56*(6), 465–472.

Bass, S. A., Caro, F. G., & Chen, Y.-P. (Eds.). (1993). *Achieving a productive aging society*. Westport, Conn.: Auburn House.

Biggs, S., & Haapala, I. (2015). Age friendly environments. *Journal of Social Work Practice, 29* (1), 1–3.

Biggs, S., Phillipson, C., Money, A., & Leach, R. (2006). The age-shift: Observations on social policy, ageism and the dynamics of the adult lifecourse. *Journal of Social Work Practice, 20* (3), 239–250.

Bindels, J., Baur, V., Cox, K., Heijing, S., & Abma, T. (2014). Older people as co-researchers: A collaborative journey. *Ageing and Society, 34*(06), 951–973.

Buffel, T., McGarry, P., Phillipson, C., De Donder, L., Dury, S., De Witte, N., et al. (2013). Developing age-friendly cities: case studies from Brussels and Manchester and implications for policy and practice. *Journal of Aging and Social Policy, 26*(1–2), 52–72.

Buffel, T., Phillipson, C., & Scharf, T. (2012). Ageing in urban environments: Developing "age-friendly" cities. *Critical Social Policy, 32*(4), 597–617.

Dagenais, C., Ridde, V., Laurendeau, M.-C., & Souffez, K. (2009). Knowledge translation research in population health: Establishing a collaborative research agenda. *Health Research Policy and Systems, 7*(28) (Online). Available from http://doi.org/10.1186/1478-4505-7-28. Accessed May 30, 2015.

Fitzgerald, K. G., & Caro, F. G. (2014). Collaborative partnership in age-friendly cities: Two case studies from Quebec, Canada. *Journal of Aging & Social Policy, 26*(1–2), 1–18.

Fitzgerald, K. G., & Caro, F. G. (2016). *International perspectives on age-friendly cities*. N.A.: Routledge (Online). Available from https://www.routledge.com/products/9780415720717. Accessed June 5, 2015.

Garon, S., Paris, M., Beaulieu, M., Veil, A., & Laliberté, A. (2014). Collaborative partnership in age-friendly cities: Two case studies from Quebec, Canada. *Journal of Aging and Social Policy, 26*(1–2), 73–87.

Hills, M., & Mullett, J. (2000). Community-based research: creating evidence-based practice for health and social change. (Online). Available from http://www.leeds.ac.uk/educol/documents/00001388.htm. Accessed May 30, 2015.

Honneth, A. (1995). *The struggle for recognition. The moral grammar of social conflicts*. Cambridge: MIT Press.

Hudson, R. B. (2015). Making a Home in the City: The age-friendly community movement. *Public Policy & Aging Report, 25*(1), 1–3 (Online). Available from http://doi.org/10.1093/ppar/pru057. Accessed June 5, 2015.

Keating, N., Eales, J., & Phillips, J. E. (2013). Age-friendly rural communities: Conceptualizing "best-fit". *Canadian Journal on Aging/La Revue Canadienne Du Vieillissement, 32*(4), 319–332.

Lassen, A. J. (2014). Keeping disease at arm's length—how older Danish people distance disease through active ageing. *Ageing and Society*, 1–20 (Online). Available from http://doi.org/10.1017/S0144686X14000245. Accessed May 30, 2015.

Lassen, A. J., & Moreira, T. (2014). Unmaking old age: Political and cognitive formats of active ageing. *Journal of Aging Studies, 30*, 33–46 (Online). Available from http://doi.org/10.1016/j.jaging.2014.03.004. Accessed May 30, 2015.

Lefebvre, H. (1968). *Le droit à la ville*. Paris: Anthropos.

Lui, C., Everingham, J., Warburton, J., Cuthill, M., & Bartlett, H. (2009). What makes a community age-friendly: A review of international literature. *Australasian Journal on Ageing, 28*(3), 116–121.

Marhankova, J. H. (2011). Leisure in old age: Disciplinary practices surrounding the discourse of active ageing. *International Journal of Ageing and Later Life, 6*(1), 5–32.

Menec, V. H., Means, R., Keating, N., Parkhurst, G., & Eales, J. (2011). Conceptualizing age-friendly communities. *Canadian Journal on Aging, 30*(3), 479–493.

Moulaert, T. (2012a). *Gouverner les fins de carrière à distance. Outplacement et vieillissement actif en emploi*. Bruxelles: Peter Lang.

Moulaert, T. (2012b). Pourquoi les francophones préfèrent-ils la sociologie du vieillissement à la gérontologie critique ? *Gérontologie et Société, 142*(3), 81–99.

Moulaert, T., & Biggs, S. (2013). International and European policy on work and retirement: Reinventing critical perspectives on active ageing and mature subjectivity. *Human Relations, 66*(1), 23–43.

Moulaert, T., & Garon, S. (2015). Researchers behind policy development: Comparing "age-friendly cities" models in Quebec and Wallonia. *Journal of Social Work Practice, 29* (1), 1–13.

Moulaert, T., & Paris, M. (2013). Social policy on ageing: The case of "active ageing" as a theatrical metaphor. *International Journal of Social Science Studies, 1*(2), 113–123.

Moulaert, T., & Viriot Durandal, J.-P. (2013) (Eds.). Production et rapport aux normes contemporaines du vieillissement. L'injonction au «bien vieillir» en question. *Recherches Sociologiques et Anthropologiques, 44*(1).

Paris, M., Garon, S., & Beaulieu, M. (2013). La lutte pour la reconnaissance de la vieillesse: Un regard critique sur l'estime sociale des aînés. *McGill Sociological Review, 3*, 5–17.

Phillipson, C. (2010). Ageing and urban society: growing old in the "century of the city". In C. Phillipson & D. Dannefer (Eds.), *The sage handbook of social gerontology* (pp. 597–606). Thousand Oaks: Sage.

Phillipson, C. (2011). Developing age-friendly communities: New approaches to growing old in urban environments. In S. Settersten & J. Angel (Eds.), *Handbook of sociology of aging* (pp. 279–293). New York: Springer.

Scharlach, A. (2009). Creating aging-friendly communities. *Generations, 33*(2), 5–11.

Scharlach, A. E., & Lehning, A. J. (2013). Ageing-friendly communities and social inclusion in the United States of America. *Ageing and Society, 33*(1), 110–136.

UN. (2002). *Madrid international plan of action on ageing*. New York: United Nations.

United Nations Evaluation Group. (2015). *Evaluation changes lives: Realizing evaluation potential to inform the global sustainable development goals*. New York: United Nations.

Walker, A. (2002). A strategy for active ageing. *International Social Security Review, 55*(1), 121–139.

Walker, A. (2007). Why involve older people in research? *Age and Ageing, 36*(5), 481–483.

WHO. (2007). *Global age-friendly cities: A guide*. Geneva: WHO.

Part I
Age-Friendly Cities and Communities: From Ideas to Practices

Chapter 2
A Critical Review of the WHO Age-Friendly Cities Methodology and Its Implementation

Louise Plouffe, Alexandre Kalache and Ina Voelcker

2.1 Introduction

In 2007, the World Health Organization released the report: *Global age-friendly cities*: *A Guide* (World Health Organization 2007a) to stimulate the creation of accessible and inclusive urban environments in order to promote active aging as defined by the WHO (2002). The Guide identified the key features of an age-friendly city from the perspective of older persons and service providers to serve as a reference for cities in developing as well as in developed countries (Plouffe and Kalache 2010; World Health Organization 2007a). The WHO initiative has rapidly spread and gained recognition as a global movement (Barusch 2013; Golant 2014; Liddle et al. 2014; Lui et al. 2009). The WHO Global Network of Age-Friendly Cities includes 209 cities and communities in 26 countries (WHO, Personal communication 2014) in addition to 10 affiliated country- or state-level programs. In Canada alone, over 850 municipalities now participate in age-friendly initiatives promoted by municipal, provincial and federal, governments (Plouffe et al. 2012).

The methodology developed for the Guide and subsequently for communities eager to become age-friendly using this approach was the Vancouver Protocol (Plouffe and Kalache 2010; World Health Organization 2007b). The foundation of

L. Plouffe (✉)
International Longevity Center Canada, University of Ottawa, K1N6N5, J8N 5J2 Ottawa, Canada
e-mail: louise.plouffe@gmail.com; louise.plouffe@ilccanada.org

A. Kalache · I. Voelcker
International Longevity Center Brazil (ILC-Brazil), Avenida Padre Leonel Franca 248, Gavea, Rio de Janeiro, RJ 22451-000, Brazil
e-mail: akalache@nyam.org

I. Voelcker
e-mail: ina.voelcker@ilcbrazil.org

© Springer International Publishing Switzerland 2016

T. Moulaert and S. Garon (eds.), *Age-Friendly Cities and Communities in International Comparison*, International Perspectives on Aging 14,
DOI 10.1007/978-3-319-24031-2_2

the Vancouver Protocol was the WHO (2002) concept of active aging, defined as: "the process of optimizing opportunities for health, participation and security over the life course in order to enhance quality of life as people age" (World Health Organization 2002, p. 12). Building on this definition and on a thorough review of elder-friendly community initiatives at that time in North America, the protocol proposed a qualitative assessment of the built, social, and service environments in eight domains: outdoor spaces and public buildings; transportation; housing; social participation; respect and social inclusion; civic participation and employment; communication and information; community support and health services. The Vancouver Protocol was finalized at an international workshop organized by the WHO in association with the Public Health Agency of Canada and the Government of British Columbia and with representatives from 12 cities (of the eventual 33 cities) that had already committed themselves to conduct the multicentric international study. Specific elements of the age-friendly cities assessment methodology included: documentation of the profile of the city and the older adult population and an inventory of programs and services for older persons; semi-structured focus groups with older persons differing by age and socio-economic status, and, where possible, differing along other characteristics that represent their diversity, as well as with a group of caregivers of older persons unable to participate in focus groups owing to disabilities and a group of service providers from the public, private, and voluntary sector.

The WHO Guide (2007a) offered the added benefit of a checklist of features of an age-friendly city for cities to use as a tool to guide municipal self-assessment. Emerging from the 'bottom-up' consultation in the 33 cities that participated in the WHO study, the checklist was intended to provide a standard set of features with which a city could compare the findings from its assessment in consultation with its older residents.

In the online instructions to join the Global Network of Age-Friendly Cities and Communities, WHO (n.d.) continues to emphasize that older persons should be closely involved in all phases of age-friendly assessment, planning, and action. Applicants are required to include the eight domains in their assessment, but they have flexibility in the assessment method in light of the diversity of cities and communities; WHO does not specify whether or how to use the checklist.

The aim of this chapter is to review how the Vancouver Protocol and checklist of age-friendly city features have been applied in diverse municipalities, regions, states, and countries in order to determine the 'lessons learned' both from experience and from further scrutiny by gerontologists. Locations listed in table below were chosen in North and South America, Europe and Australia to represent a variety of municipalities, regions, states, and countries (Table 2.1).

Published reports, online materials, and gray literature written in English, French, Portuguese, and Spanish were consulted. Firsthand accounts from persons who have led age-friendly initiatives are included as well. The initiatives were selected from those with which the authors are most familiar, are in languages they know, and/or for which program materials are accessible. Several lines of enquiry have guided the exploration.

Table 2.1 Age-friendly cities and communities initiatives reviewed

Country	Location	Jurisdiction
Australia	South Australia	State
Brazil	Sao Paulo	State
Canada	Canada	National
	Québec	Provincial
	Ontario	Provincial
	Montreal	Municipal
	Ottawa	Municipal
	Toronto	Municipal
France	Lyon	Municipal
Northern Ireland	Belfast	Municipal
Republic of Ireland	Ireland	National
	Louth	County
	Kilkenny	County
	Fingal	County
Spain	Basque Autonomous Region	State
USA	New York City	Municipal
	Portland, Oregon	Municipal

- How have the eight domains stood up to critical examination and application in different contexts?
- What methods have been used to assess a community's baseline age-friendliness, and what role has the checklist played?
- What population groups have been included in community assessments and have they represented the diversity of older adults, including those from low-income groups?
- What has been the role of older persons in the age-friendly cities' process?

In assessing these issues, the review also considers critical analyses of the WHO methodology that has appeared in the gerontological literature.

2.2 Are the Eight Domains of Age-Friendliness Adequate?

Gerontologists using the Vancouver Protocol have sought to determine the extent to which the eight domains are supported by empirical research on the environmental and social factors associated with well-being in older age. With some slight divergences, these reviewers have confirmed the consistency of the domains with gerontological research (Liddle et al. 2014; Menec et al. 2011; The New York Academy of Medicine n.d.). A recent study by Menec and Nowicki (2014) also provides empirical evidence linking six of the original domains to life-satisfaction outcomes among older persons in rural communities: physical environment (outdoor spaces and buildings), transportation, opportunities for participation (which includes social as well as civic participation and employment), communication and information, and community support and health services.

The area of *respect and social inclusion* has been the subject of criticism by Menec et al. (2011), who regard it as a basic value of the initiative rather than as a separate domain. Instead, these authors propose that variables of economic inequality and social disorder that lead to social exclusion be part of a new domain of *social environment*. While they agree that respect and social inclusion is an underlying value orientation, Liddle et al. (2014) accept it as a distinct area in view of the importance of identifying and addressing ageism and the exclusion of older persons from community decision processes. The literature review for the Age-Friendly New York program (n.d.) also supports social inclusion as an age-friendly feature and further documents the research evidence supporting *Respect*, understood as the absence of age discrimination.

The Vancouver Protocol did not give adequate coverage of social networks and social support and of protection from harm (physical safety and crime protection). The domain of Social participation considered activities, but did not specifically include unstructured contact with family, friends, and neighbors. Menec et al. (2011), the New York Academy of Medicine (n.d.) and the Basque Regional Government/Matia Foundation age-friendly municipalities assessment questionnaire (Barrios and Sanchez 2013) highlight the importance of having regular contact with family and friends as an age-friendly community feature, apart from participation in formal social activities. Access to informal support from family and friends was not considered in the domain of Community Support and Health Services, but was signaled for inclusion by Menec et al. (2011) and added to municipal assessments by Barrios and Sanchez (2013) in the Basque County and by the City of Lyon (2011).

Harm protection is featured in items within the WHO domains of outdoor spaces and buildings, transportation and housing. However, a few municipal and state projects have identified this as a distinct domain, notably the cities of Toronto (2013) and Montreal (2013) and County Fingal (n.d.). The Basque Region program examined crime security only (Barrios and Sanchez 2013).

Implementing the age-friendly city initiative has shown that some features are less relevant for older persons in less developed countries. One example relates to the availability of housing options and to community support and health services. In the municipality of São Paulo, experts responsible for conducting the age-friendly city assessment told the authors that questions about housing options were not understood in the consultations. In Brazil, the vast majority of older persons have no choice but to remain in their current dwellings or live with family because adapted, assisted-living, or congregate dwellings in both the public and the private sector are extremely scarce. They could not criticize the lack of possibilities that they did not know could exist. Similarly, because support and care of older persons is by law the responsibility of the family in Brazil (Presidência da República 2003), publicly funded support services are virtually nonexistent and private services are limited to the privileged minority. Thus, for older Brazilians, the domain of

Community support and health services is understood only in terms of services that are publically available, i.e., primary health care and hospital care. These observations underscore the requirement to complement older persons' perceptions with a community profile that documents the existing services and the views of experts and informed advocates and with an inventory of services and policies to fully assess the community's strengths and gaps.

Further modifications have been made to the WHO domains in various locations. One is to group Social Participation and Civic Participation and Employment into a single category, e.g., 'Opportunities for Participation' (Liddle et al. 2014; Menec et al. 2011), which highlights the dimension of participation itself, in all forms, as a vector for social integration and individual fulfillment. Alternatively, both the Province of Québec (Government of Québec 2013) and the City of Lyon (2011) distinguish Civic participation from other forms of participation to place particular value on the engagement of older persons as citizens and to their contribution as volunteers to formal and informal community support. County Kilkenny (n.d.) identified employment and income in order to capture needs for financial security within the scope of community services (e.g., financial literacy, banking services, and protection against fraud).

Overall, the original WHO dimensions of age-friendliness have been supported in their application in diverse locations and by gerontological research, with some additions and refinements along the way. To the question of how many dimensions are needed to assess age-friendliness, Menec et al. (2011, p. 482) reply pragmatically that "including more domains can emphasize aspects of the community environment that otherwise might not be considered". Empirical association between age-friendly community domains and the well-being of older persons, and, eventually, outcome evaluations of age-friendly community interventions in various domains may provide more conclusive answers. At the leading edge of this research is the study by Menec and Nowicki (2014) in 29 Canadian rural communities that showed that seniors' life-satisfaction was positively associated with their assessments of six dimensions of their community's age-friendliness (physical environment, social environment, opportunities for participation, informal and community supports and health care services, transportation and communication, and information) while associations with seniors' self-perceived health were found for four of the above dimensions.

Rather than focusing on which domains are the 'right' ones, Menec et al. (2011) and Liddle et al. (2014) caution us against using a model with fixed and discrete features. These authors note the interdependence among the features and the need to regard them as a dynamic whole. The interactive and dynamic nature of the domains was noted frequently in the consultation reports upon which the *Guide* is based. In distilling the 'essential' features of each domain to create the checklist, however, this interdependence was attenuated. Similar quantitative questionnaire approaches to assessing age-friendliness with discrete items for each domain may fail to capture the interrelationships that show what policy areas need to be considered jointly for effective action. Combining quantitative and qualitative methods is more effective to get both the breadth of views about the city's various features, as well as a more holistic perspective that captures their interrelationships. The WHO *Age-Friendly Cities Guide* (2007a) emphasizes the strong interconnections among the different domains, referring to them as "integrated and mutually enhancing" (p. 73) and calls for coordinated policy action, for instance, considering housing policy in connection with policies for the built environment and for opportunities for participation. Accordingly, intersectoral representation on steering or planning groups figures prominently in the actual or recommended implementation of age-friendly initiatives, for instance: Ireland (n.d.), Québec (2013), Sao Paulo (2013), and South Australia (2012a). In Sao Paulo, the decree establishing Age-Friendly Sao Paulo (AFSP) created a committee with representatives from all State Secretariats (ministries) and an executive secretariat based at the Secretary of Social Development assisted by Alexandre Kalache, as the senior consultant. The decree also instructed each of the State Secretaries to launch an 'emblematic' project, which would signal the Governor's resolve and commitment to ASFP, further ensuring the indispensable intersectoral nature of the program.

2.3 International Approaches to Assessing Age-Friendliness

The checklist of age-friendly features has led to the categorization of the WHO age-friendly city model presented in the *Guide* as being a 'top-down' approach to planning, directed by local authorities to achieve preestablished criteria (Barusch 2013; Lui et al. 2009). The checklist has received criticism on a few counts. The first is that it fails to recognize the diversity of older persons and of communities and the ongoing changes in living environments. A second, and somewhat contradictory charge, is that the checklist is unrealistically broad in the range of needs identified. And the third is that it subverts citizen engagement by older persons.

Menec et al. (2011) and Keating et al. (2013) contend that what makes a community age-friendly is having a good fit between the older person and his/her living environment, not conformity with a standard and fixed set of features. The fact that there are many diverse groups of older persons with different needs implies that there are correspondingly different 'age-friendly' environments with matching resources. Moreover, places evolve over time; a city that is age-friendly at one time may become unfriendly at another; thus, becoming or remaining age-friendly is an ongoing process.

Golant, on the other hand, makes the case that the age-friendly cities initiative, as reflected in the checklist, tries to reach too many groups with diverse needs in the community. Arguing that public resources are too limited to meet all needs, this author advises that "communities should rely on carefully tuned diagnostic and evaluative methods to prioritize the implementation of their programs" (Golant 2014, p. 11).

Starting from the perspective that an age-friendly city is one where older persons enjoy 'urban citizenship', i.e., the right to benefit from all the resources offered in the city and to participate in decisions about these resources, Phillipson (2010), Buffel et al. (2012), and Liddle et al. (2014) argue against the imposition of an arbitrary checklist of an 'ideal city' by authorities. Rather, using a "strictly bottom-up" approach, older persons themselves should be the main protagonists in defining the "actual opportunities and constraints in cities for maintaining quality of life as people age" (Buffel et al. 2012, p. 601).

In practice, different diagnostic approaches have been adopted or recommended. Approaches that rely entirely on local consultations referring to the WHO domains, but not specifically to the checklist are common. This is the case for Québec's *Municipalités-amies des aînés* (2013), the Pan-Canadian Age-Friendly Communities (Public Health Agency of Canada n.d.), and Ireland's Age-Friendly Counties (n.d.). The municipal age-friendly action plans developed by the cities of Toronto (2013) and Montreal (2013) reflect the issues identified by internal analysis and in public consultations, not the WHO checklist. A number of sites have used the checklist items in a survey questionnaire to complement qualitative data from forums or focus groups. This is evident, for instance, in the assessment materials developed for the City of Belfast (McClean 2013) and for the Basque Autonomous Region (Barrios and Sanchez 2013; Basque Government and Matia Foundation n.d.).

The Province of Ontario's *Age-friendly community planning guide* (2014) aims to marry local person-environment fit with a tailored selection of various evidence-based community assessment guidelines, including the WHO checklist. The Guide recommends that age-friendly stakeholders first consult widely within the community, via meetings, interviews, and focus groups, to identify a collective age-friendly vision and goals for the community and to choose priorities from and within the WHO age-friendly domains. Having identified the unique realities and aspirations of the community, stakeholders create a customized needs-assessment survey, drawing from existing age-friendly community and quality of life questionnaires, to address the issues of interest in the priority domains. The tailored instrument should contain questions that address individual needs (the 'person') from quality of life questionnaires with corresponding questions touching on community resources ('the environment') from the age-friendly survey instruments. The WHO checklist is one among the age-friendly questionnaires within the inventory of instruments suggested. Some, all, or none of the checklist items may be selected, depending on local priorities. With a wide distribution of the instrument in the community, the custom questionnaire data can be disaggregated at the neighborhood level to obtain even more fine-grained information about baseline age-friendliness. This guide is a new and innovative approach to harness standard instruments to respond flexibly and strategically to particular circumstances. Repeated use of this approach over time and in many communities will shed light on both the similarities and differences between communities with respect to person-environment fit, possibly contributing to community typologies (Keating et al. 2013). It will also reveal what instruments and items are the most useful in assessing age-friendliness. Finally, it could respond to requirements of communities to target age-friendly strategies to distinct groups within the older population.

The states of Sao Paulo and of South Australia have each developed programs that are modeled on the WHO checklist to achieve state-wide policy objectives, but which also rely on 'bottom-up' engagement of older persons, although in different ways.

The Age-Friendly Sao Paulo program was established through a process spanning four years until it was officially launched as a state program in 2012. The age-friendly initiatives that preceded its launch were aimed at creating a critical mass, with several municipalities adopting age-friendly approaches lead by a variety of partners: local municipal governments, academic institutions, civil society, and even private organizations. From 2008 to 2012, in close partnership with the Coordinator of the Program on health of older persons, State Secretary of Health, Marilia Louvison, Alexandre Kalache visited some 40 municipalities, making presentations, organizing seminars, and workshops or simply visiting municipal authorities with the objective of disseminating the principles behind the age-friendly cities approach, in particular, describing the WHO Active Aging Policy Framework. As a result, a number of initiatives were established at a local level, always espousing the binomial 'top-down/bottom-up' approach with government commitment to act based on views and opinions given by older people on their actual experience of growing older where they lived. A number of such local age-friendly initiatives can be found in the book edited by Rosa et al. (2013).

In a public event in November 2011, these various 'bottom-up' experiences were offered to the Governor as the foundations for a state-wide program. In May 2012, by an executive decree, the Age-Friendly State of Sao Paulo (AFSP) was launched at the Governor's Palace in the presence of over 1000 individuals, most of them older persons composing delegations representing a wide range of municipalities. The committee charged with developing the State program was chaired by the President of the State Council of Older People, to underscore the centrality of older persons' voices in developing the State initiative.

On one hand, AFSP is a highly structured, state-led initiative. It adopted a strong focus on the three pillars of the WHO Active Aging Policy Framework (2002): health; participation and security to which a fourth pillar of life-long learning has been added (Kalache 2013). It was also structured to achieve the policy obligations in the federal *Statute for Older Persons* (2003) and the *National Health Policy for Older Persons* (Ministério da Saúde 2006). The *Statute for Older Persons* affirms the rights of older persons in Brazilian society and obliges public authorities at federal, state, and municipal levels to provide a wide range of services and benefits in the areas of Health, Transportation, Housing, Labor, Social Welfare, Income Support, and Education/Sport/Recreation/Culture. The *National Health Policy for Older Persons* (2006) further mandates specific programs for older persons in disease and injury prevention and health promotion, as well as for enhancement of professional education in gerontology and geriatrics. This federal legislative framework is encompassing, ambitious, and legally binding, but implementation has been uneven and fragmentary (Kalache 2012). Thus, the implementation of the

AFSP program was an opportunity for government to fully implement these policies at the State level. On the other hand, AFSP also adopted a highly participatory approach by giving a central role to the Councils of Older Persons (*Conselhos dos Idosos*). These councils are required by federal law and are expected to be active at all government levels: federal, state, and municipal. Their main function is to oversee the implementation of the Statute within their jurisdiction and to ensure an active voice of older persons in policy development: according to the dictum "nothing about us without us."

The Age-Friendly Sao Paulo Program (2013) identifies 40 policy actions for participating municipalities, all of which are aligned with the eight WHO domains of age-friendliness and checklist. Ten of them are mandatory and must be achieved in the first year of the program. The first required action is to establish and/or strengthen the Council of Older Persons, again reinforcing the critical importance of older persons' voices in shaping age-friendly actions. In the two following years, municipalities choose four from among 30 elective policy measures, based upon local consultations with older persons and with public sector managers. To maintain full status in the program in subsequent years, municipalities must measure implementation of actions taken, renew the cycle of consultations and select further elective actions. The protagonist role of older persons through the Councils and consultation groups is again a key feature of this evaluation process.

The WHO checklist has been used to develop an 80-item questionnaire, with 10 items for each age-friendly domain to assess how and where the Statute for Older Persons is achieved or lacking. Each question is discussed fully in workshops with representatives of older persons' organizations to collectively determine a score for the item on a three-point scale (yes, partly, no). The pattern of strengths and weaknesses across the domains become the 'bottom-up' input of older citizens in the selection of municipal priorities within the predetermined requirements of the *Statute*.

In South Australia, the state-wide program was designed to develop in the opposite direction: first, the production by the government of a set of guideline documents based on the WHO checklist, and then extensive consultations with older persons and their organizations at the municipal level to decide on local priorities. *South Australia's Communities for All: Our Age-Friendly Future* (2012b) consists of three age-friendly guidelines booklets which apply the WHO checklist to orient action at the level of the state, local government, and residential development projects, respectively. The booklets map existing guidelines, standards, plans, and requirements against the checklist in the eight domains. For each domain, the booklets support implementation of the checklist items, which are identified as the 'WHO criteria', by providing corresponding guidelines, good practices, and a list of resources and technical references. Within the age-friendly guidelines as the range of potential actions, local governments have the opportunity to establish local priorities by conducting a self-assessment. Consultation with internal and external stakeholders, and with the community at large, using the WHO Vancouver Protocol was the next step. Council on the Aging (COTA), the national organization present in all states that represents older peoples' interests in policy, actively participated in the consultations that were held. However, not long after the release of the program, there was a change in government, and the state support for the community consultations on the guidelines reports has waned (for further developments on the Australian case, consult the chapter of Brasher and Winterton).

Both Sao Paulo (2013) and South Australia (2012b) made explicit use of the WHO checklist as a vehicle for identifying areas for action. In Sao Paulo, legislative requirements were the key drivers for adopting the checklist as the model, while in South Australia, the checklist served as an anchor and stimulus to advance current policy directions and existing standards of good practice. In both instances, however, the top-down action at the state level was balanced and complemented by processes (at least initially) to ensure the active engagement of older persons in local decision-making.

2.4 Capturing Diversity: Whose Voices Are Heard in the Assessment of Age-Friendliness?

Contrary to the critiques that the WHO initiative fails to acknowledge the need for a person-environment fit (Keating et al. 2013; Liddle et al. 2014; Menec et al. 2011), the age-friendly assessment method developed by WHO in the Vancouver Protocol does recommend reflecting the diversity of older persons in the community in order to capture and respond to their heterogeneity. At a minimum, the Protocol advises that assessments should include older persons varying by age, by socio-economic status and by ability level, and ideally, assessments should be conducted at a neighborhood level, considering their diversity within a city. A line of enquiry therefore was the extent to which this advice has been followed in age-friendly initiatives undertaken in different locations.

Many local initiatives have focused on large public events. Some have focused on getting views from as many older persons as possible; others have included older as well as younger persons and stakeholder groups with shared interests, such as disability advocates. Web-based and paper-and-pencil consultations also have been made open to the general public. This approach to inclusion is to be expected, given that municipal governments are concerned with meeting a broad range of population needs at the same time and with getting support for initiatives from a critical mass.

Yet some cities have made extensive efforts to be inclusive of diversity, using different consultation approaches to reach out to older persons who are less visible to decision-makers. Efforts have been made as well to broaden the age-friendly consultations to include other generations and interest groups with potentially similar issues, in particular, disability groups.

Consultations for Age-Friendly New York City (2008) systematically assured diversity with respect to geography, race, culture and language, socio-economic level, and disability as well as LGBT. Town hall meetings have been conducted in different community centers and in different languages to draw in people reflecting this diversity. Immigrants were interviewed individually as required. Focus groups were held with marginalized persons, such as persons who lived in poverty, who were previously homeless or HIV positive. Age-Friendly Portland (Orca Planning 2013) also used a range of approaches to reach a similarly diverse older urban population. The City of Ottawa (2012) used focus groups to capture the diversity of voices, including First Nations, and persons with developmental intellectual disabilities. The City of Toronto (2013) translated its age-friendly city consultation workbook into 11 languages to reach a highly multi-cultural population. Within the City of Sao Paulo, the diversity of neighborhoods (*bairros*) was respected with specific consultations in, for example, Vila Clementina (Costa Rosa et al. 2013), Mooca (Casado 2013), and the Zona Norte (Uehara et al. 2013). The City of Lyon (2011) organized focus groups with older persons in all neighborhoods, and then classified older persons into distinct groups based on expressed needs and profiles. In Belfast (McClean 2013), peer facilitators held focus groups to reach persons at the risk of social exclusion, in particular, blacks and minorities, isolated and less mobile persons in sheltered accommodation, older caregivers and older men. In this city as well, disability groups included persons who were deaf or blind. In Québec (2013) the provincial program guide specifically recommends municipal groups to ensure the participation of vulnerable persons by having them well-represented in community consultations. The Pan-Canadian Guide (Public Health Agency of Canada and Mount St Vincent University 2013) recommends that different groups in the population be represented on the project advisory committee that steers the local initiatives. These examples show that for many age-friendly programs, identifying the diversity of needs in the community is important.

Does inclusion of diversity in consultations result in action plans that respond to different needs? That is, do age-friendly initiatives attempt to provide urban resources tailored to varying needs? From the initiatives reviewed here, five cities do address diversity issues. *The Toronto seniors' strategy* (2013) adopts a general diversity lens to municipal planning rather than formulating actions targeted to specific needs. For example, commitments include ensuring that: "diverse older adults are fully involved in the development of programming"… (p. 45); "reduce financial barriers to its programming for older adults" (p. 52); and "training City staff to effectively serve all older adults including those who are vulnerable or have special needs" (p. 34). In *Age-Friendly NYC: Enhancing our city's livability for older New Yorkers* (The New York Academy of Medicine 2009) the city articulated actions targeted to the needs of specific vulnerable groups, such as improving processes for creating low-income housing and providing taxi vouchers for people unable to use public transportation. Belfast (McClean 2013) was unique in that it mentioned having compared the priorities among the WHO age-friendly domains selected by at-risk older persons with those chosen in general consultation events. The priorities were similar for both groups. Several of the city's actions focus on reducing social exclusion and disadvantage in concrete ways. In its 2012 *Older Adult Plan*, the City of Ottawa systematically identifies "target groups to benefit" from the age-friendly actions (City of Ottawa 2012). The last example is the *Lyon, Ville amie des aînés* report (2011) where the recommended actions in each domain respond to the needs of different group profiles identified.

A dissonant voice on the matter is Golant (2014), who cautions against over-ambitious age-friendly agendas which try to meet the needs of all groups in the community. Claiming this shotgun approach to be unrealistic and at times redundant with other similar urban initiatives, he advises targeting age-friendly initiatives to reach the large group of "modest or moderate income" older persons who are not so well-off that they can purchase the resources to meet their own needs nor who are so poor that their basic needs are met by a public social safety net. To date, no city has yet adopted this focus for its age-friendly city initiative. If it does occur, it will be interesting to compare the actions taken, outcomes and sustainability of the program with those of communities who target diversity.

The potential of age-friendly cities to reduce health and social inequalities at the local level has been highlighted by Kendig and Phillipson (2014) and Scharf (2013). The explicit inclusion of older persons from low-income neighborhoods in the WHO Vancouver Protocol was intended to promote this goal and several items in the checklist deal with economic barriers. Initiatives in some places have targeted income inequality, as illustrated in the examples above. In Sao Paulo, many of the actions created by the State for municipalities target low-income older persons, both because they constitute the majority of the population, and because they are the principal beneficiaries of public policy in many areas, e.g., housing, public transportation, community support, and health services.

In Europe, Scharf (2013) notes that there is little evidence of use of the age-friendly cities initiative as a community development tool in disadvantaged communities. Golant (2014) makes the same observation for North America and suggests that a lack of community leadership in disadvantaged communities is a key explanation. The importance of having leadership for successful age-friendly community development is supported by the implementation evaluations conducted in the provinces of British Columbia (Gallagher and Mallhi 2010) and Manitoba (Menec et al. 2014). While Golant (2014) and others, including the present authors, advocate for public investment in age-friendly development in disadvantaged communities, it is difficult to imagine that efforts would be sustainable or successful without including leadership development as a significant component of an implementation program.

2.5 How Does the Age-Friendly City Assessment Process Engage Older Persons?

Critics of the WHO Age-friendly cities initiative have characterized it as a 'top-down' model of intervention in which change is directed by government authorities and experts, with older people involved only in providing information to decision-makers (Barusch 2013; Buffel et al. 2012; Lui et al. 2009). Buffel et al. (2012) in particular, strongly advocate for an approach that promotes the active participation of older persons:

> Involving older people in the development and maintenance of age-friendly environments represents a crucial goal for social policy. Achieving this {…} will require a radical shift from producing urban environments *for* people to developing neighborhoods *with* and *by* older people (Buffel et al. 2012, p. 609).

For these authors, as for Liddle et al. (2014), participation of older persons in decisions about the community in which they live is inherent to the goal of enhancing social inclusion, which defines age-friendliness.

Possibly because the WHO initiative is summarily seen as a 'top-down' checklist, criticism of the WHO Guide (2007a) has overlooked its emphatic call for the active engagement of older persons in age-friendly community development:

> The same principle followed in creating the Guide applies to using it: that is: involve older people as full partners at all stages. {…} They will provide suggestions for change and they may participate in implementing improvement projects. {…} In the follow-up stages of 'age-friendly' local action, it is imperative that older people continue to be involved in monitoring the city's progress and acting as age-friendly city advocates and advisers (World Health Organization 2007a, p. 11).

The WHO Global Network has maintained this original position by requiring that groups wishing to join the Network must establish mechanisms to involve older persons through all phases of the initiative. The WHO 'model' was created as, and remains, a 'top-down' and 'bottom-up' approach, requiring the joint engagement of local decision-makers with older persons.

In practice, initiatives that have been undertaken vary considerably in their involvement of older persons. The Sao Paulo State program began with strong participation from older persons in voicing their issues and with the intention to involve them in making decisions and monitoring government action through the Councils of Older Persons. However, their involvement has been made ineffective by government control of the Council of Older Persons at the state level, and inevitably at municipal levels as well. This situation has arisen because, within the Brazilian system, presidency of the Older Persons Councils alternates every two years between the government and the civil society sectors. Unfortunately, this has led to a slowing of the program's implementation and is effectively denying older persons their right to have the critical role in monitoring it. In South Australia, following a change in government, the process of consultations with older persons in age-friendly decision-making has been eclipsed. In five initiatives in Europe and North America, involving older persons as full partners respects the principles of social inclusion and participation, which are keys to age-friendliness. The Louth Age-Friendly County Strategy Document (n.d.) announces that it is "A plan developed with older people, not for them". The Québec Guide (2013), developed with the ongoing experience of over 600 communities engaged in the initiative, advises that one of the winning conditions of the initiative is the active participation of older persons in every stage of the process and the contribution of the entire community. Age-Friendly New York City's (The New York Academy of Medicine 2012) program to create Aging Improvement Districts (that is, age-friendly neighborhoods) mentions that "leadership by older people" is one of the four things a community needs to succeed. In all these initiatives, participation of older persons on the local advisory committee is a requirement. Describing the activities undertaken in each municipality, the Basque County initiative (n.d.) shows how older persons' organizations have been involved in publicizing the initiative, garnering public support and generating wide participation in the age-friendly assessment process, and in implementing proposed age-friendly actions, and in monitoring projects undertaken.

2.6 Discussion

With the publication of the Age-Friendly Cities Guide (2007a), WHO advanced an initial model of an age-friendly city, based on the Vancouver Protocol and the checklist of frequently mentioned features in the eight areas of the Protocol (2007b). The Guide (2007a) strongly recommends the involvement of older persons in all phases of action and the adoption of an intersectoral approach to cities wishing to become more age-friendly. Since then, the WHO initiative has been implemented in varying social contexts in many locations. The methodology has undergone adaptations, and the eight original domains have been edited, adding some elements that were not sufficiently present, in particular informal social networks and social support and safety and security concerns.

The initiative has received criticism as well. One issue deals with recognizing and addressing the dynamic diversity of communities and of older persons. In this context, it is preferable to view the WHO checklist of age-friendly features as an inventory of commonly identified characteristics, a starting point for cities which may be enriched, pared down or modified, depending on the particular context in which it is used. Use and adaptation of the checklist in local questionnaires, sometimes in conjunction with other instruments, shows that it is a practical tool used flexibly by communities, but not a norm. Similarly, a flexible approach has been taken to the WHO domains, inviting people in consultations to choose their priorities from among the eight areas rather than 'imposing' all of them. To find the right fit between environments and persons within age-friendly initiatives has led to the inclusion of diverse population groups regarding their needs. Some cities have developed action plans that explicitly target diversity as well as mainstream needs, or have introduced neighborhood-level consultations and initiatives.

The initiative has been characterized as well as an arbitrary model, a checklist, with no expectation of real engagement by citizens, and especially, of older persons. As this review amply demonstrates, this view is inexact. Engagement of local older persons in age-friendly assessment and action, as recommended by the WHO Guide (2007a) and the current WHO Network of Age-Friendly Cities and Communities, is present in many programs, although the type and extent of engagement vary.

With growing practical experience in various settings, critical reflection and emerging evaluation research evidence, the WHO Age-friendly Cities and Communities initiative has perhaps reached a level of maturity that would justify a revision of the original approach by the WHO Global Network. The areas for community assessment could be modified, and different instruments proposed for conducting the assessments. Clearer guidance regarding the 2007 checklist of Age-Friendly Features could encourage its use by communities as a suggested 'starter' list based on the results of the original WHO consultation research, but not a replacement for a thorough assessment and prioritization of the needs of older persons in individual communities.

2.7 Conclusion

Buffel et al. (2012) identified three issues requiring particular attention in the further development of age-friendly communities: the need to recognize the diversity of cities and their inhabitants; the requirement for a life-course perspective that takes into account needs of residents of all ages; the need to actively involve older persons in community planning and development actions. As shown here, these issues are being addressed in practical ways in various local initiatives on different continents. These issues continue to be relevant as the WHO Global Network of Age-Friendly Cities and Communities spreads. They are keys as well to the sustainability of the program—to favor 'mainstreaming' it in policy and planning and to maintain a critical mass of support within the population that plays heavily in political decisions. The sustainability of the Age-Friendly Cities initiative may be strengthened as well if it contributes to the policy priority of reducing social inequalities within living environments.

References

Barrios, E., & Sanchez, M. (2013). *Cuestionario amigabilidade*. Euskadi lagunkoia: Matia Foundation.

Barusch, A. S. (2013). Age-friendly cities: A social work perspective. *Journal of Gerontological Social Work, 56*(6), 465–472.

Basque Government and Matia Foundation. (n.d.). *Euskadi lagunkoia. Desarollo en municipios.* Unpublished presentation. Personal communication from the Matia Foundation.

Buffel, T., Phillipson, C., & Scharf, T. (2012). Ageing in urban environments: Developing "age-friendly"cities. *Critical Social Policy, 32*(4), 597–617.

Casado, R. (2013). Subprefeitura da Mooca amiga dos idosos:ações com base no envelhecimento ativo. In T. E. D. C. Rosa, Á. E. S. Barroso & M. C. P. Louvison (Eds.), *Velhices: experiências e desafios nas políticas do envelhecimento ativo* (pp. 149–160). Sao Paulo: Instituto de Saúde.

City of Lyon. (2011). *Lyon, Ville amie des aînés* [Online]. Available: www.lyon.fr/page/solidarite/ville-amie-des-aines.html. Accessed January 14, 2013.

City of Montreal. (2013). *Plan d'action municipal pour les aînés 2013–2015* [Online]. Available: www.ville.montreal.qc.ca/…/Plan_action_municipal_aines. Accessed July 20, 2014.

City of Ottawa. (2012). *Older Adult Plan 2012–2014.* Ottawa [Online]. Available: http://ottawa.ca/cs/groups/content/@webottawa/documents/pdf/mdaw/mjg1/~edisp/cap290801.pdf. Accessed July 20, 2014.

City of Toronto. (2013). *The Toronto seniors strategy: Towards an age friendly city* [Online]. Available: http://joshmatlow.ca/ward-22/delivering-results/1268-toronto-seniors-strategy-creating-an-age-friendly-city.html. Accessed July 20, 2014.

Costa Rosa, T. E., et al. (2013). Implantação do bairro amigo do idoso da Vila Clementino do município de São Paulo/SP. In T. E. Da Costa, A. E. S. Barroso & M. C. Prado Louvison (Eds.), *Velhice: experiência e desafios nas políticas do envelhecimento ativo* (pp. 137–148). Sao Paulo: Instituto de Saúde.

County of Fingal. (n.d.). *Fingal age-friendly county* [Online]. Available: www.agefriendlycounties.com/resource_library/afc_strategies. Accessed July 20, 2014.

County of Kilkenny. (n.d.). *Kilkenny age-friendly county* [Online]. Available: www.agefriendlycounties.com/resource_library/afc_strategies. Accessed July 20, 2014.

County of Louth. (n.d.). *Louth age-friendly county* [Online]. Available: www.agefriendlycounties.com/resource_library/afc_strategies. Accessed July 20, 2014.

Gallagher, E., & Mallhi, A. (2010). *Age-friendly British Columbia. Lessons learned from October 1, 2007–September 30, 2010. Victoria, British Columbia.* Report submitted to the Seniors Healthy Living Secretariat, British Columbia Ministry of Health Services.

Golant, S. M. (2014). Age-friendly communities: Are we expecting too much? *IRPP Insight, 2*(5).

Government of Ireland. (n.d.). *Ireland's age friendly cities and counties programme* [Online]. Available: www.agefriendlycounties.com/. Accessed May 16, 2014.

Government of Ontario. (2014). *Finding the right fit. Age-friendly community planning* [Online]. Available: www.seniors.gov.on.ca/en/resources/AFCP_Eng.pdf. Accessed May 16, 2014.

Government of Québec. (2013). *Programme municipalités amies des aînés* (2nd ed) [Online]. Available: www.aines.gouv.qc.ca/mada/12-830-21W_BR.pdf. Accessed May 16, 2014.

Government of Sao Paulo. (2013). *Como se tornar um município amigo do idoso: Programa São Paulo Amigo do Idoso* [Online]. Available: www.desenvolvimentosocial.sp.gov.br/a2sitebox/arquivos/documentos/518.pdf. Accessed May 16, 2014.

Government of South Australia. (2012a). *Age-friendly neighbourhoods guidelines and toolkit for local government* [Online]. Available: http://www.sahealth.sa.gov.au/wps/wcm/connect/public+content/sa+health+internet/resources/age-friendly+neighbourhoods+guidelines+and+toolkit+for+local+government+2012. Accessed May 16, 2014.

Government of South Australia. (2012b). *South Australia's communities for all: Our age-friendly future.*

Kalache, A. (2012). Prefácio. In M. V. Berzins & M. C. Borges (Eds.), *Políticas públicas para um país que envelhece* (pp. 5–8). Sao Paulo: Martinari.

Kalache, A. (2013). *The longevity revolution: Creating a society for all ages. Adeleide thinker in residence 2012–2013.* Adeleide: Government of South Australia.

Keating, N., Eales, J., & Phillips, J. E. (2013). Age-friendly rural communities: Conceptualizing "best-fit". *Canadian Journal on Aging/La Revue Canadienne Du Vieillissement, 32*(4), 319–332.

Kendig, H., & Phillipson, C. (2014). Building age-friendly communities: New approaches to challenging health and social inequalities. In *If you could do one thing... Nine local actions to reduce health inequalities* (pp. 102–111). British Academy: London.

Liddle, J., Scharf, T., Bartlam, B., Bernard, M., & Sim, J. (2014). Exploring the age-friendliness of purpose-built retirement communities: Evidence from England. *Ageing and Society, 34*(9), 1601–1629.

Lui, C., Everingham, J., Warburton, J., Cuthill, M., & Bartlett, H. (2009). What makes a community age-friendly: A review of international literature. *Australasian Journal on Ageing, 28*(3), 116–121.

McClean, M. (2013). *Age-friendly Belfast. Additional consultation with lesser heard older people.* Unpublished report prepared by AgeNI for Belfast City Council and Healthy Aging Strategic Partnerships.

Menec, V. H., Means, R., Keating, N. C., Parkhurst, G., & Eales, J. (2011). Conceptualizing age-friendly communities. *Canadian Journal on Aging, 30*(3), 479–493.

Menec, V. H., Novek, S., Veselyuk, D., & McArthur, J. (2014). Lessons learned from a Canadian province-wide age-friendly initiative: The Age-friendly Manitoba initiative. *Journal of Aging and Social Policy, 26*(1–2), 33–51.

Menec, V. H., & Nowicki, S. (2014). Examining the relationship between communities 'age-friendliness' and life satisfaction and self-perceived health in rural Manitoba, Canada. *Rural and Remote Health, 14*(2594).

Ministério da Saúde. (2006). *PORTARIA Nº 2.528 DE 19 DE OUTUBRO DE 2006.* Política Nacional de Saúde da Pessoa Idosa. Brasília: Ministério da Saúde [Online]. Available: http://portal.saude.gov.br/portal/arquivos/pdf/2528aprova%20a%20politica%20nacional%20de%20saude%20da%20pessoa%20idosa.pdf. Accessed May 16, 2014.

Orca Planning. (2013). *Toward an age-friendly Portland.* Portland [Online]. Available: https://www.portlandoregon.gov/bps/article/425455. Accessed May 16, 2014.

Phillipson, C. (2010). Globalisation, global ageing and intergenerational change. In M. Izuhara (Ed.), *Ageing and intergenerational relations. Family reciprocity from a global perspective* (pp. 13–28). The Policy Press: Bristol.

Plouffe, L., Garon, S., Brownoff, J., Eve, D., Foucault, M.-L., Lawrence, R., Lessard-Baupré, J.-P., & Toews, V. (2012). Advancing age-friendly communities in Canada. *Canadian Review of Social Policy/Revue Canadienne de Politique Sociale*, 2(68–69). [Online]. http://pi.library.yorku.ca/ojs/index.php/crsp/article/view/34389. Accessed May 16, 2014.

Plouffe, L., & Kalache, A. (2010). Towards global age-friendly cities: Determining urban features that promote active aging. *Journal of Urban Health, 87*(5), 733–739.

Presidência da República. (2003). *Estatuto do Idoso, Lei Nº 10.741, de 1º de outubro de 2003.* In S. para & A. J. Casa Civil (Eds.), Brasília/DF: Presidência de República [Online]. Available: http://www.planalto.gov.br/ccivil_03/leis/2003/l10.741.htm. Accessed May 16, 2014.

Public Health Agency of Canada. (n.d.). *Age-friendly communities* [Online]. Available: http://www.phac-aspc.gc.ca/seniors-aines/afc-caa-eng.php. Accessed May 16, 2014.

Public Health Agency of Canada and Mount St Vincent University. (2013). *Age friendly communities in Canada: Community implementation guide and toolbox* [Online]. Available: http://www.phac-aspc.gc.ca/seniors-aines/publications/public/afc-caa/guide/index-eng.php. Accessed May 16, 2014.

Rosa, T. E. D. C., Barroso, Á. E. S., & Prado Louvison, M. C. (Eds.). (2013). *Velhices: experiências e desafios nas políticas do envelhecimento ativo.* Sao Paulo: Instituto de Saúde.

Scharf, T. (2013). *Unequal ageing and age-friendly communities.* Paper presented at the Presidential symposium: international Perspectives on age-friendly communities. Gerontological Society of America Annual Scientific Meeting, Nov, 23.

The New York Academy of Medicine. (n.d.). *Respect and Social Inclusion: How this domain relates to active ageing.* New York: The New York Academy of Medicine.

The New York Academy of Medicine. (2008). *Toward an age-friendly New York City: A findings report.* New York: The New York Academy of Medicine [Online]. Available: http://www.health.ny.gov/diseases/aids/providers/conferences/docs/nyam_age_friendly_report.pdf. Accessed May 16, 2014.

The New York Academy of Medicine. (2009). *Age Friendly NYC: Enhancing our city's livability for older New Yorkers.* New York: The New York Academy of Medicine [Online]. Available: http://www.nyc.gov/html/dfta/downloads/pdf/age_friendly/agefriendlynyc.pdf. Accessed May 16, 2014.

The New York Academy of Medicine. (2012). *Creating an age-friendly NYC: One neighborhood at a time.* New York: The New York Academy of Medicine [Online]. Available: http://www.nyam.org/agefriendlynyc/docs/Toolkit_Report_0321-VA-new.pdf. Accessed May 16, 2014.

Uehara, C. A., Nicoletti Braga, D., Magalhães, A. C., Brumini, C., & Miguel, D. F. (2013). *O processo de implantação do bairro amigo do idoso na Zona Norte – São Paulo/SP.* In M. C. Rosa, T. E. D. C., Barroso, Á. E. S. & Prado Louvison (Eds.), *Velhices: experiências e desafios nas políticas do envelhecimento ativo* (pp. 161–168). Sao Paulo: Instituto de Saúde.

World Health Organization, Personal communication. (2014).

World Health Organization. (n.d.). *Application form to join the WHO global network of age-friendly cities and communities* [Online]. Available: http://www.who.int/ageing/application_form/en/. Accessed May 13, 2014.
World Health Organization. (2002). *Active ageing: A policy framework*. Geneva: WHO.
World Health Organization. (2007a). *Global age-friendly cities: A guide*. Geneva: WHO.
World Health Organization. (2007b). *WHO Age-friendly cities project methodology: Vancouver protocol*. Geneva: WHO.

Chapter 3
The WHO Global Network of Age-Friendly Cities and Communities: Origins, Developments and Challenges

Lisa Warth

3.1 Introduction

Thibauld Moulaert *Lisa Warth, you have coordinated the WHO Global Network of Age-friendly Cities and Communities since 2012. While Louise Plouffe, Alexandre Kalache and Ina Voelcker present a critical perspective on the original methodology of the WHO Age-friendly Cities Project in their chapter, we would like you to give us a comprehensive perspective on the network. Where it has come from and where is it heading…*

3.2 The Origins of the Network

Thibauld Moulaert *What motivated WHO to establish a network of age-friendly cities and communities?*

Lisa Warth The WHO Global Network of Age-friendly Cities and Communities (GNAFCC) was set up in 2010 to connect local governments that are using the WHO's age-friendly cities and communities approach in their engagement with older people, to identify areas for improvement and implement policies and measures to address these with the aim of making their city or community more age-friendly. The creation of GNAFCC was inspired by the age-friendly cities

The author is a staff member of the World Health Organization. The author alone is responsible for the views expressed in this publication and they do not necessarily represent the decisions or policies of the World Health Organization.

L. Warth (✉)
Department of Ageing and Life Course, World Health Organization, Geneva, Switzerland
e-mail: warthl@who.int

© Springer International Publishing Switzerland 2016

T. Moulaert and S. Garon (eds.), *Age-Friendly Cities and Communities in International Comparison*, International Perspectives on Aging 14,
DOI 10.1007/978-3-319-24031-2_3

project ten years ago that led to the publication of the Global Age-friendly Cities Guide in 2007 (WHO 2007). In this project, older people, care givers and service providers around the world had identified features of age-friendly cities. The Guide sparked off a lot of interest in the concept of an age-friendly city and there was an enthusiasm to use it to promote change at the local level.

TM *What are the conceptual foundations of the network?*

LW The Age-friendly Cities Guide operationalized the Active Ageing Policy Framework (WHO 2002) that had been published by the WHO a few years earlier and showed ways in which active and healthy ageing could be facilitated through concrete measures at the local level, adapting the structures and services to the diverse capacities and needs of older persons.[1] To this date, the Guide and its comprehensive framework for physically accessible and socially inclusive environments have remained the centrepiece of WHO's age-friendly cities approach. Building on the conceptual foundation provided by the United Nations (UN) principles for Older Persons, the social determinants of health, the Active Ageing Policy Framework and the Age-friendly cities and communities Guide, the WHO age-friendly cities and communities approach uses participatory age-friendly assessments to inform policy making at the local level. Local governments assess, plan, implement, monitor and evaluate their progress in a "continual cycle of improvement". The network connects the growing number of cities and communities that are using this approach to assess the age-friendliness of their cities and communities and to guide their policy planning and implementation.

TM *What does an age-friendly city/community look like?*

LW It is not a question of what an age-friendly city looks like, but what it enables older persons to do and how well it caters to their needs. Age-friendly cities and communities foster the health and well-being of people as they age. This necessitates barrier-free built environments as well as supportive and enabling services and attitudes. For example, the lack of affordable public transport isolates older people who can no longer drive and can make participation outside their immediate neighbourhood difficult. When transport is available and adapted to the needs of older persons, both in terms of scheduling and destinations, it enhances mobility and facilitates social participation and a sense of belonging in one's community.

An age-friendly city enables older persons to remain active and involved in all spheres of life. This requires dismantling existing barriers to this participation as well as proactively fostering older people's inclusion and involvement. Cultural offers and entertainment of interest for older people, opportunities for volunteering or civic engagement contribute to a fulfilling and enjoyable older age and continued meaningful engagement.

The provision of health care, social support and security is equally important. Health and social services that are coordinated and people-centred should promote

[1]Note from the editors: for further details on these origins, see the chapter with Alexandre Kalache.

good health and prevent or delay the onset of functional decline as far as possible. They should strive for recovery of function and—where this is not possible—compensate for its loss, so that people can continue to do the things that are important to them even when they start to experience sensory loss, reduced mobility, or cognitive impairments.

Both physical structures and services need to be designed to maximize the safety and security of all. This becomes particularly important to older persons experiencing functional decline that makes them more vulnerable to falls and abuse and increases their subjective concern for security. Security also applies to income and food security, which is particularly important for older people who are no longer able to work and ensure their own livelihood.

3.3 The Age-Friendly Cities and Communities Approach

TM *How can cities and communities start to work towards this ideal of an age-friendly city and put these ideas into practice?*

LW The needs and preferences of older people and the culture and the sphere of influence of local governments are highly context-specific: just as there is no one-size-fits-all solution to meeting the needs of older people, there is no one-size-fits-all recipe for creating age-friendly environments. The WHO age-friendly cities and communities approach, therefore recommends a flexible process that can help local actors to understand the needs of older persons and stakeholders in the community and identify ways in which these could be better met within the scope of their possibilities and within their locally specific context. It provides the key "ingredients", but remains flexible enough to be adapted to the local realities that local governments operate within. Those key ingredients are political commitment, a participatory approach involving older people and key stakeholders, coordinated action and collaboration across sectors, evidence-informed and results-oriented planning, as well as monitoring and evaluation of progress.

1. **Political commitment**

Leadership and political commitment are important for any policy agenda. It is, however, particularly important for policy initiatives such as the creation of age-friendly environments as these require a whole-of-government effort. Leadership at the highest level of government is therefore important. This does not, however, imply a purely top-down approach to policy making.

2. **A participatory approach**

In addition to being driven by political commitment, the process of creating age-friendly environments should also be participatory involving those intended to benefit (older persons and their families) as well as those needed to achieve success. The age-friendly cities and communities process recommends a highly participatory approach that engages not only older people in a meaningful way throughout the process, but also seeks alliances across government and with key stakeholders across all sectors of the society.

The participation of older people is both the goal of age-friendly environments and important in the process of creating them. A participatory approach helps to ensure that interventions meet the needs and expectations of citizens, but also to foster ownership. As key beneficiaries of age-friendly environments, they are best placed to identify the areas of needs and priorities for action, to monitor and to evaluate progress. With their firsthand experience, they present an invaluable source of expertise to decision-makers and involving them throughout all stages of research, planning, implementation, monitoring and evaluation is likely to result in measures that are better oriented at the needs of older people than services designed on their behalf without their involvement.

3. **Coordination and collaboration across sectors**

The Global Age-friendly Cities Guide highlighted that coordination and collaboration between all government sectors are key to creating age-friendly outcomes, in which accessible structures work seamlessly with accessible services that are oriented at the various needs, preferences and diverse capacities of older persons. A cross-cutting and coordinated approach across sectors is important to avoid "pockets of good practice" that are not joined up and therefore may not achieve the desired outcomes. For example, when senior housing is not connected to accessible public transport, older people may be able to "age in place", but not to continue to go out and about and engage with the community, increasing their risk of social isolation.

The age-friendly cities and communities approach often provides the opportunity for local policy makers to gather actors across government departments and sectors around one table for the first time to think jointly around the challenges and opportunities that an ageing population presents and how they can best address these in the city. In New York City, for example, the Office of the Mayor and the New York City Council asked all city departments to consider how they can improve the way they integrate and serve older adults through their work. This review led to 59 initiatives to improve the quality of life of older persons in the city.[2]

Age-friendly environments cannot be created by one group of actors alone. Collaboration with the private sector and civil society enables a more comprehensive initiative for change. In the Age-friendly Ireland initiative, for example, creating an age-friendly city or county alliance that will lead the programme is a requirement for all participating local governments.[3] The city or county alliances are chaired by the Chief Executive of the local authority and comprise the most senior decision-makers from the key public, private and not-for-profit agencies

[2][Online]. Available: http://www.nyam.org/agefriendlynyc/initiatives/current/city-government.html [Accessed December 23, 2014].

[3][Online]. Available: http://agefriendlyireland.ie/programme/age-friendly-structure [Accessed December 23, 2014].

involved in providing supports and services to older people. The alliances further include representatives from the business and service provider forums and the older people's council established as part of the age-friendly city or county initiative.

Jointly, multiple stakeholders, including older persons, can plan around age-friendly outcomes that could not be achieved by one sector alone.

4. **Evidence-informed and results-oriented planning**

Evidence-informed and results-oriented planning involves assessing the status quo to understand what is going well and where the areas of need are, what priorities a community wishes to tackle most urgently. The WHO Age-friendly Cities Guide proposes eight different domains of intervention covering transportation, housing, accessible public places and buildings, opportunities for social participation, employment and civic participation, respectful attitudes and social inclusion, information and communication and health and social services.

An important objective of evidence-informed planning is to obtain information about the diverse needs and capacities of the older population in a city or community and to grasp the heterogeneity among older persons of different sex (and sexual identity) education and income level, ethnic or minority background and neighbourhood to name, but some of the categories to consider. One of the main reasons for such a differentiated analysis is to identify existing inequities that should be addressed.

Evidence gathered to inform policy making should therefore provide a clear picture of the diversity of and inequities among the older population. A more comprehensive analysis would equally include a comparison of the situation of older persons compared to younger cohorts.

Results-oriented planning that sets goals and concrete targets, and includes a plan for monitoring and evaluating progress over time is important in terms of transparency and accountability to citizens. The age-friendly cities and communities approach proposes to cities and communities to create an action plan with concrete targets and indicators against which they can monitor their progress over time.

5. **Monitoring**

Monitoring and evaluation of progress over time enables all actors involved in age-friendly city and community initiatives to celebrate successes and to identify areas that have fallen short of expectations and need increased attention. Assessing developments at regular intervals also helps to build knowledge on the situation of older persons in the city or community and their (changing) needs over time.

3.4 The Mission of the Network

TM *What is the mission of the GNAFCC network?*

LW It supports the development of age-friendly cities and communities: the GNAFCC seeks to do this by **connecting** cities and communities worldwide to facilitate the exchange of information, experiences and lessons learnt; **inspiring** change by showing what can be done and how it can be done; and **supporting** cities to find **solutions** through innovative and evidence-based technical guidance.

TM *Connecting cities and communities: who can join the network?*

LW The focus is on age-friendly policy making at the local level. A requirement for membership is to have the authority to act on the priorities identified in the course of the age-friendly city process. The size of the city or community does not matter in this regard: membership currently includes small rural villages as well as mega-cities such as New York City, Seoul, and Tehran. In order to accommodate smaller communities, including rural areas, the network's name was changed from age-friendly cities to age-friendly cities and communities.

TM *Is this diversity not a problem?*

LW Not in the sense that participation in the network is based on the process applied rather than the outcome achieved and therefore caters for the diversity among members. In other words, cities and communities join with the commitment to becoming more age-friendly using the WHO age-friendly cities approach in their work, including adhering to the key principles of involving older people throughout the process and seeking alliances across sectors. WHO is not prescriptive as to how this should be done so every member is free to apply the principles to their possibilities. Clearly, a mega-city will have different means and might be able to dedicate many resources to the effort than smaller municipalities, but they will also face challenges at a different scale. Urban/rural differences also need to be acknowledged: small communities in rural areas will face different challenges, particularly with regard to infrastructure and services, than small communities in urban areas.

TM *Are age-friendly cities and communities in the GNAFCC already age-friendly when they join?*

LW Cities and communities join the GNAFCC with a commitment to becoming more age-friendly. The objective of the network is to support cities as they are working towards this goal. The concept of an age-friendly city is an ideal that we strive for. The name of the network might lend itself to the misinterpretation that membership certifies a status of age-friendliness that is already achieved—this is not the case.

TM *The network currently has twelve network affiliates. What are network affiliates?*

LW In addition to cities and communities that are members in the network, WHO collaborates with a growing number of age-friendly cities or communities networks and with organizations that promote the creation of age-friendly environments. These *network affiliates* play a significant role in supporting the network's mission through advocacy, the advancement of knowledge and action on age-friendly environments. They are catalysts and multipliers at national, regional, and international level.

TM *The network wishes to inspire change at the local level. How is it working?*

LW By showing what can be done and how it can be done, all cities and communities in the network share the commitment to create urban and rural environments that foster healthy and active ageing, and between them possess a wealth of innovative policy solutions to concrete local challenges. The network provides a platform on which these experiences—and achievements—are shared between members of the network and beyond. We have created a web portal on age-friendly cities and communities featuring the initiatives, resources and age-friendly practices developed and implemented by the network's members and affiliates. Age-friendly World (www.agefriendlyworld.org) is a participatory website for and by the network which we hope to develop over time to become a one-stop-shop for inspiration, learning and guidance on how we can make cities and communities more age-friendly.
We are in the process of building a *Global Database of Age-friendly Practices*. It features concrete measures developed by local actors, aimed at creating more supportive social and accessible physical environments to enable older people to actively participate in community life and maintain their autonomy and independence.

TM *How can the network help cities and communities find solutions to their unique challenges?*

LW There are three main ways in which the network supports cities and communities in finding solutions. First,—and this after all is the primary purpose of a network—GNAFCC provides members with **access to hundreds of other practitioners** who might have faced similar challenges in the past and can provide guidance and mentorship or simply share their experience.

Second, the network is building a growing **reference library on age-friendly cities and communities**. Network members share the results of their age-friendly assessments, priorities identified, action plans, lessons learnt and achievements with each other and thus contribute to building new knowledge based on challenges and solutions developed at the local level. Cities wishing to learn from others how they designed their action plan, what indicators they used to monitor progress or how they conducted their age-friendly assessment can find solutions by browsing the resources section on Age-friendly World (www.agefriendlyworld.org).

Last but not least, the network provides innovative **guidance and tools** related to the age-friendly cities and communities process—developed by WHO and by network affiliates and members. One example is the core set of **age-friendly city indicators** developed by the WHO Kobe Centre[4] in close collaboration with network members and affiliates. The core indicators cover age-friendly environment outcomes both with regard to the physical environment (such as neighbourhood

---

[4][Online]. Available: http://www.who.int/kobe_centre/ageing/age_friendly_cities/en/ [Accessed December 23, 2014].

walkability and the accessibility of public spaces and buildings, transportation and housing) and the social environment (such as engagement in volunteering and employment, attitudes towards older people and availability of services), as well as the impact on well-being of the older population. Importantly, it also includes equity measures. The Guide offers structured guidance on how to select, define and measure indicators. It is not prescriptive, but designed to be adaptable so that indicators can be selected and measured in a way that is locally relevant and appropriate.

3.5 Achievements to Date

TM *Taking stock after the first five years, what do you see as the key achievements to date?*

LW Over the past five years, membership in the network has grown to over 280 cities and communities across 33 countries worldwide that are engaged in age-friendly city initiatives. Together these cities and communities cover over 113 million people.

Global interest in age-friendly cities and communities has grown. The Age-friendly Cities Guide has been translated into many languages. Regional and country level networks have been created in a number of countries including in Canada, France, Ireland, Slovenia, Spain, the Republic of Bashkortostan in the Russian Federation, the United Kingdom and the United States of America.

Network members and affiliates have contributed to raising awareness on the age-friendly cities and communities framework by sharing their work in publications and at a large number of national, regional and international meetings on ageing and age-friendly cities and communities that have raised awareness on age-friendly environments and provided opportunities for face-to-face exchange and networking. The European Year on Active and Healthy Ageing in 2012 and the European Innovation Partnership on Active and Healthy Ageing that has been active since then has contributed to increased collaboration around age-friendly environments in Europe for example.[5] We have seen two successful International Conferences on Age-friendly Cities and Communities hosted by network affiliates, in Dublin, Ireland in 2011 and Québec City, Canada in 2013 and the International Federation on Ageing Global Conferences have dedicated conference streams on age-friendly cities and communities.

With **Age-friendly World**, we have created an information-sharing platform that has the potential to become a one-stop-shop for information on age-friendly cities

[5]Note from the editors: for further details on the European development of AFCC, see the chapter of Parent and Wadoux.

and communities. The network itself is an engine of knowledge production. As the age-friendly city and community initiatives in the network mature, we gain increasing insights into innovative policy solutions that local actors are developing around the world to better meet the needs of older citizens—often in the context of constrained public budgets. The Global Database of Age-friendly Practices is a tool that will facilitate the collection and dissemination of age-friendly measures which can form the basis for more advanced case-study and evaluation work in the future.

Working closely with network members and affiliates, WHO has developed a number of new tools building on the WHO Age-friendly Cities Guide and complementing it. They draw on local expertise and insight and aim to assist local actors in their work: a new guide on age-friendly environments in Europe, guidance on action plan development, monitoring and evaluation and indicators to assess progress towards age-friendliness that were mentioned earlier.

3.6 Challenges and Outlook

TM *What are the challenges for the network?*

LW One of the challenges for the network is sustainable growth. What was started as a small network of cities five years ago has grown considerably with over 280 members to date, and many more enrolled in affiliated networks. A future challenge will be how to ensure that WHO, with its limited resources, will have governance and management structures for the network that are fit for purpose to serve a membership that might grow to become significantly larger than it is today. The Strategic Advisory Group and Executive Committee of the GNAFCC are already thinking about this and are reviewing how to manage membership and governance to prepare the network for the future.

Another challenge relates to the question of whether we can maintain the network free of charge to its members. We wish to create a space for the co-production of knowledge on age-friendly environments at the local level. Membership is free of charge but in turn members are expected to contribute and take an active part in building a global knowledge and evidence-base on age-friendly cities and communities for the benefit of all—by actively sharing their progress, experience and lessons learnt as well as the results of research and evaluation work they engage in.

This experiment is of course not without its own challenges. One of the major barriers to vibrant exchange is the **language barriers within the network which spans 33 countries and more than 20 languages!** Network affiliates play an important role in bridging these and facilitating communication between their members and the rest of the network. The Francophone Network on Age-friendly Cities for examples was created in 2012 to facilitate exchange between French-speaking members in France, Québec, Belgium and Switzerland.

Another challenge will be to address the regional bias in membership. The network has spread unevenly across the world's regions, with most membership growth in Europe and North America, which is largely due to active network affiliates in these regions. Going forward, we would like to intensify the network's outreach to countries in Latin America, Africa, the Eastern Mediterranean region and South East Asia.

Last but not least, we wish to strengthen the knowledge and evidence base on age-friendly cities and communities through research and policy. There is potential for collaboration between network members and the research community interested in the role of the environment in healthy and active ageing. As the Network matures, and cities and communities advance in their age-friendly city and community initiatives, there is growing interest in the question of what impact policy interventions have on the health and well-being of older people. To date, many interventions are based on common sense but it is unclear whether they do achieve their intended outcomes.

Many cities and communities in the network are already collaborating with universities and research centres to conduct base-line assessments and monitor progress over time. We hope to see growing interest among the research community to address some of the research questions that age-friendly cities and communities are raising: for example about the impact of age-friendly policy interventions on the health and well-being of older people. The network certainly holds unused potential for stronger collaboration between academia and local governments that could be leveraged.

References

World Health Organization. (2002). *Active ageing: A Policy framework*. Geneva: WHO.
World Health Organization. (2007). *Global age-friendly cities: A guide*. Geneva: WHO.

Chapter 4
Population Ageing from a Global and Theoretical Perspective: European Lessons on Active Ageing

Alan Walker

4.1 Introduction

Active ageing is now established as the leading global policy strategy in response to population ageing. This pre-eminence was assured by its promotion by international governmental organizations (IGOs) such as the World Health Organization (WHO), United Nations (UN) and Organization for Economic Co-operation and Development (OECD). In practice, however, the term 'active ageing' often serves merely as a convenient label for a wide range of contrasting policy discourses and initiatives concerning ageing and demographic change. A key theme of this chapter is that this lack of clarity about precisely what active ageing consists of is a serious barrier to its widespread adoption as a policy strategy. There are other barriers too and these are also examined in the chapter. It is argued that the over-emphasis on productivity and the labour market in active ageing discourses has detracted from the major potential of this approach to promote much wider well-being. Thus the chapter concludes with an outline of the steps necessary to realize this potential. The starting point is the demographic context which has given rise to active ageing and, in particular, the changing nature of the political discourses concerning this major social transformation and policy challenge. The main geo-political reference point for this chapter is Europe but it begins with a global overview.

A. Walker (✉)
Department of Sociological Studies, University of Sheffield, Sheffield S10 2TU, United Kingdom
e-mail: a.c.walker@sheffield.ac.uk

© Springer International Publishing Switzerland 2016

T. Moulaert and S. Garon (eds.), *Age-Friendly Cities and Communities in International Comparison*, International Perspectives on Aging 14,
DOI 10.1007/978-3-319-24031-2_4

4.2 An Ageing World

In this century advanced ageing will become a global phenomenon. In the twentieth century it was the global north that saw the most rapid ageing whereas, in the twenty-first century it is the turn of the southern hemisphere. The main drivers of this unique historical transformation are declining fertility and improved health and longevity. Soon people aged 65 and over will outnumber children under the age of 5 for the first time in human history. Within this ageing trend the numbers of people aged 80 and over, the 'oldest old', are rising particularly rapidly: 230 % to 2040 compared with 160 % for the population aged 65 and over and 33 % for the total population of all ages (Kinsella and He 2008). This demographic transition is currently in overdrive as a result of the high fertility levels after the Second World War (the 'baby boomers') and the reduced death rates at older ages.

Ageing is a function of economic and social development, which is why it has, for a century, been associated mainly with the industrialized countries of Europe and North America, as well as Japan in Asia. As economic growth, education and incomes rise in the hitherto less developed countries, fertility declines and longevity increases. Three-fifths of the world's population aged 65 and over now live in less developed countries (62 % or 313 million people, in 2008). By 2040 this share is expected to be more than three-quarters, or over one billion people.

Two specific features of ageing in the global south require mention. First the less developed countries are ageing much more rapidly than their northern counterparts. For example it took France more than 100 years to grow its older population from 7 to 14 %, the United States (US) 69 years, Canada 65 years and the United Kingdom (UK) 45 years. In contrast this transition, in UN terms from an ageing to an aged society, will take around 20 years in many less developed countries: China 26 years, Brazil 21 years and South Korea 18 years (Kinsella and He 2008, p. 14). This means that the time available to adjust policy and practice regimes to respond to population ageing is much shorter in the south than was available to the north. Second, the countries of the south are undergoing ageing and development simultaneously whereas, generally, in the north the former followed the latter. Thus the resources available to respond to ageing will be constrained by the economic growth priority.

Western Europe is the world's oldest region and, in global comparative terms, it is only Japan that matches its ageing profile. Both of them also have the most prominent demographic ageing trends over the next 20 years. The ageing of the post-war baby boomer generations started to boost this ageing process in Europe from 2005. While this sketch of the quiet demographic revolution is familiar to many, other dimensions of it are less well known. For example, in many of the world's advanced industrial countries (Australia, Ireland, Japan, The Netherlands, New Zealand, Norway, Sweden and Switzerland) a linear increase has been observed since 1840! This means that there is no reason to expect the rise in life expectancy will level off in the foreseeable future (Oeppen and Vaupel 2002). The remarkable persistence of life expectancy increases—2.4 years on average each decade for women and 2.2 years for men—has not been understood until very recently and this helps to explain why earlier population forecasts were prone to error. Moreover, the sheer scale of the demographic changes still taking place and their far-reaching social policy implications have not been grasped by the policy community let alone more widely. A mid-range projection for the European Union (EU) shows that, by 2050, the proportion aged 65 and over will have risen by 77 % while the working age population will have fallen by 16 % (Economic Policy Committee 2005). It is as a result of the longevity revolution that the segment of the older European population that is growing most rapidly is the very elderly: the proportion aged 80 and over will more than double by 2050 (Eurostat 2011). Not surprisingly for such a large and diverse continent Europe's ageing has not progressed uniformly nor will it in the future. A large number of sub-regions in Europe saw their populations cease to grow and age rapidly before the end of the last century and this will extend to the majority, some of which will experience population reductions, before 2030. With the additional impact of migration some parts of France, Germany, Italy and Spain will have average ages between 44 and 50 by 2015 (Walker and Maltby 1997).

By any standards, historical or global, these are truly remarkable changes which have been felt in every corner and at every level of the developed world—from families and communities to the state. These far reaching consequences are being felt in the global south and will be so increasingly over the rest of this century. Despite its major economic and social implications, however, the issue of population ageing became a 'European' policy one only very recently. This might be contrasted with the global stage where the UN launched its first plan of action on ageing as early as 1982. The next section tracks its origins and then emergence on the European stage and, in particular, the framing of a specific policy discourse around 'active ageing' as part of a new politics of ageing in Europe.

4.3 Policy Discourses on Ageing—Old and New

Current policy discourses on ageing echo the past as well as contemporary social and economic changes. In particular the close relationship between ageing and public policy was forged in Europe, as in other developed countries, in the so-called 'golden age' of welfare state construction following Second World War. This association between older people and the welfare state produced both positive and negative outcomes for this group: summarizing drastically, on the one hand it raised their living standards substantially in most Western European countries but, on the other hand, it contributed to their social construction as dependent in economic terms and encouraged popular ageist stereotypes of old age as a period of both poverty and frailty, this being particularly studied in Europe (Walker 1980; Townsend 1981, 1986; Binstock 1991). Although many stereotypes have some tangential relationship to reality, in this case the high levels of poverty in old age in most European countries in the 1950s and 1960s (one in three in the UK and one in five in Germany), they tend to exert a disproportionate influence on public perceptions and discourses even long after the grain of truth has disappeared. Thus age discrimination is a universal feature of European countries even though the standard of living experienced by older people differs widely and, in a few, poverty has been virtually eradicated (Naegele and Walker 2009). A similar story may be told for other developed regions, such as North America and Australasia. However in less developed countries poverty in old age is still deep-seated and persistent.

The close relationship between older people and the welfare state is also a powerful factor in determining the current nature of ageing discourses in the developed world. All welfare states originated, to some extent, in provision for old age and public pension systems. Now they constitute not only the largest items of national social expenditures, which pose sharp policy questions about future sustainability, but also they comprise the heart of the particular welfare regime found in each country. This means that the character of a country's retirement pension system tends to have a major role in determining the rest of the welfare regime (Esping-Andersen 1990; Walker 2003). Notwithstanding variations in regime types between European countries, however, there was a common stereotype of older people as, essentially, passive recipients of pensions and, in the case of those reliant on social assistance, welfare dependents, albeit deserving ones. In other words, under contrasting flat-rate Beveridge (UK) and earnings-related Bismarck (Germany) pension systems, similar discourses of dependency were found although they were, and are, contingent on the broad social legitimacy of a national pension scheme. These discourses, even the compassionate ones, are expressions of the ageism that, over the past six to seven decades, has become institutionalized in all developed countries (Macnicol 2006). Institutional ageism appears to be no less significant in the developing world (Lloyd-Sherlock 2010).

This generally negative social construction of old age and older people was reinforced by their exclusion from the political and policy making systems of most developed countries (Binstock 1991). The key development here was superannuation: the expectation that older people would leave the labour force at fixed (arbitrary) ages, exchange wages for pensions and disengage themselves from formal economic activity (Phillipson 1982; Walker 1980). Retirement also operated as a process of social and political exclusion: older people were simultaneously detached from paid work together with the main sources of political consciousness and channels of representation. This exclusion contributed to the popular perception of older people as being politically, as well as economically, inactive. This then fed into age discriminatory stereotypes that portrayed older people as passive, acquiescent, family orientated and disinterested in social and political participation.

Other factors also operated to reinforce these stereotypes of passivity. For example, age was less significant than it is today: there were fewer older people; they were less healthy; and retirement acted as an efficient regular of labour force exit. Also, in political terms, old age was less salient because it was not yet perceived as an economic threat and, in any case, attention in many developed countries was directed at rebuilding their physical infrastructures following the Second World War and constructing the major institutions of the modern society, including the welfare state.

4.3.1 Emergence of a New Politics of Ageing: The European Case

Summary accounts such as the preceding and following ones, are prone to sweeping generalizations. They are, nonetheless, necessary here to convey the important broad transformations that have taken place in the public policy discourses accompanying population ageing (for fuller accounts see Walker and Naegele 1999; Walker 2006). The tide in public discourses on ageing began to turn in the 1970s. At the macro level policy makers began to reject the welfare state consensus and to question, more openly than hitherto, the cost of population ageing. Underlying this tidal shift in policy discourses were macroeconomic developments: first the fiscal crisis of the 1970s following the Yom Kippur War and, then in the 1980s, the rising influence of neoliberalism with its inherent critique of public welfare (an ideology that was soon to be globalized—see below). Because pension systems are the keystones of welfare states they came under close scrutiny from this new doctrinal perspective. Moreover, this change in economic and political ideology coincided with the maturation of some national pension schemes and the beginning of the present concern with projections of the future costs of long term care.

It was in this period too that Europe witnessed a huge growth of early exit from the labour force (Kohli et al. 1991). This was in part the product of rising affluence in later life and also the result of public policies aimed at encouraging older workers to leave the labour market in the (vain) hope that their jobs would be taken by the young unemployed whose numbers were growing rapidly as a result of neoliberal deflationary measures. What happened, in fact, was not job substitution but job destruction. The expansion of early retirement, often subsidized by social protection schemes, exacerbated the financial consequences of population ageing for national budgets. All European countries began to focus on the ageing issue at the highest levels of policy making. Few took significant action however, which emphasizes the essential role of political/economic ideology in shaping policy responses to demographic change as well as the urgency of their introduction. Thus there were big variations between European countries at this stage, ranging from substantial reductions in public pensions in the UK (under the neoliberal Thatcher Governments) to the maintenance or improvement of existing pension systems in many other countries (Pierson 2000; Scharpf and Schmidt 2000). Despite the negative consequences of the public policy measures encouraging early exit all of them remained in operation apart from the UK's. It was not until the latter part of the following decade that a different policy perspective emerged in the majority of European countries one which, as is outlined below, portrayed an active role for older people.

Globalization was yet to have its later decisive impact on the spread of neoliberal ideas including those concerning ageing and its economic consequences (Estes and Phillipson 2002; Walker and Deacon 2003). The early signs of what would become common ground among the IGOs appeared in the late 1980s in the form of two OECD (1988a, b) reports. These set out, and were followed by others in the same mould, a 'burden of ageing' discourse and advocated policy prescriptions that involved a reduction in public pay-as-you-go and private/occupational defined benefit pension schemes and an increase in private, defined contribution ones (World Bank 1994; OECD 1998). The IGOs made much, some would argue too much, of the 'ageing crisis' and, in doing so, reinforced negative perceptions of ageing and older people (Quadagno 1989; Walker 1990; Vincent 1996). The idea of the public burden of ageing was not accepted by all policy makers nor were the neoliberal prescriptions that accompanied it. There remained substantial variations across the world in general and in the European continent in particular: in the west only the UK went along this policy route wholeheartedly while, in the east, the key roles played by the IGOs in advising the ex-communist bloc countries meant that the neoliberal prescriptions were followed more closely (Ferge 2002).

There is a common but officially unrecognized 'structural lag' between social and cultural changes and institutional ones (Riley 1992). So, while the public policy discourses were dominated by either the deserving model of ageing or the public burden one, or sometimes a mixture of both, within some European countries new grassroots discourses were emerging. For example, in the late 1980s and early 1990s, there was a growth in direct political participation among older people. Such action is invariably a minority pursuit but, nonetheless, new or reconstituted movements of older people were seen in Denmark, Germany and the UK while, in 1992, the Italian pensioner party, the oldest of its kind in Europe, had its first representative elected to the regional government in Rome. A year later seven pensioner representatives were elected to the Dutch parliament (Walker and Naegele 1999). Similar developments were also seen in North America spearheaded by American Association of Retired Persons (AARP) in the US (Binstock 2005). The characters of the political and policy discourses emanating from these social movements were, of course, fundamentally different from the public policy ones. In contrast to the latter the grassroots movements emphasized human (including welfare) rights, participation, social inclusion and fiercely opposed age discrimination regardless of whether it was blatant or benign.

These new social movements of civil society often reflected the close relationship between ageing and the welfare state in two separate ways. On the one hand the success of Western European welfare provision meant that increasing numbers of older people were not only surviving longer than previously but, also, were doing so in better health. On the other hand the negative impact of the changes in economic and political ideology discussed earlier had a mobilizing effect and led to direct action in the form of protests against cuts in pensions, health and social services. Policy makers in several countries have responded to this new politics by, for example, establishing advisory boards of older people at the local level. Non-governmental organizations (NGO) have also often supported the activities of older people engaged in this new, more direct, politics of ageing (Walker and Naegele 1999).

Intertwined with this growth in social movements concerned with ageing issues has been the cultural shift in society usually labelled as the transition from modernity to late modernity/post modernity (Harvey 1989). The key dimension of this transition, for this account, is the rise of individualistic consumerism. Its effects are apparent in both the state and the market sectors (and of course, its spread is closely related to neoliberal economic globalization). As far as the state is concerned the pressures for more individually tailored services and for a participating voice by service users has led to new, more flexible forms for service provision, such as individual payments in Germany and the UK in lieu of services, and the establishment of user groups to represent the interests of older users, as in Denmark. On the market front, the emergence of the 'older consumer' and the 'silver economy' over the last decade can be seen across Europe and is even stronger in the United States. Indeed one comparison between Germany and the UK concluded that differences are greater between age cohorts within the two countries that between them (Pongratz et al. 2009). In other words, new generations of younger older people who, as a result of continuous employment and pension scheme maturation, are more affluent than their forebears, also display higher levels of hedonism concerning consumption and life style. This finds its expression in a wide variety of forms, from anti-ageing medicines to 'silver travellers' (Pongratz et al. 2009).

4.4 From Successful to Active Ageing

Although the term 'active ageing' is of relatively recent origin, its roots stretch back to the 1950s and 1960s when the activity perspective in gerontology was developed. This was derived from the empirical observation of the connection between different forms of physical activity and well-being (Blau 1973). This approach was a reaction to the first major theory of social gerontology, 'disengagement', which argued that old age is an inevitable mutual period of withdrawal from roles and relationships (Cumming and Henry 1961). From a much sounder empirical vantage point the activity perspective argued that the key to 'successful ageing' (Pfeiffer 1974; Rowe and Kahn 1987) was the maintenance in old age of the activity patterns and values typical of middle age (Havighurst 1954, 1963; Havighurst and Albrecht 1953). In short successful ageing was to be achieved by denying the onset of old age and by replacing those relationships, activities and roles of middle age that are lost, with new ones in order to maintain life satisfaction and well-being.

Later Rowe and Kahn elaborated their initial model of successful ageing to focus on three main components: low probability of disease and disease-related disability, high cognitive and physical functional capacity and active engagement with life. In the US this idea became a reference point for public and political discourses on ageing and made an important contribution to the case for rejecting the negative notion that older age is an inevitable succession of losses (Boudiny 2013). It also attracted scientific interest to research the factors that determine ageing well and clinical practitioner interest to develop preventative measures (Villar 2012). In essence there was a subtle shift in the research and practice focus from those 'doing poorly to those doing well' (Strawbridge et al. 2002).

While successful ageing derived from the broad activity perspective, however, the adjective 'successful' brought a negative judgment to the concept which has proved to be its major weakness (Foster and Walker 2013). For one thing it placed an unrealistic expectation on ageing individuals themselves to maintain levels of activity and to defeat the causes of disease. Overlooked were not only the biological or anatomical limitations but also the economic and social structures that frequently inhibit or prevent people from remaining active—enforced retirement and age discrimination being obvious examples (Walker 1980, 1981). In moral terms the adjective 'successful' implies that there are necessarily winners and losers in the ageing process. Of course this is true, because ageing is unequal everywhere (Cann and Dean 2009) but the fault is less often with individuals than society. Moreover it is stigmatizing to label someone 'unsuccessful' because they have a disease or disability. Finally, even if a person suffers from such limitations, they may still engage in a range of activities and experience a relatively high quality of life (Tate et al. 2003; Bowling 2005). Despite its continuing currency then, the idea of successful ageing tends to be exclusionary and discriminatory and lacks a clear single definition.

Partly in reaction to the deficiencies of successful ageing, from the same intellectual source the concept of active ageing began to emerge in the 1990s, under the influence of the WHO, which, not surprisingly, emphasized the vital connection between activity and health (Butler et al. 1990, p. 201) and the importance of healthy ageing (WHO 1994; see also WHO 2001a).[1] Given the link with health and the influence of the European Union on its development, this approach to active ageing has focused on a broader range of activities than those normally associated with production and the labour market, and has emphasized health and the participation and inclusion of older people as full citizens (see for example Walker 1993, 1994). The thinking behind this new approach is expressed perfectly in the WHO dictum "years have been added to life now we must add life to years". This suggests a general lifestyle strategy for the preservation of physical and mental health as people age rather than just trying to make them work longer. Thus the essence of the emerging modern concept of active ageing is a combination of the core element of productive ageing but with a strong emphasis on quality of life and mental and physical well-being (European Commission 1999; Cabinet Office 2000). The WHO (2001a, b), for example, sees active ageing in terms of health, independence and productivity of older people and defines it as "the process of optimizing opportunities for health, participation and security in order to enhance quality of life as people age" (WHO 2002, p. 12).

[1]Note from the editors: the influence of healthy ageing is also discussed in the presentation of Kalache and in the chapter of Moulaert, Boudiny and Paris.

4.5 Barriers to Active Ageing

Despite the radical promise of the WHO formulation of active ageing and its rhetorical prominence in Europe and across the globe the actual realization of this promise, in terms of fully formulated and implemented strategies, is incredibly limited. Any radical policy proposal faces barriers to acceptance and take-up because institutions and professions are inherently conservative and, through their bureaucratic rules and training and supervision processes, they reproduce this conservatism. But, in the case of active ageing, while the concept is at least superficially appealing to policy makers, there are additional barriers that have impeded thorough acceptance and implementation. Five such barriers are identified here.

First and perhaps foremost there are two political barriers. On the one hand there is a simple confusion about aims and purposes while, on the other hand, there is a purposeful and sometimes cynical hijack of the concept which betrays its original intentions.

The confusion arises out of the array of cognate terms that are employed simultaneously to describe ageing well: as listed above, successful ageing, healthy ageing and so on. Of course there is always a risk attached to the transfer of scientific terms into policy and popular discourses that their original meanings and intentions will be forgotten, purposely or otherwise. However, scientists must take some of the blame here for both using terms as synonyms and for failing to at least try to correct incorrect usage in the policy domain. As indicated in the previous section successful ageing and active ageing are two fundamentally different concepts and should be recognized as such. The continuing tendency for US academics to favour 'successful ageing' or 'productive ageing' means that they are speaking a different language to their European counterparts who favour active ageing. In policy terms the implications of these two formulations are very different. Similarly the term often conjoined with active ageing is healthy ageing. The latter is an important idea and goal but it is not the same as active ageing. Briefly, because this point is discussed at length by Moulaert, Boudiny and Paris in their chapter, healthy ageing concerns health and health interventions and, therefore, is monodimensional, it tends to be institutional in focus (health services) and thereby top-down and it therefore favours professional perspectives. This ambivalence may be an explanation for the ambivalence about healthy/active ageing inside the WHO organization itself, as illustrated by Moulaert, Boudiny and Paris. In contrast active ageing is multidimensional and demands a joined-up approach, which includes health; it favours wide stakeholder engagement and, therefore, is inclusive rather than exclusive. In practice healthy ageing should be treated as an important subset of active ageing but this is rarely the case. The EU for example frequently employs the couplet 'active and healthy ageing' and its strategic framework for research and innovation includes a major initiative on active and healthy ageing (European Commission 2010; Walker and Maltby 2012).[2]

---

[2]Note from the editors: The WHO (2002) Policy Framework on active ageing has therefore joined participation and security to its three pillars of active ageing next to the health dimension. In Age

The second political barrier is ideological and hence much more fundamental than the issue of nomenclature. Under the influence of neoliberalism, promoted by IGOs such as the OECD, policy makers have come to see the concept of active ageing as a narrowly productivist one. Thus an idea that is intended to embrace the whole life course with a focus on human development has become a policy instrument almost exclusively concerned with encouraging, enabling and even forcing older people to work longer. This tendency is particularly marked in Europe and its evolution has been discussed elsewhere (Walker 2009a; Walker and Maltby 2012). The working longer priority is dominant at national level in the EU and also, at European level, it has had high political prominence. For example, the Lisbon Agenda which set the strategic framework for EU policies between 2000 and 2010, included the key targets to increase to 50 % the employment rate of those aged 55–64 and the average retirement age by 5 years—targets incidentally that few EU countries achieved in the case of the former and none achieved in the case of the latter (Walker 2009a; Zaidi and Zolyami 2011).

It would be wrong to suggest that the narrow productivist interpretation of active ageing as working longer is so dominant that it excludes other interpretations and clearly there are competing ones within the European Commission itself. Thus we may contrast these following statements. The first is from a demographic report and the second is from the announcement that 2012 would be the European Year for Active Ageing and Solidarity between Generations.

> Active ageing constitutes in itself a comprehensive and sustainable approach which must employ a range of tools beyond retirement reforms (European Commission 2006, p. 9).The European Year 2012 aims to help create better job opportunities and working conditions for the growing number of older people in Europe, help them to take an active role in society and encourage healthy ageing (European Commission 2010).

At the WHO level, employment has been associated with 'civic participation' as one of the eight domains gathered into the Vancouver Protocol. Nevertheless, in local implementations, it has rarely appeared as a central theme according to Quebec's data (see the chapter by Garon et al. in this book); following Plouffe et al. in their chapter, only "County Kilkenny identified Employment and Income in order to capture needs for financial security within the scope of community service".

Faced with, at best, policy makers' ambivalence and, at worst, the constraints imposed by neoliberal ideology, it is not difficult to see why Europe has not succeeded in developing a concerted strategy for active ageing which joins-up all of the potentially influential policy domains. In its absence the neoliberal reduction of active ageing to working longer remains the main policy goal.

(Footnote 2 continued)

Friendly Cities developments, such multidimensional approach is illustrated by the 8 domains suggested to be taken into account by the Vancouver Protocol (for a detailed review on the Vancouver methodology and its implementation, see the chapter of Plouffe, Kalache and Voelcker).

The second barrier to active ageing is cultural. Of highest importance here are misleading and often damaging stereotypes. The most common active ageing stereotype is of a super-fit pensioner who does extraordinary feats of gymnastics or athletics. Such stereotypes severely distort the meaning of active ageing by transforming it from a mass pursuit to an exclusive minority one. Although there is no evidence to support this contention, these misleading stereotypes are likely to deter anyone other than the fit young-old from believing that active ageing has any relevance to their lives. Moreover they always emphasize physical prowess and rarely focus on mental capacity.

Although not with the same global impact as ageist stereotypes there is evidence of a semantic barrier created by the term ‘active ageing’, in one region at least. In the central European, Eurasian, Commonwealth of Independent States (Russian Federation, Georgia, Ukraine, Kazakhstan and so on) active ageing has two negative connotations. On the one hand it means *accelerated* ageing at the individual level because of the impact of harsh living and working conditions leading to premature death. At the societal level, on the other hand, it means *rapid* ageing due to low fertility and high mortality among the young (Sidorenko and Zaidi 2013). In this, albeit limited case, it is clear that an alternative term, such as healthy ageing, is essential (see chapter by Moulaert et al.).

The third barrier is bureaucratic. As indicated active ageing requires a holistic approach but governments, local and national, are not geared to respond to such strategic needs. Instead, everywhere (even in state socialist China) responsibilities are divided between ministries and departments. This division of labour encourages silo thinking and militates against the implementation of an effective active ageing approach. In the same vein this division supports the reduction of the strategic potential of active ageing to older workers or older people rather than emphasizing the full life course. Thus, in governments everywhere, older people and children are usually represented by ministries but no one is responsible for ageing. In Age Friendly Cities (AFC) application, the will to work behind silos is a clear application of such an attempt to defeat this barrier (Garon et al. 2014).

The fourth barrier, or set of barriers, is societal. This includes the age segmentation that predominates in thought and practice. As illustrated below the traditional paradigm segments the life course into three major stages. Although the life course and working life have been transformed over recent decades—the former for example by increased longevity and the latter by the replacement of careers for large sections of the working population by discontinuous employment—social institutions and popular discourses still operate as if the traditional model is the dominant one. This embeddedness encourages silo thinking in policy and practice: active ageing is for the retired and so on. In contrast the age integrated paradigm opens the door to a life course active ageing approach (Fig. 4.1).

Equally important are the barriers created by age discrimination or ageism. These can include direct discrimination, when older workers are excluded from jobs or vulnerable older patients are abused, but also encompasses less direct, more insidious, stereotyping for example when older people are described as a ‘burden’ or accused of robbing resources from the young. Discrimination has two

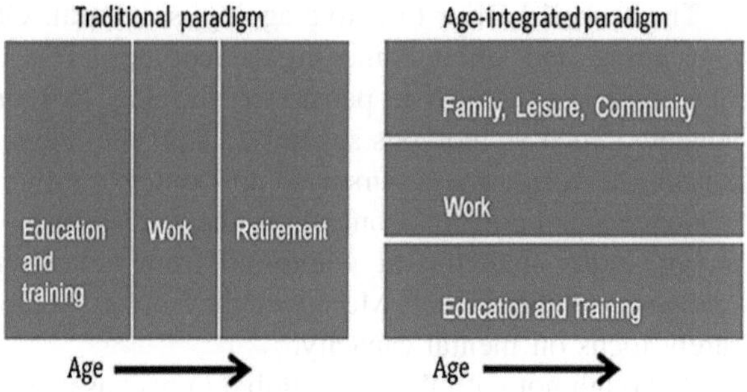

Fig. 4.1 Life course segmentation. *Source* Walker and Maltby (2012)

unfortunate effects. On the one hand it excludes and stigmatizes older people, particularly frail older people and, on the other hand, it encourages younger people to ignore later life, to push it to the back of their minds. Both effects limit the potential of active ageing policies. For example among older people there is often a resigned stoicism—'at my age what can you expect?'—which militates against active engagement.

The fifth barrier is unequal ageing—the deep-seated inequalities that exist both between older people and over the life course. Within countries there are inequalities between different groups of older people, for example based on social class, gender and race, which segment the experience of later life. These inequalities are usually created not in old age but at earlier stages of the life cycle (Walker 2009b). Then there are substantial inequalities in ageing and later life between countries at similar levels of development. In the EU for example there are substantial differences between member states in healthy life expectancy—10 years between Denmark and Estonia (Jagger et al. 2009). Then there are huge disparities between rich and poor countries—the global north and south. These three aspects of unequal ageing make the task of implementing an active ageing strategy more difficult than it already is because they cry out for flexibility in the design and implementation of such a strategy when the preference of policy makers is closer to 'one size fits all'. It is particularly well illustrated in this book by the different applications of Age Friendly Cities in both developed and less developed countries.

As indicated these five barriers help to account for the fact that active ageing is not yet reaching its full potential in policy terms. So, what steps are necessary to confront these barriers and enable the comprehensive approach that is called for? The next section considers this question.

4.6 Towards Active Ageing

As indicated in the previous section there are formidable barriers confronting a comprehensive active ageing approach of the kind espoused by the WHO. Some of these stem directly from ideology and must be considered beyond the scope of this chapter and book. The fight for social justice is taking place elsewhere but be in no doubt about its importance for the active ageing agenda (Walker et al. 2011). Here my concern is with the nature of the active ageing strategy itself, adopting science's public role to ensure that policy makers and wider society are adequately equipped to pursue specific goals. Thus it is essential to start by clarifying what an active ageing strategy should look like, including the principles upon which it should be based.

Seven key principles have been proposed as the basis for a strategy on active ageing to ensure that it is both comprehensive and consistent (Walker 2002).

First of all, 'activity' should consist of all meaningful pursuits which contribute to the well-being of the individual concerned, his or her family, local community or society at large and should not be concerned only with paid employment or production. Thus, in terms of active ageing, volunteering should be as highly valued as paid employment. Second, it should be primarily a preventative concept. This means involving all age groups in the process of ageing actively across the whole of the life course. Third, active ageing should encompass *all* older people, even those who are, to some extent, frail and dependent. This is because of the danger that a focus only on the 'young-old' will exclude the 'old-old' and the fact that the link between activity and health (including mental stimulation) holds good into advanced old age (WHO 2001b). There is also an important gender aspect to this principle in that most of the very old are women. Thus this strategy is framed to be gender-sensitive not gender neutral (Foster and Walker 2013). Fourth, the maintenance of intergenerational solidarity should be an important feature of active ageing. This means fairness between generations as well as the opportunity to develop activities that span the generations. Fifth, the concept should embody both rights and obligations. Thus the rights to social protection, life-long education and training and so on should be accompanied by obligations to take advantage of education and training opportunities and to remain active in other ways. Again, from a gender perspective, this requires support to enable women to participate. Sixth, a strategy for active ageing should be participative and empowering. In other words, there must be a combination of top-down policy action to enable and motivate activity but, also, opportunities for citizens to take action, from the bottom up, for example in developing their own forms of activity. Seventh, active ageing has to respect national and cultural diversity. For example there are differences in the forms of participation undertaken between the north and the south of Europe, therefore value judgments about what sort of activity is 'best' are likely to be problematic (European Commission 2000). Within some EU countries, such as Belgium, there are major cultural variations that require a flexible approach to the implementation of an active ageing strategy. Indeed this cultural diversity and the unequal ageing discussed in the previous section suggest that 'flexibility' should be an eighth principle (Foster and Walker 2015).

These principles suggest that an effective strategy on active ageing will be based on a *partnership* between the citizen and society. In this partnership the role of the state is to enable, facilitate and motivate citizens and, where necessary, to provide high quality social protection for as long as possible. This will require interrelated individual and societal strategies. As far as individuals are concerned they have a duty to take advantage of lifelong learning and continuous training opportunities and to promote their own health and well-being throughout the life course. As far as society is concerned the policy challenge is to recognize the thread that links together all of the relevant policy areas: employment, health, social protection, social inclusion, transport, education and so on. A comprehensive active ageing strategy demands that all of them are 'joined-up' and become mutually supportive. The primary discourse behind this strategic vision of active ageing is the UN's one of a *Society for all ages* (http://www.un.org/esa/socdev/iyop/iyopcfo.htm).

With regard to the scope of the actions necessary to achieve such a comprehensive strategy the WHO has highlighted eight main determinants of active ageing: culture and gender (both of which are cross-cutting), health and social service, behavioural, the physical environment, the social environment, economic determinants and those related to the person concerned (such as biology, genetics and psychology) (WHO 2002). In EU policy terms this would mean linkage between policy domains that have hitherto been separated: employment, health, social protection, pensions, social inclusion, technology, economic policy and research. At the same time, in line with the WHO's call, there is a need to mobilize all stakeholders to "popularize the term 'active ageing' through dialogue, discussion and debate in the political arena, the education sector, public fora and media such as radio and television programming" (WHO 2002, p. 55). The basis for such a comprehensive approach exists already in some EU member states but appears to be stymied by the huge challenge of transcending traditional departmental boundaries and changing deeply entrenched reactive policies into preventative ones. A case in point is the UK, which has had a comprehensive strategy on active ageing since 2005 which has yet to be thoroughly implemented (Department for Work and Pensions 2005).

Of course the key stakeholders are not dormant while they wait for the perfect strategic framework to be assembled. Thus there are countless examples of local community and grass roots level initiatives by older people, NGOs and municipalities aimed at raising the participation and well-being of this group (Walker and Naegele 1999). In some countries there are national programmes to encourage healthy ageing such as 'FinnWell' in Finland. There is plenty of evidence too that some employers, albeit a minority, have developed a variety of age management measures designed to retain, recruit and maximize the potential of an ageing work force (Walker 1999; Naegele and Walker 2006). What is lacking at present, however, is a comprehensive strategy on active ageing which includes the sharing of the many examples of good practices between EU countries.

Research and development have a critical role to play in advancing the active ageing agenda and, especially, in providing the evidence base for policy. European research, under Framework Programmes 5, 6 and 7, has already added considerably to this knowledge base and the future research, priorities have been mapped by, for example the ETAN initiative, coordinated actions like FORUM and ERA-AGE (http://era-age.group.shef.ac.uk/) and the Road Map project (FUTURAGE 2011). At EU level, these are also initiatives organized by Age Platform (see the chapter of Parent and Wadoux that describes the vision of this organisation).

4.7 Conclusion: Active Ageing or Age Friendly?

This chapter set out to provide an overview of both the social policy discourses following in the wake of the unprecedented global ageing and the leading response from policy makers, namely active ageing. While the science is set from a global perspective the specific illustrations are mainly European. This is for reasons of convenience and pertinence. My research base is Europe and I have studied policy development in that region. Happily for students of policies on ageing Europe is the oldest region in the world and, therefore, there are rich and varied experiences in the application of these policies at local and national levels, and also uniquely at regional level. Cases in point are active ageing and related age friendly city indicatives. Therefore, while not suggesting that policy can be simply transported from one region to another, there are at the very least some important cautions in the European experience, not least the gap between rhetoric and action with regard to active ageing. This gap emphasizes the need for a strategic approach to the achievement of active ageing and the application of agreed principles designed to ensure that this policy is both comprehensive and that no group is excluded.

Of course active ageing should also be an integral part of the Age Friendly Cities (AFC) network. In practice, however, the relationship is not clear or consistent. In fact very often the comprehensive, all ages, aspect of active ageing is ignored in AFC policies in favour of an old age focus. This is apparent in the key source for the initiatives, which acknowledges active ageing but then reduces the focus to older people:

> Informed by WHO's approach to active ageing, the purpose of this guide is to engage cities to become more age-friendly so as to tap the potential that older people represent to humanity (WHO 2007, p. 1).

Of the eight topic areas listed in the WHO Guide to AFCs almost half of them explicitly refer to older people (and disabled people). There is a risk therefore that 'age friendly' instead of 'ageing friendly' becomes the dominant focus. Even in the case of the world's leading example of an AFC governmental plan as proposed in Quebec in 2011, *Ageing and Living Together*, the focus on ageing, rather than older people, is not always clear. However in the Quebec Action Plan there is no ambivalence: the emphasis is firmly on ageing as recalled by Garon et al. in their chapter.

The Quebec approach may serve as a model for other AFC initiatives. The key lesson, to my mind, is in the integration of active ageing and AFC policies and, especially, in the application of a principled approach to active ageing. Thus, if all AFC initiatives truly embodied a comprehensive active ageing model, this movement could have the potential to transform the ageing experience globally.

References

Binstock, R. (1991). Aging, politics, and public policy. In B. Hess & E. Markson (Eds.), *Growing old in America* (pp. 325–340). New Brunswick: Transaction Publishers.

Binstock, R. H. (2005). The contemporary politics of old age policies. In R. Hudson (Ed.), *The new politics of old age policy* (pp. 265–293). Baltimore: John Hopkins University Press.

Blau, Z. S. (1973). *Old age in a changing society*. New Viewpoints.

Boudiny, K. (2013). 'Active ageing': From empty rhetoric to effective policy tool. *Ageing and Society, 33*(6), 1077–1098.

Bowling, A. (2005). *Ageing well: Quality of life in old age*. Maidenhead: Open University Press.

Butler, R., Oberlink, M., & Schecter, M. (Eds.). (1990). *The promise of productive aging*. New York: Springer.

Cabinet Office. (2000). *Winning the generation game*. London: The Stationery Office.

Cann, P., & Dean, M. (2009). *Unequal Ageing*. Policy Press.

Cumming, E., & Henry, W. (1961). *Growing old—The process of disengagement*. New York: Basic Books.

Department for Work and Pensions. (2005). *Opportunity age*. London: DWP.

Economic Policy Committee. (2005). *The impact of ageing on public expenditure*. Brussels: European Commission.

ERA-AGE. Available http://era-age.group.shef.ac.uk/ [Accessed May 16, 2014].

Esping-Andersen, G. (1990). *The three worlds of welfare capitalism*. Cambridge: Policy Press.

Estes, C., & Phillipson, C. (2002). The globalisation of capital, the welfare state and old age policy. *International Journal of Health Services, 32*(2), 279–297.

European Commission. (1999). *Towards a Europe for all ages*. COM (1999) 221, Brussels: European Commission.

European Commission. (2000). *Social agenda 2000–2005*. Brussels: European Commission.

European Commission. (2006). *The demographic future of Europe—From challenge to opportunity*. Brussels: European Commission.

European Commission. (2010). *2012 to be the European year for active ageing*. DG Employment, Social Affairs and Inclusion website. Available http://ec.europa.eu/social/main.jsp?langId=en&catId=89&newsId=860 [Accessed May 16, 2014].

Eurostat. (2011). *The greying of the babyboomers: A century-long view of ageing in European populations*. Statistics in Focus, Eurostat, Statistical Office of the European Union, 23/2011.

Ferge, Z. (2002). European integration and the reform of social security in the accession countries. *European Journal of Social Quality, 3*(1/2), 9–25.

Foster, L., & Walker, A. (2013). Gender and active ageing in Europe. *European Journal of Ageing, 10*(1), 3–10.

Foster, L. & Walker, A. (2015). Active and successful ageing: A European policy perspective. *The Gerontologist*, forthcoming.

FUTURAGE. (2011). *A road map for European ageing research*. Available http://www.futurage.group.shef.ac.uk/roadmap.html [Accessed May 16, 2014].

Garon, S., Paris, M., Beaulieu, M., Veil, A., & Laliberté, A. (2014). Collaborative partnership in age-friendly cities: Two case studies from Quebec, Canada. *Journal of Aging & Social Policy, 26*(1–2), 73–87.

Harvey, D. (1989). *The condition of postmodernity. An inquiry into the origins of cultural change*. Oxford: Blackwell.

Havighurst, R. (1954). Flexibility and the social roles of the retired. *American Journal of Sociology, 59*, 309–311.

Havighurst, R. (1963). Successful aging. In R. Williams, C. Tibbitts, & W. Donahue (Eds.), *Process of Aging 1* (pp. 299–320). New York: Atherton.

Havighurst, R., & Albrecht, R. (1953). *Older People*. London: Longmans.

Jagger, C., Gillies, C., Moscone, F., Cambois, E., Van Oyen, H., & Nusselder, W. et al. (2009). Inequalities in healthy life years in the 25 countries of the European Union in 2005: A cross-national meta-regression analysis. *Lancet, 372*, 2124–2131.

Kohli, M., Rein, M., Guillemard, A.-M., & Gunsteren, H. (Eds.). (1991). *Time for retirement*. Cambridge: Cambridge University Press.

Kinsella, K., & He, W. (2008). *An Ageing World: 2008*. Washington DC: US Census Bureau.

Lloyd-Sherlock, P. (2010). *Population ageing and international development: From generalisation to evidence*. Policy Press.

Macnicol, J. (2006). *Age discrimination. An historical and contemporary analysis*. Cambridge: Cambridge University Press.

Naegele, G., & Walker, A. (2006). *A guide to good practice in age management*. Dublin: European Foundation for the Improvement of Living and Working Conditions.

Naegele, G., & Walker, A. (Eds.). (2009). *Ageing and social policy: Britain and Germany compared*. Houndmills: Palgrave.

OECD. (1988a). *Ageing populations—The social policy implications*. Paris: OECD.

OECD. (1988b). *Reforming public pensions*. Paris: OECD.

OECD. (1998). *Maintaining prosperity in an ageing society*. Paris: OECD.

Oeppen, I., & Vaupel, J. (2002). Broken limits to life expectancy. *Science, 296*, 1029–1031.

Pfeiffer, E. (Ed.). (1974). *Successful aging: A conference report*. Durham, NC: Duke University.

Phillipson, C. (1982). *Capitalism and the construction of old age*. London: Macmillan.

Pierson, P. (2000). *The new politics of the welfare state*. Oxford: Oxford University Press.

Pongratz, W., Gross, T., & Hilbert, J. (2009). The silver economy—Purchasing power and the quest for quality of life. In G. Naegele & A. Walker (Eds.), *Ageing and social policy: Britain and Germany compared* (pp. 82–105). United Kingdom: Palgrave Macmillan.

Quadagno, J. (1989). Generational equity and the politics of the welfare state. *Politics and Society, 17*, 353–376.

Riley, M. (1992). Cohort Perspectives. In E. Borgatta & M. Borgatta (Eds.), *The encyclopedia of social sciences* (pp. 52–65). New York: Macmillan.

Rowe, J., & Kahn, R. (1987). Human aging: Usual and successful. *Science, 237*, 143–149.

Scharpf, F., & Schmidt, V. (Eds.). (2000). *Welfare and work in open economies*. Oxford: Oxford University Press.

Sidorenko, A., & Zaidi, A. (2013). Active ageing in CIS countries: Semantics, challenges and responses. *Current Gerontology and Geriatrics Research, 1*. Available: doi:10.1155/2013/261819 [Accessed July 20, 2014].

Strawbridge, W. J., Wallhagen, M. I., & Cohen, R. D. (2002). Successful aging and well-being self-rated compared with Rowe and Kahn. *The Gerontologist, 42*(6), 727–733.

Tate, R., Leedine, L., & Cuddy, T. (2003). Definition of successful aging by elderly Canadian males: The Manitoba follow-up study. *The Gerontologist, 43*, 735–744.

Townsend, P. (1981). The structured dependency of the elderly: The creation of social policy in the twentieth century. *Ageing and Society, 1*(1), 5–28.

Townsend, P. (1986). Ageism and social policy. In C. Phillipson & A. Walker (Eds.), *Ageing and social policy* (pp. 15–44). Aldershot: Gower.

Villar, F. (2012). Successful ageing and development: The contribution of generativity in older age. *Ageing and Society, 32*, 1087–1105.

Vincent, J. (1996). Who's afraid of an ageing population? *Critical Social Policy, 16*(2), 3–44.

Walker, A. (1980). the social creation of poverty and dependency in old age. *Journal of Social Policy, 9*(1), 49–75.

Walker, A. (1981). Towards a political economy of old age. *Ageing and Society, 1*(1), 73–94.

Walker, A. (1990). The economic 'burden' of ageing and the prospect of intergenerational conflict. *Ageing and Society, 10*(4), 377–396.

Walker, A. (1993). *Age and attitudes*. Brussels: EC.

Walker, A. (1994). Work and income in the third age—An EU perspective. *The Geneva Papers on Risk and Insurance, 19*(73), 397–407.

Walker, A. (1999). *Managing an ageing workforce: A guide to good practice*. Dublin: European Foundation for the Improvement of Living and Working Conditions.
Walker, A. (2002). A strategy for active ageing. *International Social Security Review, 55*(1), 121–140.
Walker, A. (2003). Securing the future of old age in Europe. *Journal of Societal and Social Policy, 2*(1), 13–32.
Walker, A. (2006). Aging and politics—An international perspective. In R. Binstock & L. George (Eds.), *Handbook of aging and the social sciences* (pp. 339–359). Amsterdam: Elsevier Academic Press.
Walker, A. (2009a). The emergence and application of active ageing in Europe. *Journal of Ageing and Social Policy, 21*, 75–93.
Walker, A. (2009b). Why is ageing so unequal? In P. Cann & M. Dean (Eds.), *Unequal ageing* (pp. 141–158). Bristol: Policy Press.
Walker, A., & Deacon, B. (2003). Economic globalisation and policies on ageing. *Journal of Societal and Social Policy, 2*(2), 1–18.
Walker, A., & Maltby, T. (1997). *Ageing Europe*. Milton Keynes: Open University Press.
Walker, A., & Maltby, T. (2012). Active ageing: A strategic policy solution to demographic ageing in the European Union. *International Journal of Social Welfare, 21*(s1), S117–S130.
Walker, A., & Naegele, G. (1999). *The politics of old age in Europe*. Open University Press.
Walker, A., Sinfield, A., & Walker, C. (Eds.). (2011). *Fighting poverty, inequality and injustice*. The Policy Press.
WHO. (1994). *Health for all: Updated targets*. Copenhagen: WHO.
WHO. (2001a). *Health and ageing: A discussion paper*. Geneva: WHO.
WHO. (2001b). *Active ageing: From evidence to action*. Geneva: WHO.
WHO. (2002). *Active ageing: A policy framework*. Geneva: WHO.
WHO. (2007). *Global age friendly cities: A guide*. Geneva: WHO.
World Bank. (1994). *Averting the old age crisis: Policies to protect the old and promote growth*. Oxford: Oxford University Press.
Zaidi, A., & Zolyomi, E. (2011). *Active Ageing Research Note 7/2011*. Produced by the Social Situation Observatory, for the European Commission, DG for Employment, Social Affairs and Inclusion, Brussels.

Chapter 5
Active Ageing and Age-Friendly Cities—A Personal Account

Alexandre Kalache

Thibauld Moulaert and Suzanne Garon *Doctor Kalache, you directed the Ageing and Life Course Department of the World Health Organisation (WHO) for twelve years. You brought a focused 'active ageing' approach to the organization and initiated the WHO Age Friendly Cities program that you first instigated during your keynote speech at the 2005 Congress of International Association of Gerontology and Geriatrics (IAGG) in Brazil. Firstly could you recall how you became interested in ageing?*

Alexandre Kalache My professional interest in ageing began in the mid-1970s when I went to the UK to study for a Masters Degree in Social Medicine. A footnote in a Lancet article caught my attention. It alluded to the fact that over 80 % of all geriatricians working in Britain at the time were foreign born—the vast majority originating from South Asian countries such as India, Pakistan, Sri Lanka and Bangladesh. Coming from a background in medical education and having recently worked in the field on a WHO project in my home city of Rio de Janeiro, the reference intrigued me. Why were British medical schools failing to attract students to a specialty area that was so clearly relevant to its aged society? My curiosity lead me to design and then conduct two research projects in order to try to shed light on the reasons.

The first study surveyed all British Geriatric Society (BGS) members—that is, doctors who had already committed to Geriatric Medicine as their specialty—in an attempt to learn the factors that influenced that decision. It soon became clear from the large group of around 800 respondents, that they were fundamentally divided over satisfaction with their specialty choice. Regardless of nationality or where they had studied medicine, those who were content with their specialty choice were much more likely to have lived with their grandparents in their childhood and/or adolescence. The second group lacked such familiarity with older people at the beginning of their lives. The clear message was that positive attitudes toward older

A. Kalache (✉)
International Longevity Center-Brazil (ILC-BRAZ), International Longevity Center Global Alliance (ILC-GA), Avenida Padre Leonel Franca 248, Gavea, Rio de Janeiro, RJ 22451-000, Brazil
e-mail: kalache@ilcbrazil.org; akalache@nyam.org

© Springer International Publishing Switzerland 2016

T. Moulaert and S. Garon (eds.), *Age-Friendly Cities and Communities in International Comparison*, International Perspectives on Aging 14,
DOI 10.1007/978-3-319-24031-2_5

persons are mostly formed early in life and are critical for job satisfaction among those who work with them.

The second study measured medical students' attitudes towards ageing, old age, older people, end of life issues, geriatrics as a vocation and other related subjects. It was conducted in one of the medical schools of the University of London, the Royal Free Hospital. In that year (1976), geriatric medicine had been inserted into the curriculum for the very first time. Most of the students lacked any real familiarity with older people, even within their own families. They had, by and large, been raised in nuclear families and only saw their older relatives occasionally within the rather formal context of special family events. Many of them had never spent a full half an hour talking to an older person. At medical school they were placed in the context of a 'forced' relationship with older people for the very first time. And these were older people in very particular circumstances—often confused, depressed, neglected, with multiple pathologies, dying—not positive circumstances to start relationships with older people.

The findings of these two studies irretrievably changed my career path and I made a firm commitment to focus my public health training on ageing. There was a certain predictability about the decision because the most influential figures throughout my own childhood and youth had unquestionably been the towering personalities of older relatives within my extended family.

TM/SG *Why did you decide to focus on public health?*

AK Even before and certainly throughout my medical training, I knew that public health was going to be my eventual route. Growing up in a country with the characteristics of Brazil at the time meant that it was difficult not to recognise and feel activist about the importance of the social determinants to health. I came to medicine through sociology and community work. The choice of the UK for my post-graduate studies was largely because of that country's pre-eminence in what was then called 'community health'.

I was first based at the University of London but, in 1977, I was invited to join the University of Oxford Department of Community Medicine where I spent the next seven years working as Clinical Lecturer and simultaneously studying for my doctorate. Ageing became my professional north star with epidemiology as my compass and I was privileged to work alongside several of the most celebrated epidemiologists in the world. The knowledge that I acquired during these formative years consolidated my focus on public health.

TM/SG *Then you came into contact with WHO. How did it happen?*

AK In fact, my professional life became intertwined with WHO very soon after medical graduation. I worked for the Organisation for two years in Brazil as Research Development Officer at the Pan American Health Organisation (PAHO) Latin American Centre for Education, Technology and Health. My UK Masters Degree was made possible through a WHO fellowship. From the mid-1970s onwards I was a regular consultant at WHO technical meetings and I spent two separate six month secondments to the WHO global programme on ageing in the 1980s.

As a member of the WHO Scientific Group on the Epidemiology of Ageing, I participated in a pivotal meeting of experts in Geneva (January 1983) on the uses of epidemiology in the study of older age (WHO 1984). The main recommendation of that meeting was a call for more training on the subject with a special emphasis on health professionals from the developing world. It was an appeal that held great personal resonance—combining as it did my interests in education, epidemiology, ageing and the developing world. With this mandate, I returned to the London School of Hygiene and Tropical Medicine (LSHTM) where I had studied for my Master Degree and established the Epidemiology of Ageing Unit (1984). From this platform, over the following years, I launched a series of WHO supported international courses on the Epidemiology of Ageing. As increasing numbers of people came to recognise that worldwide ageing was the demographic imperative of our times, post-graduate students came from more than fifty countries to attend the courses. Using the London-based courses as a template, I was able to replicate the training model to many other countries, particularly throughout Latin America but also within Europe, Africa and Asia—again, typically, with WHO financial support.

TM/SG *As an academic working in the field of ageing and public health and with such an association with WHO, how did you view the then 1st World Assembly on Ageing, held in 1982 in Vienna?*

AK The 1st World Assembly on Ageing very clearly reflected the belief at the time that population ageing was an issue of concern to mostly European and other similarly developed countries. The decision to physically locate the WHO *Global* 'Programme on Health of the Elderly' (PHE) within the WHO European *Regional* Office in Copenhagen was further illustration of that mindset. The world community in 1982 was manifestly not yet ready to accept the rapid demographic transformations beginning to emerge in such countries as Brazil. There were very few voices calling for a truly global focus at the meeting in Vienna. It is important to also recall that, even later, when I was seconded to PHE (1987), the preoccupation was almost entirely disease-orientated and the Programme was strongly partnered with the Programme on Disabilities—conveying the firm message that health in older age was synonymous with disease, decline and disability. It was not a perspective that I could share. Significantly, more influential on my thinking than the Vienna Assembly was the WHO Ottawa Charter on Health Promotion (1986). I became an enthusiastic supporter on my first read and I eagerly began to incorporate the principles behind the new Health Promotion rationale into ageing. On my return to LSHTM, I readily applied the approaches to all aspects of my work and initiated the ground work to later launch and coordinate the first ever Master Degree on Health Promotion (1991).

TM/SG *While you continue pedagogical support to Health Promotion, the 1990s changed your life-trajectory. Could you explain what happened?*

AK I do not think that my life-trajectory changed to any significant extent. I view my recruitment by WHO to direct the then labelled 'Programme on Health of the Elderly' (PHE) very much as a continuation. I had already worked in and around that office for a decade or so and there was a solid intellectual continuity. I simply took the view that the new role added a larger platform. A source of satisfaction in my professional life is that it has been remarkably consistent from very early. A shift of paradigm however, was underway. In line with a gradual realisation that population ageing was indeed a world-wide process, a political decision had been taken to relocate PHE to the WHO global headquarters in Geneva.

Inspired by the Ottawa Charter, my first action was to submit a technical paper to the WHO Executive Board requesting an official change of name as well as clearance for a total reorientation of the programme. No longer was the focus going to be on diseases. The emphasis would now be increasingly placed on the determinants of healthy ageing and related principles. The newly branded 'Programme on Ageing and Health' (designated as AHE) would no longer stress 'the elderly' but 'ageing'. It would adopt a strong life course approach and remind everyone, including colleagues working in other WHO departments, that good health in older age is everybody's business.

TM/SG *Do you mean that colleagues regarded the changes with suspicion? How did you set about changing that perception?*

AK All changes provoke resistance. Most medical professionals have been schooled in the notion that health is simply the absence of disease. Moreover, to this day, most medical education remains heavily focused on child and reproductive health in spite of the new demographic imperatives. The medical profession, among others, has still not undergone the necessary recalibration for the realities of the twenty-first century. I was acutely aware that a significant part of my role, as head of the programme, was to raise the level of awareness about ageing in all settings.

In 1997, WHO dedicated the 50th year edition of the Organisation magazine to 'Active Ageing' (WHO 1997). As its editor, I commissioned a number of papers intended to add elucidation to the concept of active ageing. It was the first time that the label was publicly referenced to best describe the new thinking in the WHO work on ageing.[1] Working within the constraints of language, my view was that 'active' conveyed a more inclusive message than alternative labels such as 'healthy', 'successful' or 'positive'. The intention was to reference a continued participation in social, economic, cultural, spiritual and civic affairs—not simply the capacity to be physically or economically active.

TM/SG *However, inside the magazine, the emphasis is still on 'healthy ageing'. How do you explain this?*

AK 'Active ageing', both as a concept and a policy tool, did not arrive assembled and ready to use. Its definition and application has evolved and continues to evolve

[1]Note from the editors: in their chapter, Moulaert, Boudiny and Paris showed that the cover of this document was indeed labelled with 'active ageing' but that inside the publication, the authors continue to refer to 'successful ageing'. This is not contradictory with what is exposed here; it simply stresses, from another perspective, how social and medical gazes are intertwined since the WHO's launch of 'active ageing'.

within the context of a shifting political and social landscape. From the mid 1990s onwards WHO embarked on the process to identify the features and to refine the language of a fresh ideological approach to ageing that would have global resonance. Understandably, health was a large focus of that process but it was not to the exclusion of all other aspects. The concept of active ageing achieved widespread currency in this period, largely owing to the efforts of the WHO (Walker 2002).

The vocabulary within the 1997 WHO Active Ageing publication may have lacked a consistency but the content was part of the conversation towards a formalised Active Ageing strategy and its eventual world-wide application—principally through the Age-friendly cities and communities' movement that continues to actualise it on multiple fronts to this day. I co-authored the central article in the publication with Ilona Kickbush, the intellectual mentor behind the Ottawa Charter and then Director of the Health Promotion Division in which AHE was anchored. As showed in Fig. 5.1, the emphasis is strongly placed on a life course approach to the concept of functional capacity:

> For a number of functional capacities (for example, respiratory capacity, muscular strength, and cardiovascular performance) individuals reach a maximum early in adulthood. From there on, there is a decline. However, this decline can be faster or slower depending on a number of characteristics often associated with living conditions and behaviour (Kalache and Kickbush 1997, p. 5).

This life course perspective, with a large health promotion component that considers both the environmental as well as the social circumstances, forms the fundamental building block of active ageing as later formally defined and applied by WHO.

TM/SG *Were you aware, at the time, that the US Department of Health had commissioned the Organisation for Economic Co-operation and Development (OECD) for a similar comprehensive new vision of ageing, a vision that later became limited to 'productive ageing', according to* Moulaert and Biggs (2013)*?*

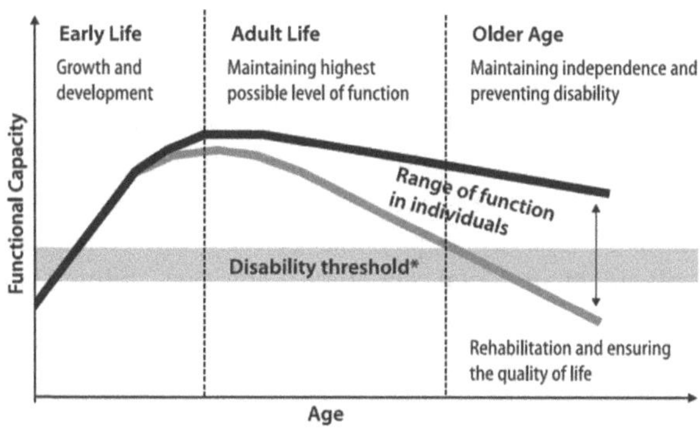

Fig. 5.1 Maintaining functional capacity over the life course graph. *Sources* Kalache and Kickbush (1997, p. 5), WHO (2002, p. 14). Reproduced, with the permission of the publisher, from WHO (2002)

AK I was aware at the time that many institutions and researchers, particularly in those countries most immediately impacted by ageing, principally within the OECD, were waking up to the economic implications of their ageing populations. However, I was wary about any policy approaches that over-emphasised the monetary aspect and championed longer working lives regardless of the social context. Furthermore, I was mindful of the growing need for relevant guidance and policy tools that related to the vast majority of the world's ageing population—most of whom already lived beyond the borders of the developed economies. I was also conscious of WHO's almost unique power of global convocation. The vision that we were beginning to articulate, in line with our organisational mandate, had the ambition to have a very broad reach and one that was firmly rooted in a 'rights-based' approach that shifted strategic planning away from a 'needs-based' approach.

TM/SG *What was the route to consolidate this vision?*

AK I was sensitive to the potential criticism that WHO was limiting its work on ageing to the promotion of physical activity in older age. It is certainly true that we vigorously promoted work in this area from 1995–2000. For instance, for over a year AHE engaged in a process (in collaboration with academic institutions and civil society organisations from both developed and developing countries) through seminars, workshops and the commission of papers that culminated in the launch of the WHO Heidelberg Guidelines for Promoting Physical Activity Among Older Persons (WHO 1996). Firmly based on empirical evidence, it stimulated training programmes as well as public campaigns.[2] The climax of such campaigns, however, was reached in 1999, the International Year of Older Persons (IYOP).

It was symbolic that the United Nations chose to dedicate the last year of the twentieth century to, arguably, the most enduring achievement of the previous 100 years: ageing not as the privilege of the few but increasingly as the expectation of most. Throughout this last century, some 30 years had been added to life expectancy at birth. It was an extraordinary accomplishment and celebration was in order. In 1997, WHO put into motion a process that would culminate with the 'Global Embrace on Active Ageing' on 1st October, International Day of Older Persons (IDOP), within the International Year of Older Persons (IYOP). Arguably, the largest single health promotion event in history, the world-wide walk event of the Global Embrace involved well over five thousand communities and in excess of two-million walkers and was magnified to a much larger audience by extensive media coverage (Chodzko-Zajko and Schwingel n.d.). WHO, however, did not limit its definition of 'active' to physical activity but referenced within it the 'continuing involvement in social, economic, spiritual, cultural and civic affairs' (WHO 2001).

---

[2]Note from the editors: some authors, like Lassen and Moreira (2014) rightly interpreted WHO as focusing on lifestyle; however, they only focus on this 'physical' perspective and do not take into account the WHO evolution towards Age-Friendly Cities and Communities that enlarges the sole focus on such physical activity.

Although by this time WHO had become inexorably wedded to an Active Ageing approach much remained to be done in order to build solid policy around it. The announcement of the 2nd World Assembly on Ageing (WAA) in Madrid (2002) provided the impetus. It was clear that the international community expected WHO to adopt a strong protagonism—health being the most important single determinant to quality of life in older age. The Organisation was an active participant at the expert and drafting meetings for WAA. It provided scientific evidence, particularly in relation to Health which is one of the three principles of the Madrid International Plan of Action on Ageing (MIPAA), and lent to it a strong life course perspective.

TM/SG *At this point you launched the so often-cited document, "Active ageing: a Policy Framework"* (WHO 2002). *You have stated that the intention was to give population ageing a global focus. How did you manage to embrace all parts of the world?*

AK Extensive discussions, not only within WHO but also with outside experts, governments and non-governmental organisations, were held for more than two years before my now renamed 'Ageing and Life Course Programme' (ALC) launched the definitive document 'Active Ageing: a Policy Framework' (WHO 2002) at the Madrid Assembly. Particularly, valuable were a series of joint workshops with the UK Faculty of Community Health. The process was immensely facilitated by the Canadian Government who provided resources for meetings and by the secondment of a Canadian Health Promotion writer, Peggy Edwards, to assist as rapporteur alongside my WHO ALC colleague, Irene Hoskins. Workshops were conducted in a number of countries,[3] experts from all WHO regions were appointed to provide evidence and models of good practice and drafts were circulated for feedback from academics and non-governmental organisations from both developed and developing countries. Eventually, at a conclusive meeting of experts at the WHO Kobe Centre, the Active Ageing Policy Framework was finalised.

TM/SG *Could describe the final WHO Active Ageing Policy Framework?*

AK The WHO Policy Framework document, defined Active Ageing as: 'The process of optimizing opportunities for health, participation and security in order to enhance quality of life as people age' (WHO 2002, p. 12).

The Framework was designed to apply to both individuals and population groups. Its intention was to assist people to realise their potential for physical, social and mental well-being throughout the life course, thus enabling them to participate in society according to their needs, desires and capacities. The document additionally emphasises, however, the importance of adequate protection, security and care for individuals as they age.

---

[3]Workshops were conducted in Argentina, Brazil, Chile, Jamaica, Lebanon, Jordan, the Netherlands, Spain, South Africa, Botswana, Hong Kong, Thailand and Malaysia.

Active ageing depends on a variety of influences or 'determinants' that surround individuals, families, communities and nations. Understanding the available evidence about these determinants must inform the design of effective policies and programmes. In the document, WHO discusses in some detail how the broad determinants of health affect the process of ageing, with an emphasis on the health and quality of life of older persons. The Framework called attention to the fact that more research was—and still is—needed to identify and to clarify the role of each determinant and the interaction between them, in the active ageing process. A substantial body of evidence about health outcomes suggests that all of these determinants—and the interplay between them—are good predictors of how well both individuals and populations age. The determinants of Active Ageing diagram (Fig. 5.2) shows that while each group of them can individually influence Active Ageing, they are all interconnected: behavioural determinants, personal, economic, those related to the physical as well as to the social environment and access to health and social services—all embedded on a background of the crosscutting determinants of gender and culture.

The final chapter of the Policy Framework focuses on the Policy Response and calls for action on the three basic pillars[4]:

> Health: "when the risk factors (both environmental and behavioral) for chronic diseases and functional decline are kept low while the protective factors are kept high, people will enjoy both a longer quantity and quality of life; they will remain healthy and able to manage their own lives as they grow older; fewer older adults will need costly medical treatment and care services. For those who do need care, they should have access to the entire range of health and social services that address the needs and rights of women and men as they age" (WHO 2002, p. 45).Participation: "when labour market, employment, education, health and social policies and programmes support their full participation in socioeconomic, cultural and spiritual activities, according to their basic human rights, capacities, needs and preferences, people will continue to make a productive contribution to society in both paid and unpaid activities as they age" (WHO 2002, p. 46).Security (Protection): "when policies and programmes address the social, financial and physical security needs and rights of people as they age, older people are ensured of protection, dignity and care in the event that they are no longer able to support and protect themselves. Families and communities are supported in efforts to care for their older members" (WHO 2002, p. 46).

TM/SG *However, the WHO 'Active Ageing Policy Framework' of 2002 was very theoretical. The same observation has been made about the academic definition by*

---

[4]Life-long Learning (LLL) or 'continuing education throughout the life-course' was an intrinsic component of the 2002 WHO Framework as an aspect of 'Participation' (WHO 2002, p. 51) but it's importance has subsequently been given much greater prominence. The case for LLL to be incorporated into the 'fourth pillar' of active ageing has been strongly articulated. Support for this change initially came from an international conference on active ageing hosted by the Andalusian Government in 2010 which I attended as keynote speaker. The idea became firmly embedded in the active ageing thinking following the publication of a State's White Book on Active Ageing in the same year. The current conception of active ageing is presented in a major revision of the WHO Active Ageing Policy Framework published by the International Longevity Centre-Brazil in July 2015.

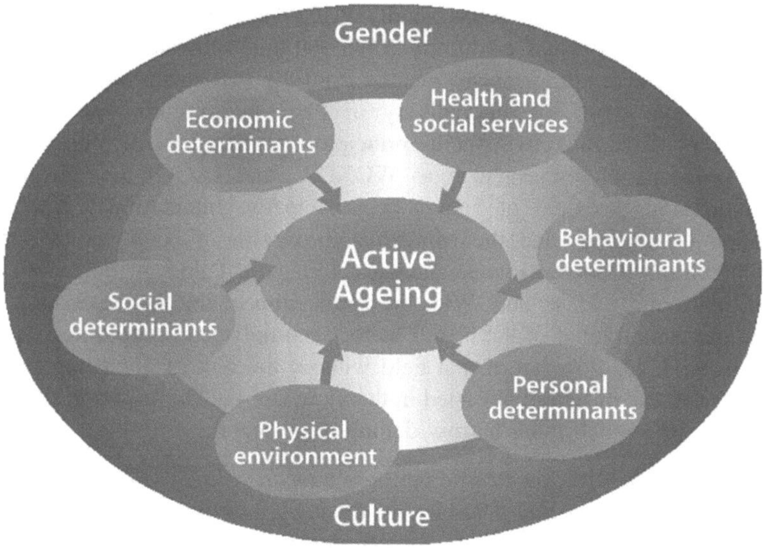

Fig. 5.2 The determinants of active ageing diagram. *Source* WHO (2002, p. 19). Reproduced, with the permission of the publisher, from WHO (2002)

Alan Walker from the same period (Walker 2002, 2009). *What were the steps then taken to put the theories into practice?*

AK Very soon after the launch of the Active Ageing Policy Framework, I quickly recognised the need to translate the Active Ageing theory into practice. The ambition was consolidated by the challenge posed by the United Nations (UN) to adopt a bottom up approach in implementing the Madrid International Plan of Action on Ageing (UN 2003). It is when the idea of 'age friendly approaches' started to resonate more loudly inside me.

The initial step in that direction was to establish in 2003 a project focused specifically on 'Age Friendly Primary Health Care' (PHC)—inevitably incorporating an Active Ageing perspective. WHO had long recognised the critical role that PHC centres play in the health of older people worldwide and the need for these centres to be accessible and adapted to the needs of older populations. No real momentum, however, had been established. Supported by the Australian Government, ALC set up a series of workshops to define a common methodology that could then be applied in simultaneous studies in Australia, Brazil, Canada, Jamaica, Malaysia, the Philippines, Singapore, Thailand and Spain. Older persons and their providers of PHC were asked to describe and define the barriers to care and give suggestions for change (bottom up approach). The results of these qualitative studies, reinforced by background research and a consensus meeting of experts, led to the development of a set of Age-Friendly principles for PHC (WHO 2004). Subsequent work based on the generated principles led to the launch of WHO toolkits (piloted and evaluated in some of the above mentioned countries) showing how to implement Age-Friendly PHC. They were based on three main directions, (i) the physical environment; (ii) attitudes towards ageing and older people and (iii) core competencies needed from the PHC team (WHO 2004).

This seminal work on the delivery of PHC incorporating an age-friendly approach embedded in Active Ageing principles was subsequently endorsed by the leading professional bodies (such as WONCA[5] and IAGG), non-governmental organisations in the field of ageing (e.g. HelpAge International, International Federation of Ageing) and the Ministries of Health of countries as varied as Australia, Brazil, Canada, France, Jamaica, Lebanon, Pakistan, Singapore, South Africa, Spain and Switzerland. With the application of the Active Ageing framework to the central tenet of the health sector accomplished, I could now apply the Active Ageing 'ideology' further a field without the risk of alienating my WHO colleagues—some of whom persisted in the view that the Organisation should limit its activities to a much more narrow definition of health.

TM/SG *Consequently, you steered between a medical and a more comprehensive approach to ageing. In doing so, 'active ageing', with its prominent social component, might be understood as a conversation (with its implicit conflicts) with a traditional 'healthy ageing' formula.*[6] *What was the next 'social' step?*

AK The Ottawa Charter on Health Promotion had taught me that 'Health is created and lived by people within the settings of their everyday life; where they learn, work, play and love' (WHO 1986, p. 4). For Active Ageing to truly reflect all its determinants, it was required to go beyond the constraints of the health sector and to embrace a truly multi-sectorial perspective that would instigate the necessary responses from a myriad of other players from other societal arenas.

In 2005, the Congress of the International Association of Geriatrics and Gerontology (IAGG) provided the perfect stage to launch the response that I had long felt was missing. With the invitation to give the keynote address at the opening session, the organisers requested that I initiate a big idea that would galvanise the media to engage with gerontology, the Congress and the importance of ageing for modern societies. The opportunity was ripe to launch the concept of an 'Age Friendly City'. I based my presentation on actual data collected in informal and improvised focus groups with older persons living in Rio de Janeiro's Copacabana district—a dense urban environment with the highest concentration of older people in Brazil. It was a provocative choice as, despite the demographic realities, for many, Copacabana Beach retains an association with the youthful vigour of its past.

---

[5]Founded in 1972, World Organisation of Family Doctors (WONCA) is a global not-for-profit professional organisation representing family physicians and general practitioners from all regions of the world. WONCA is an organisation in official collaborative relations with the WHO representing family doctors and family medicine.

[6]Note from the editors: this perspective is presented in the Chapter of Moulaert, Boudiny and Paris.

Not only did the media respond to the 'appeal' but international participants came in large numbers to ask for details to see if they could replicate what had been done in such an improvised way.

TM/SG *Researchers in gerontology thus started to invest in 'Age Friendly Cities'. Authors like Chris Phillipson had indeed called attention to the link between ageing and the environment* (Phillipson 2004). *How did researchers come to work on this theme with WHO?*

The global conversation on ageing was entering a new phase in this period. Researchers in several locations, most notably in Canada, were indeed starting to look at ageing more contextually, including within the urban frame. This was inevitable as, early into the new millennium it was becoming increasingly evident to more and more people that population ageing and urbanisation were amongst the greatest legacies of the previous century and in turn, were going to become major challenges for the new century.[7]

WHO is not a research-based organisation per se. Although it incorporates research advances and establishes collaborations with study centres and universities, it has an entirely different *raison d'être* to that of academia. Its role is to highlight directions in health domains, to identify competent partners, to stimulate and to coordinate targeted research and to formulate and disseminate globally relevant protocols based on firm empirical evidence. As it had done with Active Ageing and other initiatives before it, ALC employed advocacy as its most effective tool to generate the necessary momentum towards defining an Age-friendly cities methodology that could be adopted and applied in the same way around the world.

Following the IAGG Congress, I put considerable effort into galvanising support towards the establishment of a formalised global project on Age-Friendly Cities. Once again, cognisant of the almost unique convocation powers of WHO and emboldened by the worldwide impact of the earlier Global Embrace movement, I lobbied hard for resource support. Such support materialised through the Canadian Government. The Director of the Division of Aging and Seniors within the Canadian Public Health Agency (PHAC), Margaret Gills, approved a grant that enabled us to organise a meeting in Vancouver in March 2006 where we could assemble the representatives of the first twelve cities that had expressed a commitment to conduct the relevant research. Crucially, PHAC additionally released their top policy person on ageing, Louise Plouffe, to join our small team in Geneva where she remained for close to three years. The Vancouver Meeting was preceded by a thorough literature review on age-friendly (or similarly named) initiatives. That research formed the basis for identifying the eight topic areas eventually adopted by the Vancouver Protocol. The broad lines of the methodology were defined in consultation with a small group of advisers with expertise in policy, community action or qualitative research and who were familiar with the social context of both developing and developed countries.

---

[7]I would also add 'migration', both internal and external, as the other great legacy—one that remains still largely neglected by ageing researchers.

The first twelve cities were soon joined by others. By August 2006, ALC was coordinating the qualitative research that was being conducted in parallel in 33 cities that represented various urban realities—from mega-cities such as Tokyo, New Delhi, Mexico City and Rio de Janeiro to mid-sized cities such as La Plata (Argentina), Geneva, Kingston (Jamaica) to a number of smaller cities such as Melville (Australia), Ponce (Puerto Rico) and Sherbrooke (Canada). WHO awarded small research grants to non-governmental organisations or research centres in the developing world to facilitate their participation.

TM/SG *While the AFC Guide* (WHO 2007) *has been an application of the comprehensive vision of WHO on 'active ageing'* (WHO 2002)*, another application failed to meet the same success—that is,* WHO (2008) Older Persons in Emergencies: An Active Ageing Perspective. *How do you explain this failure?*

AK Rather ironically, owing to the WHO mandatory retirement policy, I left the Organisation in 2008 shortly after launching the Older Persons and Emergencies publication. I moved to the United States to take up a position at the New York Academy of Medicine (NYAM) where I advised on the implementation of that City's Age-friendly initiative. In addition, my collaborator Louise Plouffe had to return to her home position at PHAC. Neither of us remained at ALC to oversee the work to apply this particular WHO active ageing framework to emergency planning and response for older persons. The document nevertheless remains as valid today as it did in 2008. Emergency preparedness and a response that takes into account older persons' needs and contributions however, is among the characteristics of Age-Friendly Cities (WHO 2007) and some cities have gone further than others in implementing it. In the wake of Hurricane Sandy, that left thousands of older Americans stranded without electricity, food, water and medication, an older adults disaster preparedness and response initiative was prepared and the Age-Friendly New York City programme, which engages all municipal sectors, is the platform that presents such recommendations (Goldman et al. 2014).

TM/SG *Finally, what do you consider are the main challenges for 'active ageing' and 'Age Friendly Cities and Communities'?*

AK Since I departed WHO, I have worked strenuously to actualise Active Ageing through Age-friendly projects in many regions of the world. I am involved in on-going initiatives on Age-friendly States (Sao Paulo, South Australia, Andalusia, Queensland, the Basque Country), Age-friendly Hospitals, Age-friendly Universities and Age-friendly Businesses among others. I maintain great interest in the expanding WHO Age-friendly Cities and Communities Network, and I continue to sit on its Strategic Advisory Group but I additionally have watched the ideology spread and develop its own life-force.

The first challenge is not to lose sight of the importance of the Active Ageing thinking to inform and guide Age-friendly. Without it, Age-friendly, in my opinion, is hollow and the results will be tokenistic and short-lived. Second, more effort must be made to sustain the political support of any initiatives. Continuity can only be achieved by establishing an ownership as wide as possible so that the 'top down' interventions go beyond a single elected administration. Third, while an Age-friendly City or Community or any other Age-friendly project is, by nature, friendly to ***all*** ages, the defined constituency of older persons must be kept at its core. Without the specificity of this particular population group to instruct, nurture and defend its development, the Age-friendly movement would not have thrived and it will not be able to move successfully forward. It is the upward flow of the 'bottom up' component of the structure that feeds it. All too often, age-friendly actions have not lived up to their promise either as a result of this lack of on-going political commitment or as a result of an over predominance of 'top down' interventions and a neglect of the all-important 'bottom-up' protagonism.

References

Chodzko-Zajko, W., & Schwingel, A. (n.d.). Active aging history. Available http://www.humankinetics.com/active-aging-history [Accessed December 23, 2014].

Goldman, L., Finkelstein, R., Schafer, P., & Pugh, T. (2014). *Resilient communities: Empowering older adults in disasters and daily life*. New York, NY: The New York Academy of Medicine. Available http://www.nyam.org/news/docs/pdf/Resilient_Communities_Report_Final.pdf [Accessed December 23, 2014].

Kalache, A., & Kickbush, I. (1997). A global strategy for healthy ageing. *World Health, WHO Magazine, 50th Year* (4), 4–5.

Lassen, A. J., & Moreira, T. (2014). Unmaking old age: political and cognitive formats of active ageing. *Journal of Aging Studies, 30*, 33–46.

Moulaert, T., & Biggs, S. (2013). International and European policy on work and retirement: Reinventing critical perspectives on active ageing and mature subjectivity. *Human Relations, 66*(1), 23–43.

Phillipson, C. (2004). Urbanisation and ageing: towards a new environmental gerontology. *Ageing and Society, 24*, 963–972.

UN. (2003). *United Nations General Assembly 58th Session, Follow-up to the International Year of Older Persons: Second World Assembly on Ageing*. New York, NY: United Nations.

Walker, A. (2002). A strategy for active ageing. *International Social Security Review, 55*(1), 121–139.

Walker, A. (2009). Commentary: The Emergence and Application of Active Aging in Europe. *Journal of Aging & Social Policy, 21*(1), 75–93.

WHO. (1984). *Report of the Scientific Group on the Epidemiology of Ageing*. WHO, Technical Report Series 706.

WHO. (1986). *Ottawa Charter for Health Promotion*. WHO.

WHO. (1996). *The Heidelberg guidelines for promoting physical activity among older persons—Guidelines series for healthy ageing I*. WHO.

WHO. (1997). Active Ageing. *World Health, WHO Magazine, 50th Year* (4).

WHO. (2001). *Health & Ageing: a Discussion Paper*.

WHO. (2002). *Active ageing. A policy framework World Health Organization to the Second United Nations World Assembly on Ageing*. Geneva: WHO.

WHO. (2004). *Towards Age-friendly Primary Health Care*. WHO, Active Ageing Series.

WHO. (2007). *Global age-friendly cities: A guide*. Geneva: WHO.

WHO. (2008). *Older persons in emergencies: An active ageing perspective*. Geneva: WHO.

Part II
Age-Friendly Cities and Communities: One Model, Many Programs

Chapter 6
It's About the People...: Seniors' Perspectives on Age-Friendly Communities

John L. Lewis and Arlene Groh

6.1 Introduction

By now, most urban planning and design professionals are acutely aware of the changing demographic character of urban North American communities and of the challenges that this change represents for community services and physical infrastructure. In brief, the retirement of the baby boomer generation has already begun, and in just over two decades it is projected that 20 % of the North American population will be 65 years or older (Alley et al. 2007; Health Canada 2002). In the province of Ontario, that proportion will reach nearly 25 % (Ministry of Finance 2011). The growing senior population in North America has been portrayed as a looming catastrophe as growing numbers of increasingly frail older people exhaust pension funds and overwhelm the capacity of public health and long term care systems (Peterson 1999). Although an aging population will represent a dramatic shift in housing choices, mobility patterns, and community support needs, the "senior surge" (Hodge 2008) or "silver tsunami" (Ottawa Citizen 2011) is likely to be more of a challenge than a crisis. The challenge rests in the observation among planning academics, professionals, and aging residents that many of the communities in which we live can be difficult environments in which to grow old (Phillipson 2011). For instance, multistoried suburban homes built on spacious lots that were originally planned for young families become increasingly difficult to maintain for aging occupants, and are often isolated from the commercial, social, and health services required by older residents who may no longer drive.

J.L. Lewis (✉)
School of Planning, Faculty of Environment, Research Institute for Aging,
University of Waterloo, 200 University Avenue West, Waterloo, ON N2L 3G1, Canada
e-mail: j7lewis@uwaterloo.ca

A. Groh
Healing Approaches to Elder Abuse and Mistreatment, Waterloo, ON, Canada
e-mail: Arlene.Groh@healingapproaches.com

© Springer International Publishing Switzerland 2016

T. Moulaert and S. Garon (eds.), *Age-Friendly Cities and Communities in International Comparison*, International Perspectives on Aging 14,
DOI 10.1007/978-3-319-24031-2_6

Whether and to what extent older residents are satisfied with a community's social and physical infrastructure, and how to assess that infrastructure, are questions of growing interest to planning professionals and policy makers. In part, this interest is a recent and perhaps belated acknowledgment of the senior surge that is occurring in both urban and rural communities. It is also a reflection of the lack of consistent theoretical frameworks and practical guidance in urban planning regarding the definition, components of or strategies for creating communities that are conducive to successful aging. Several initiatives have sought to address this deficiency, however, the most publicized is a program undertaken by the World Health Organization (WHO), which defined and developed a generalized checklist of characteristics (i.e. applicable in "widely varying countries," according to Plouffe and Kalache (2010, p. 735)) that describes the essential determinants of an age-friendly city.

Recent research suggests that a broad consensus is emerging regarding the fundamental components of an age-friendly community (Hanson and Emlet 2006; Kihl et al. 2005). While the age-friendly frameworks developed by international organizations such as the WHO are a critical and necessary first step in the development of age-friendly concepts and policy, local action plans and strategies require more than globally generalized frameworks. In effect, the characteristics of aging populations across large state and provincial jurisdictions such as Ontario can be quite diverse. Ethnic, health, housing tenure, gender, and income diversity as well as the spatial distribution of senior populations within and across urban, suburban, and rural communities implies that policy strategies will vary depending on the particular needs of each community (Alley et al. 2007). To capture this diversity of condition and needs, Phillipson calls for qualitative research that moves beyond surveys and generalized checklists to document the lived experiences of aging urban residents and "deepen our understanding about the way in which cities are changing, and about the positive and negative contributions that the changes have on the quality of daily life in old age" (2011 p. 289).

In response to the primary research question—*What are the challenges associated with aging in the City of Waterloo*?—the following chapter reports on research conducted for the City of Waterloo's Advisory Committee on Age-Friendly Cities that sought to provide this level of nuance and depth. The following section begins with an overview of the environmental gerontology literature and documents the progression of and relationship between concepts that are central to age-friendly planning and research—i.e., livable communities, aging in place, person-environment fit, environmental press, etc. The chapter then presents the results of interviews conducted with a sample of Waterloo's senior population and concludes by suggesting that while the sources and implications of environmental press are significant for the seniors interviewed, "social press" ought to be an important focus of Waterloo's age-friendly strategy.

6.2 Conceptual Background

As conceptions of the aging process evolve in social and health sciences research, it is clear that popular conception of and policy approaches to aging lag considerably behind. In popular terms, there is a tendency to view the aging experience through a medical lens (Imrie and Hall 2001; Landorf et al. 2008). In effect, aging is conceptualized as a homogeneous and predictable process through which an individual will invariably encounter diminished mental capacity, impairment, health degeneration, and increasing dependence. Ageist attitudes are remarkably common (Lewis 2011) and it "may influence decisions at all levels of human interaction and policy development" (Spencer 2009).

Not surprisingly, the policy approaches that address the aging "problem" are rooted in a philosophy that advocates professional intervention and increased dependence on retirement, nursing, and continuum of care facilities that compensate for physical, cognitive, and sensory decline and social loss.

The inconsistency within North American cultural assumptions and policy approaches toward older people is that substantial physical and financial capital is directed toward safeguarding the health and material stability of people as they age, while implicitly denying or failing to recognize that older people can be active contributors to the economic and social capital of their communities (Thomas and Blanchard 2009). The age-friendly city/community concept has recently emerged to challenge and recast the vision of older people as passive service recipients to active contributors to their communities (WHO 2002, 2007), and with it, bring an enhanced understanding of the aging process to public discourse and policy making. Although it is a relatively new concept, the idea of age-friendly communities is deeply rooted in the environmental gerontology perspective developed in the 1950s by Kleemeier (1959) and subsequently refined by Lawton and Nahemow (1973), Carp (1966), and Kahana (1982). The theoretical framework of environmental gerontology is rooted in the notion that human development over the lifespan is driven by an ongoing exchange or transaction between individuals and the social and physical environment (Hodge 2008). This exchange is expressed behaviorally, cognitively, and emotionally through person-environment *agency* (Wahl and Oswald 2010). In effect, older people strive to maintain their independence for as long as possible, as well as their sense of self-esteem, well-being, and identity, by actively or passively adapting to the demands of the physical environment. For instance, older people living at home and experiencing diminished capacity to climb stairs due to chronic health or mobility impairment can use assistive devices such as a cane or install stair lifts to maintain their independence. Alternatively, they can cognitively re-evaluate the relative value of different domestic spaces—e.g., the basement, attic, and high cupboards—that are painful, difficult, or hazardous to access and adapt their behavior accordingly (Wahl and Gitlin 2007).

Environmental gerontology's focus during the 1950s and 1960s was placed on the day-to-day contexts of aging individuals, reinforcing the importance of "natural settings" or the home and immediate home environment for empirical research.

The rationale is that older people spend approximately three-quarters of their daytime in the home (Oswald and Wahl 2005). One factor influencing the field's focus on housing was the rapid development of retirement living developments during the 1960s, which raised debates about the relative advantages of "age integrated" versus "age segregated" home environments (Kleemeier 1956; Webber and Osterbind 1960), and the degree to which older people wish to remain part of the larger community (Friis et al. 1968). Teaff et al. (1973) subsequently extended this research through their study of age-segregation's advantages for personal security, friendship formation, and access to specialized recreational facilities and health support.

During the 1970s, environmental gerontology's theoretical and empirical scope developed beyond the narrow focus on the relative merits of specific housing types, to conceptualize the ways in which the aging process may be influenced by the physical environment. The ecological theory of aging posits that old age is a critical phase in the life course of an individual and that optimal functioning in older life is a function of person-environment fit (p-e fit)—i.e., the dynamic interaction between the physical and mental competence of an individual and the "press" of his or her environment (Lawton and Nahemow 1973). Competence is portrayed as a characteristic of an individual and consists of such processes as physical health, sensory and perceptual capacities, motor skills, cognitive abilities, and mental health (Lawton 1982). Environmental press encompasses the characteristics of the physical and social environment, particularly the demands that are placed on an individual. Difficulties arise as people in varying degrees and stages of the aging process experience changing health and functioning such that their competence (i.e. cognitive, sensory, health and motor abilities) no longer matches the demands of their environment. The goal is to adjust the level of environmental press or optimize an individual's quality of life in old age through behavioral adaptations or design interventions in the home environment that address accessibility, safety, privacy, autonomy, and personal independence (Hodge 2008; Phillipson 2011; Wahl and Gitlin 2007).

Influenced in part by neighborhood satisfaction research (Parkes et al. 2002), scholars and policy makers have acknowledged that while the home environment may be adapted to accommodate the changing needs of its occupants, the p-e perspective should be broadened beyond domestic and institutional settings to acknowledge that the broader community setting is a fundamental component of maintaining older people's "{...} continuing participation in {...} social, economic, cultural and civic affairs{...}" (WHO 2002, p. 12). This broader perspective was influential in the development of ideas associated with lifetime neighborhoods (DCLG 2008), livable communities (Kihl et al. 2005), and age-friendly cities (WHO 2007). The common thread underlying these concepts is the assertion that effective support for older people requires a range of policy interventions that integrate different parts and scales of the urban system—i.e., from microlevel approaches that address housing, mesoscale strategies that address accessibility to neighborhood services and amenities, to macro or city-wide interventions that encompass the design of streets and transportation networks (Wahl and Oswald 2010). The flagship standard for age-friendly principles is centered on a program initiated by the WHO. Defining an age-friendly city as one that "encourages active aging by optimizing opportunities for health, participation and security in order to enhance quality of life as people age" (Plouffe and Kalache 2010), the program originated in 2006 with a project that sought to define qualitatively the characteristics of an urban community that would support this goal. Through focus groups conducted in 33 cities internationally, senior participants were asked to identify the advantages and challenges of urban living according to eight elements or domains (Table 6.1).

Although significant progress has been made over the past five decades toward the development of a more holistic understanding of how facets of the built and social environments can enhance or complicate the aging process, more research, and policy discussion is required. A significant concern is recognizing the considerable heterogeneity of the aging experience (i.e. the process is not the same for every individual) and the diversity of aging residents and communities. The trends toward urbanization and an aging urban population throughout North America are largely "universal." Within Ontario, however, the needs and choice of policy strategies by municipalities across the province will depend on a consideration of multiple factors such as the community's scale (i.e. urban metropolitan center, mid-sized city, town or village); location (i.e. southern urban/rural, northern urban/rural); the age, condition, and cost of the housing stock; as well as the

Table 6.1 Age-friendly assessment domains

WHO age-friendly cities checklist[a]	AARP livable communities evaluation guide[b]	Advantage initiative[c]
1. Outdoor space and buildings	1. General neighborhood assessment	1. Addresses basic needs
2. Transportation	2. Transportation	2. Promotes civic and social engagement
3. Housing	3. Housing	3. Optimizes physical and mental health and well-being
4. Respect and social inclusion	4. Caring community	4. Maximizes independence for frail and disabled
5. Community and health services	5. Health services	
6. Social participation	6. Recreation and cultural facilities	
7. Civic participation and employment	7. Walkability	
8. Communication and information	8. Safety and security	
	9. Shopping	

Age-friendly domains drawn from a sample of assessment frameworks

Sources [a]WHO (2007), [b]Kihl et al. (2005), [c]Hanson and Emlet (2006)

Reproduced, with the permission of the publisher, from WHO for AFC checklist; from AARP for Livable Communities Evaluation Guide; from The Visiting Nurse Society of New York's for AdvantAge Initiative ethnicity, health status, and income distribution of the resident older population (Alley et al. 2007). Clear guidance about how to develop age-friendly objectives and indicators of policy success that are measurable, reflect the diversity and unique requirements of different communities, and which embody the priorities and lived experience of a community's older residents are crucial steps in addressing the challenges of an aging population.

6.3 Waterloo's Age-Friendly City Initiative: Our Story

The City of Waterloo is a mid-sized city ($\sim$120,000 residents) located 70 km west of Toronto—Ontario's provincial capital and Canada's largest metropolitan center. Along with the cities of Kitchener and Cambridge and three rural townships, Waterloo is a "lower tier" municipality within the Region of Waterloo, home to two universities (the University of Waterloo and Wilfred Laurier University), and is an internationally significant center for computer technology corporations and research institutes (e.g. Blackberry, Open Text, Sybase, the Institute for Quantum Computing). Fueled in part by the Region's continuing growth in the high technology sector, westward expansion of the Greater Toronto Area and relatively affordable housing stock, Waterloo Region's population is projected to grow by 60 % over the next 25 years (Region of Waterloo 2005). While a significant portion of that growth is anticipated to be young university graduates and families seeking employment in "Canada's Technology Triangle," nearly 35 % of Waterloo's population will consist of senior residents (Ministry of Finance 2011). As such, the City of Waterloo provides an ideal context to investigate the challenges associated with a rapidly growing and aging population.

The inception of our work can be traced to the 2009 World Congress of Gerontology in Paris. A key message from this event was the need for supportive and enabling living environments to compensate for the challenges associated with aging. This precipitated the questions—could adequate policies affecting older people result in conditions that decrease the risk of ageist behavior and elder abuse? The nexus between public policy and elder abuse and the potential for an age-friendly community to breach that relationship was the impetus to approach Mayor Brenda Halloran to engage her support for Waterloo's endeavor to become an age-friendly city. Mayor Halloran enthusiastically supported our request and agreed to host a Mayor's Forum through which an Age-Friendly City Advisory Committee was established and several community volunteers were recruited.

The community driven nature of Waterloo's AFC initiative is a unique approach. First, the advisory committee receives no cash funding to support its work from the City of Waterloo (the City provides in-kind support through staff time, access to meeting space, etc.). However, the momentum to sustain our work comes from the passion of committee members, many of who are retired professionals (e.g. planners and engineers, health care professionals, educators) or dedicated volunteers from the City of Waterloo's staff. Thus far, there have been six community AFC events.

The events have confirmed that a high level of interest exists among residents to enhance Waterloo's age friendliness, and they have provided a source of baseline data to identify key areas of concern for community members.

Second, Waterloo's AFC initiative has been sustained through a partnership with the School of Planning at the University of Waterloo. One of our first tasks was to gather preliminary data from the City's senior population to address the question—*what are the challenges associated with aging in the City of Waterloo*? The research team worked throughout 2010 with seniors' centers, senior housing sites, and community organizations that serve elders to inform residents about the Advisory Committee's mandate, discuss the objectives, and invite individuals to participate in the research. A total of 76 individuals representing a range of perspectives agreed to participate. A breadth of viewpoints was sought by inviting participants from different neighborhoods within Waterloo (i.e. the UpTown core, post-war communities and neighborhoods constructed within the past 20 years), multiple seniors' organizations and centers, as well as individuals living independently in their original homes, and those residing within retirement communities (participants did not include assisted living or nursing home residents). All of the participants were either married or widowed, and had lived in Waterloo for 10 years or more. However, the participants were almost evenly split between males and females, and 20 % of the sample was either below 65 or above 85 years of age, with the remainder between 65 and 85 years. Overall, 85 % of the individuals interviewed characterized their health as "good," and approximately the same proportion indicated no "severe or significant" limitations in active daily living. However, 30 % of the participants indicated that they experience or are beginning to experience limitations with instrumental activities of daily living (i.e. housework, cooking, etc.) due to physical and/or sensory impairments.

A fundamental objective of this research was to gather accounts and understand the experience of living in the City of Waterloo from the perspective of senior residents. To this end, a largely qualitative research design was employed. Qualitative methods essentially provide descriptive data that permit a deep understanding of the perceptions and beliefs of participants situated within the context of their lived experience. Given the interest in perceptions and the experience of aging in Waterloo, a structured questionnaire accompanied by semi-structured interview questions was deemed an appropriate data collection strategy. The WHO's checklist and the American Association of Retired Persons (AARP'S) *Livable Communities Evaluation Guide* were used as the basis for the questionnaire, in large part because they are the most extensive instruments developed specifically for use by community groups to evaluate multiple dimensions of livability for seniors. Moreover, the AARP encourages researchers to modify their questionnaire in order to better reflect particular community issues and needs.

The interviews typically took 60–90 min to complete and were generally regarded by the participants as a pleasant opportunity to reflect on the changing character of Waterloo, share issues of personal salience, and dialog with younger members of the community. A pair of graduate research assistants from the University of Waterloo's School of Planning conducted each interview. The first research assistant was responsible for asking the bulk of the questions, while the second research assistant handled the digital audio recorder, compiled notes, and asked supplementary questions based on their notations. Given that a portion of the data consists of Likert-based ratings, the first data analysis task was to tabulate some basic descriptive statistics (i.e. frequencies and means) of the participants' ratings. However, the bulk of the data consisted of qualitative responses and general comments regarding each domain topic. To be conceptually meaningful, the interview recordings were transcribed and, together with field notes collected during the interview, subjected to a form of content analysis based on the Constant Comparative Method (Strauss and Corbin 1990). The data were organized into a research database by iterative readings and examined in terms of consistencies and inconsistencies among individual participants and, more particularly, themes that described the virtues and limitations of aging in Waterloo.

6.4 Results

6.4.1 General Assessment of Waterloo and Sources of Press

For several domains, the findings revealed positive news about Waterloo's livability for seniors. The participants' satisfaction with and affection for their neighborhoods was palpable during the interviews, and clearly reflected in the quantitative and qualitative results. A clear majority of participants (95 %, $n = 71$) indicated that they are satisfied with their neighborhood and home, and would like to remain living in their home environment for as long as possible. From their comments, the prevalent feeling of satisfaction was attributed to Waterloo's compact and therefore "small-town" neighborly atmosphere, combined with the commercial and cultural amenities of a larger city:

> I live in a good neighborhood with good friends and people, strong social ties, this is an average neighborhood. The one thing that I like about living in Waterloo is that it's not too big, but there's a lot to do here; it's definitely not Toronto (Similar comments expressed by 20 other participants).

A common refrain among the participants was the sense of safety and social rapport that results from living among people who share a common history and attachment to their neighborhood, and a mutual willingness to attend to one another's well-being:

> We have a nice home, nice neighborhood and neighbors. We like seeing the children playing in their yards or in the snow. Also, when we need help with snow blowing or whatnot, the people on our block are always ready to lend a hand (Similar comments expressed by 16 other participants).

Nevertheless, the challenges of aging at home are becoming increasingly salient to some participants and have compelled others to re-evaluate their living circumstances. Personal independence or the capacity to make autonomous long-term and daily living choices is perhaps the most important factor underlying participant conceptions of aging well:

> The biggest consideration for us is being able to make decisions for ourselves, leading life at our own pace and avoiding isolation" (Similar comments expressed by 24 other participants).

However, the factors that constrain senior independence stem from two sources —accessibility (both domestic and neighborhood) and social behavior. As discussed above, the degree to which the physical environment can create "press" or present challenges and barriers is a well-established concept in the environmental gerontology and disability studies literatures (Hodge 2008; Lawton 1982). In effect, as physical strength and/or sensory acuity diminishes and chronic health issues become more prevalent with increasing age, domestic features such as kitchen cabinets and stairs, as well as daily tasks (e.g. laundry, yard maintenance) that were once tacit components of daily existence become overt problems:

> It's getting tougher to manage this place, indoor and out. I'm getting to the point where I'm questioning whether I need a four-bedroom house just for one person. With my husband gone, I now have to hire people to help take care of it (Similar comments expressed by 12 other participants).

A common theme in many of the interviews was the relative tractability of the "presses" that are associated with aging. On the one hand, while the sources of environmental press and accessibility are very important and substantive concerns, there is an overt sentiment among the participants that growing old should not imply that the capacity to make independent choices and adapt personal behaviors or one's environment has vanished. For some, this was most clearly expressed through the assertion that younger people have the unfounded sense that seniors are passive victims of the aging process and tend not to appreciate their capacity to be resilient and active decision-makers:

> I may be getting older but I still have what it takes to look after myself. Sure, some things are getting harder, but I cope and try to make things work (Similar comments expressed by 18 other participants).

Ultimately, these participants acknowledge that a point is reached when various forms and degrees of chronic illness and impairment (particularly for the 'old-old' or 'oldest old' such as those over 85; Marshall and Bengston 2011) appreciably constrain an individual's competence and capacity to live independently. At that juncture, the physical environment becomes less tractable and the decision to move into an assisted living facility or reside with adult children becomes increasingly unavoidable. However, the central message from the participants posits that while older people are capable of addressing environmental presses as their minds and bodies change, the "social press" that is associated with the behavior and underlying attitudes of other community members can be the most difficult challenges to address. In effect, while environmental features can be modified to improve their accessibility, or avoided until essential modifications are made, the social presses that arise from the attitudes and manifest behaviors of public service providers, neighbors and youth, for instance, cannot be changed or readily avoided:

> I can deal with the house, it's a struggle and it may cost a penny or two, but you can't change people. I'm not saying that seniors aren't respected in Waterloo, I love living here, but there's the odd person, strangers really, who makes life pretty darned difficult. That you can't change (Similar comments expressed by 15 other participants).

6.4.2 Transportation

Transportation is a critical factor in sustaining senior independence, and older people in Waterloo employ a variety of means of traveling throughout the city to shop, visit friends or family, participate in social events, or attend to personal finances or health. For the most part, 68 % ($n = 52$) of the participants indicated that walking or driving are their preferred travel modes, and 14 % ($n = 11$) remarked that they had quit driving due to poor vision, neck or back pain, or slower reflexes and have come to rely on friends or family for rides to destinations that are outside their immediate neighborhood. Remarkably, only 9 % ($n = 7$) of the participants take public transit on a regular basis, while 78 % ($n = 59$) commented that they avoid the public transit system and prefer to hire a taxi or request a ride from a friend or relative. Among the 59 participants who avoid Waterloo's bus system, the reasons given for this strong response encompassed a variety of factors including the legibility of schedule information (4 %), fare costs (4 %), and the regularity of bus services (11 %). However, the most prevalent explanation (43 %) pointed toward the behavior of bus drivers and other transit patrons. In terms of the latter, a significant source of frustration and physical discomfort follows from the reluctance of younger passengers to offer their seats on full buses or, in some cases, deliberately occupying seats that are visibly marked for older or impaired persons:

> What really rattles my cage is when I get on the bus, and there's already a lot of other passenger on you know, and there's kids sitting in the disabled seats at the front. Sometimes they give them up when you board, but more often than not they don't (Similar comments expressed by 14 other participants).

Greater levels of frustration and the real potential for physical injury can result from the propensity of some bus drivers to move their vehicle before an older passenger has been able to seat themselves (Ottawa Sun 2011):

> Drivers don't wait for seniors to sit, they're not courteous. I have hip problems and there's times I've had to hold on for dear life while the driver pulls away from the stop and I nearly wind up on the floor. I need to take the bus to get to Conestoga [Mall], but when drivers do things like that, I'm nervous taking transit (Similar comments expressed by 19 other participants).

6.4.3 Walkability

The walkability of a city's roads, sidewalks, and paths is an important variable in carrying out daily activities, but in maintaining a physically active lifestyle as well. From prior research through which the author evaluated the walkability and physical accessibility of selected street segments in Waterloo's UpTown neighborhood, it was expected that factors such as the timing of pedestrian crosswalks, or the absence of curb extensions along wide arterial roads would surface during the interviews, particularly among those participants with mobility and sensory impairments, and chronic health issues that limit physical stamina. For the most part, coping strategies involved behavioral adaptations such as changing walking routes to road segments that are relatively easy to cross or adjusting their walking schedule to periods of the day when traffic volumes are comparatively low:

> Pavers are unlevel in many parts of UpTown and cause tripping hazards. Most of the walk signals are too fast, especially at Erb and Caroline, they don't give you enough time. You always have to hurry. It's gotten to the point where I'll cross where there isn't as much traffic, or just not leave the house during rush hour (Similar comments expressed by 12 other participants).

While these strategies may be an inconvenient but effective means of addressing walkability challenges, the greater difficulty often ensues from trying to "deal with those idiots on the road…" who can make walking a hazardous experience at any time of day or location. For instance, older pedestrians sometimes contend with drivers who cannot spare the time to allow slower walkers to move through a crosswalk:

> Crossing the road can be difficult {…} people don't stop on Father David Bauer drive to let us cross, and the drivers on King Street get impatient when I'm in the intersection. We need better crosswalks, longer stoplights or something else. The pedestrian crossing sign seems to mean nothing to people in a car (Similar comments expressed by 27 other participants).

Moreover, it is perhaps inevitable that in a smaller city with two universities and a community college the prevalence of young adults cycling to and from their respective campuses would result in pedestrian conflicts. In this regard, several participants remarked on the habit of young cyclists to travel along sidewalks and fail to signal their presence to older pedestrians, who may in turn be startled and physically strained when the cyclist's sudden appearance and rapid movement elicits an avoidance reflex:

> They're terrifying when they whip past you with no bell. Cyclists come up behind me so fast and surprise me and skateboarders are a problem too (Similar comments expressed by 18 other participants).

In part, the onus rests on cyclists to signal their approach or avoid using sidewalks and pedestrian paths, where dedicated bicycle lanes are provided. In the latter regard, some participants acknowledged that cyclists are often compelled to use pedestrian infrastructure in order to avoid aggressive motorists:

> The problem is the cyclist lanes are too dangerous to begin with so I don't blame them for using the sidewalk. The cars are moving too fast and sometimes you can see them driving in the cyclist lanes (Similar comments expressed by 5 other participants).

6.4.4 Housing

As the locus of daily activity, the home is the most immediate source of environmental press. For the interview participants, there were three general concerns with Waterloo's housing stock—its accessibility, affordability, and perceived security. First, the scale and layout of a dwelling can make the performance of routine activities more challenging—e.g., climbing stairs, carrying laundry to the basement, placing waste bins on the curb, lawn maintenance, and snow removal—because of the loss of physical strength, agility, motor or sensory loss, or the presence of chronic pain:

> If you are handicapped then it's a poor place to live, if you are well then it's a wonderful place. My shoulder though is making housework a problem in the winter, and it's getting difficult to carry the laundry basket up and down the stairs at any time of the year (Similar comments expressed by 17 other participants).

As such challenges intensify, installing accessible features such as stair-mounted chair lifts, extendable shelving, and specialized bathroom fixtures can mitigate the home's environmental press. However, a point is invariably reached when the addition of accessible features can no longer compensate for the resident's loss of competence, or building retrofits become increasingly unaffordable. In general terms, housing affordability can be a significant concern for aging Waterloo residents where residential property tax rates are among the highest in Ontario (The Record 2011):

> It's getting expensive to run a large home. There's the usual costs with taxes, heat and hydro, but it's now getting to the point where we have to arrange for snow removal, salt and grass cutting. As we get older we may not be able to do these things anymore, and paying someone else to do them begins to eat away at a fixed income (Similar comments expressed by 11 other participants).

In addition to the physical and financial challenges associated with aging at home, the behavior of neighbors can play a substantial role in the decision to relocate. For the most part, relationships with neighbors were described in generally favorable terms. In addition to friends and family, they are an important facet of older residents' social networks and, on occasion, a source of support with domestic responsibilities. However, many post-secondary students in Waterloo reside off the two university campuses and, as it is perhaps common for many university towns, the standard of care for student rental properties can be lower than their homeowner neighbors. Combined with periodic occurrences of vandalism and noise, some participants have found themselves living in areas of relative decline, both socially and in terms of the asset value of their properties:

> Things have gotten poorer here over the last couple of years. A lot of the students that have moved into the neighborhood get foolish. Every now and then, but especially Fridays, we have spill over where damage occurs to personal property, some cars have been vandalized, things have been broken (Similar comments expressed by 17 other participants).

6.4.5 Social Inclusion

Although discussions regarding the accessibility of the home environment and the availability of public services provided interesting insights into the participants' aging experience, perhaps of greater interest were the comments that centered on the social inclusiveness of community services. In brief, some participants (25 %, $n = 19$) characterized their experience of the aging process as one in which opportunities to experience new ideas and people becomes markedly constrained. For those participants who reside in retirement communities this sentiment was somewhat anticipated. For the most part, they described their living circumstances as clean, pleasant, and enjoyable, but lacking in the social and intellectual fulfillment that could be derived from interacting with people outside their residence:

> Aren't we sort of isolating ourselves as seniors, and is that a good thing? When you want to have a community of just seniors, I think the seniors are losing out on new ideas, I mean you have to keep your mind open (Similar comments expressed by 12 other participants).

While statements such as this were less prevalent among those participants who continue to live in their original homes, the feeling of increasing social isolation did surface on occasion:

> It's harder to stay connected with other people {…}. Just having younger people around with different interests and just brightening the day. I mean you don't always want to talk with your old neighbors, I really enjoy talking to young people with their ideas and ambitions (Similar comments expressed by 7 other participants).

According to these participants, the City of Waterloo does a tremendous job of providing senior focused programs, but there is a relative dearth of opportunities for seniors to mix with others outside their usual social networks and encounter the experiences and ideas that other generations possess. Although this is not a universal sentiment, for those participants who shared this mindset, the learning process does not abate or cease when an individual enters their later years. Indeed, inter generational interaction and learning can be an essential means of sustaining the feeling of excitement and fun that comes through discovery, which ultimately maintains mental fitness in later life.

In general, the challenges associated with aging in the City of Waterloo are both physical and social in nature. For older people, it matters a great deal whether commercial, social, and recreational amenities are available and within easy access, and it matters whether the built environment at home and in the public realm accommodates or places excessive press on the physical, sensory, and cognitive capacities of a senior resident. The forms of environmental press described above were characterized by the participants as more or less tractable issues to the extent that architectural interventions can be made to address the home environment; civic governments, and commercial establishments can be pressured to improve the accessibility of their buildings, open spaces, and services; and friends and extended family members can be called upon to provide transportation or assistance with domestic responsibilities. As Hodge (2008, p. 124) posits, the person-environment relationship is an "active transaction" between a senior and their environment, which is defined by changes to the person as well as the home to maintain competence and independence. The underlying message from the preceding discussion is that while elements of the physical environment are amenable to behaviors that seek to overcome environmental press, expressions of *social press* (e.g. aggressive drivers and cyclists, impatient transit drivers) are less amenable to adaptive behavior. In effect, it is easier to change proactively one's living conditions to maintain independence than it is to adapt (or adapt to) the behavior of others. The implications of these findings for future research and planning policy are addressed in the concluding discussion.

6.5 Reflections

Waterloo's age-friendly initiative has encountered considerable success. In September 2011, the WHO designated Waterloo as a member of the Global Network of Age-Friendly Cities. In June 2013, a Mayor's report was compiled and endorsed by City Council (http://www.waterloo.ca/agefriendly), which includes an action plan to address Waterloo's continued growth as an age-friendly community. In response, Waterloo staff developed a detailed action plan to address the recommendations outlined in the report. In September 2013, the action plan received unanimous approval by City Council and subcommittees of City staff and community volunteers were formed to collaborate with the City to operationalize the plan.

In general, our experience has taught us the importance of people as the keystone of an age-friendly initiative. Waterloo's initiative would not have succeeded without the support of political champions such as Mayor Brenda Halloran, the diligent work of City staff and community volunteers, and Waterloo residents who are embracing the initiative to ensure that Waterloo continues to grow as an age-friendly city. In addition, a clear theme from the research conducted in collaboration with the University of Waterloo is the importance of both the built and social environments for older people and the relationship between them. For instance, the significance of the home and neighborhood environment is evident from comments that underscored the need for affordable and accessible housing, public and commercial services available within reasonable travel time, and safe and accessible pedestrian infrastructure. Commensurate with these concerns, a community's respect for older adults and the extent to which it provides the physical and social infrastructure that allow seniors to stay connected to their social networks and involved in the life of the community are important determinants of aging well (Gardner 2008).

The impetus for undertaking Waterloo's AFC initiative was the conviction that AFC could be a strategy to address ageism and elder abuse. Elder abuse may be considered one of the most extreme forms of ageism, which is socially constructed and perpetuated (Struthers 2007). Does Waterloo's age-friendly initiative address the proposition that ageist and abusive behaviors are rooted in, among other possible causes, inadequate public policies that affect older adults? This of course requires further research and evaluation. One contribution of the City of Waterloo's research is the proposition that the environmental presses that result from person environment transactions are important considerations in community planning for the elderly. However, community planners must also recognize that the age-friendly community concept is fundamentally about *people* and the presses that ensue from social transactions. On the one hand, the interview participants acknowledged the significant policy and financial commitments made by the City of Waterloo toward senior social and cultural programs and the management of public spaces to enhance their safety and accessibility. Moreover, while financial and temporal constraints may limit a senior's capacity to age indefinitely in their home environment, the participants expressed the characteristic human desire and capacity to maintain their independence in the face of challenges that are within their locus of control. In effect, environmental press can be managed. On the other hand, social press or the ways in which other people can impinge on seniors' lives is a distinct and far less manageable issue (i.e. it is more challenging to change or anticipate troublesome behavior from others). As described above, social press results from the tacit attitudes and assumptions about aging, which lead to behaviors that can appreciably affect life satisfaction, physical, and social well-being. Ultimately, aging urban residents are less "prisoners of space" than they are prisoners of attitude and behavior (Phillipson 2011).

Although addressing social press is perhaps no more tractable for community planners than it is for seniors, there are perhaps two paths through which the social presses associated with aging could be addressed. The most common approach is to call for greater public awareness through various forms of government sponsored social marketing campaigns, public outreach, or professional training events (Lewis 2011). The underlying assumption is that the public generally expresses favorable attitudes toward seniors and people with impairments, but possess deeper unspoken beliefs and attitudes that are frequently misinformed or, on occasion, unfavorable (Daruwalla and Darcy 2005). Given the growth of social marketing efforts to address issues such as elder abuse (Seniors Canada 2009), the propensity of academic researchers and consultants to advocate for education or awareness training as a policy strategy, and the mixed pattern of findings in the educational psychology and disability studies literatures, the potential effectiveness of educational strategies merits closer examination (Lewis 2011).

A more compelling suggestion is that the presence of ageist attitudes and abusive behavior toward seniors is a consequence of our modern culture's tendency to simplify the ageing experience and conceptualize it through a medical lens that emphasizes dependence, increasing ill-health, disability, and decline (Landorf et al. 2008). This perspective has led to policies that advocate professional intervention to facilitate independence at home or, more commonly, segregation from the rest of the community in retirement/continuum of care/assisted living facilities. We posit that stereotypes, ageist attitudes, and inappropriate behaviors are likely to exist when the most prevalent response to aging is to isolate older adults and limit opportunities to remain visible, interact with the broader community, exchange perspectives and experiences, and maintain or establish new social networks. Advertently or inadvertently, our society has transitioned from a conception of older adults as "elders" or, in other words, as respected and vital contributors to family and community life. The dominant paradigm today conceptualizes aging as a debilitating, problem-ridden phase of life that is fraught with medical challenges that necessitate specialized—i.e., institutionalized—care and intervention. Planning policies that treat or *zone* retirement and assisted care facilities as "institutions" implicitly support this paradigm by removing specialized older adult residences from neighborhoods where residents may sustain or establish new social relationships. Although we have made considerable progress in extending life and managing age-related diseases through such facilities, considerable effort needs to be redirected toward ensuring that social and planning policies do not diminish the connections that seniors have with the larger community.

Rather than a dichotomous conception of aging well in terms of either aging in place or aging in self-contained (i.e. in terms of medical, recreational, cultural services) but socially isolated "villages" for seniors, some sources suggest that planning and gerontology professionals ought to refocus their efforts toward policies that enable "aging in community" (Thomas and Blanchard 2009). Approaches to aging in community include revising municipal planning policies (zoning by-laws in particular) to allow accessory dwelling units on single detached lots, co-housing developments that provide private living units within communally managed living areas, and multi-generational "urban villages" where seniors are an integral part of receiving as well as providing commercial and community services—e.g., library/reading rooms, day care, community gardens. The aging in community concept is predicated on the idea that aging should neither be seen as a crisis or a threat, but as an opportunity; where people across life's age spectrum are valued, reside within and are involved in community life, and are encouraged to share interests and sources of fulfillment (Thomas and Blanchard 2009). Whatever its manifestation, aging in community is a response to the prevailing practice of institutionalizing seniors, not only because it consumes large quantities of financial capital, but more importantly because it isolates substantial reservoirs of social capital.

Community planners, architects, urban designers, gerontology, and public health professionals ought to be able to envision new and more innovative models of living that provide more choice for seniors and fill the wide dichotomy that exists between aging at home and assisted living. The dominant planning and land development paradigm needs to shift from the creation of housing districts that support young families to the creation of inclusive and socially vibrant communities that provide access to resources, foster opportunities for older people to pass their experiences and talents onto younger generations, and live a full life regardless of age or other circumstances.

References

Alley, D., Liebig, P., Pynoos, J., Banerjee, T., & Choi, I. (2007). Creating elder friendly communities. *Journal of Gerontological Social Work, 49*(1), 1–18.

Carp, F. (1966). *A future for the aged: The residents of Victoria Plaza*. Austin, TX: University of Texas Press.

Daruwalla, P., & Darcy, S. (2005). Personal and societal attitudes to disability. *Annals of Tourism Research, 32*(3), 549–570.

Department of Communities and Local Government. (2008). *Lifetime homes, lifetime neighbourhoods: A national strategy for housing in an aging society*. London: DCLG.

Friis, H., Townsend, P., & Shanas, E. (1968). Old people in three industrial societies: An introduction. In E. Shanas, P. Townsend, D. Wedderburn, H. Friis, P. Milhoj, & J. Stenhower (Eds.), *Old people in three industrial societies* (pp. 2–18). London: Routledge.

Gardner, P. (2008). *The public life of older people: Neighbourhoods and networks*. PhD Dissertation, Graduate Department of Public Health Sciences, University of Toronto.

Hanson, D., & Emlet, C. (2006). Assessing a community's elder friendliness: A case example of the advantage initiative. *Family and Community Health, 29*(4), 266–278.

Health Canada. (2002). *Canada's aging population. Division of aging and seniors*. Health Canada, Minister of Public Works and Government Services Canada.

Hodge, G. (2008). *The geography of aging: Preparing communities for the surge in seniors*. Montréal and Kingston: McGill-Queen's University Press.

Imrie, R., & Hall, P. (2001). *Inclusive design: Designing and developing accessible environments*. London: Taylor and Francis.

Kahana, E. (1982). A congruence model of person-environment interaction. In M. P. Lawton, P. Windley, & T. Byerts (Eds.), *Aging and the environment: Theoretical approaches* (pp. 97–121). New York, NY: Springer.

Kihl, M., Breenan, D., Gabhawala, N., List, J., & Mittal, P. (2005). *Livable communities: An evaluation guide*. Washington DC: AARP Public Policy Institute.

Kleemeier, R. (1956). Environmental settings and the aging process. In J. Anderson (Ed.), *Psychological aspects of aging* (pp. 105–116). Washington DC: American Psychological Association.

Kleemeier, R. (1959). Behavior and the organization of the bodily and external environment. In J. Birren (Ed.), *Handbook of aging and the individual* (pp. 400–451). Chicago: University of Chicago Press.

Landorf, C., Brewer, G., & Sheppard, L. (2008). The urban environment and sustainable aging: critical issues and assessment indicators. *Local Environment, 13*(6), 497–514.

Lawton, M. P. (1982). Competence, environmental press, and the adaptation of older people. In M. P. Lawton, P. Windley, & T. Byerts (Eds.), *Aging and the environment* (pp. 33–59). New York, NY: Springer.

Lawton, M.-P., & Nahemow, L. (1973). Ecology of the aging process. In C. Eisdorfer & M. P. Lawton (Eds.), *Psychology of adult development and aging* (pp. 619–624). Washington DC: American Psychological Association.

Lewis, J. (2011). Student attitudes toward impairment: An assessment of passive and active learning methods in urban planning education. *Teaching in Higher Education, 16*(2), 237–249.
Marshall, V., & Bengston, V. (2011). Theoretical perspectives on the sociology of aging. In R. Settersten & J. Angel (Eds.), *Handbook of sociology of aging* (pp. 17–34). New York, NY: Springer.
Ministry of Finance. (2011). *Ontario population projections update: 2010–2036, Ontario and its 49 census divisions*. Toronto: Queen's Printer for Ontario.
Oswald, F., & Wahl, H.-W. (2005). Dimensions of the meaning of home in later life. In G. Rowles & H. Chadhury (Eds.), *Home and identity in later life* (pp. 21–46). New York, NY: Springer.
Ottawa Citizen. (2011). *Growing old at home: The silver tsunami is on the way and studies show boomers want to stay in their own homes as they age. Will builders rise up to meet the accessibility needs of the coming wave?* (p. I1). August 27, 2011.
Ottawa Sun. (2011). *Injured bus riders Sue OC for $4.2M* (p. A3). June 2, 2011.
Parkes, A., Kearns, A., & Atkinson, R. (2002). What makes people dissatisfied with their neighbourhoods. *Urban Studies, 39*(13), 2413–2438.
Peterson, P. (1999). *Gray dawn*. New York, NY: Times Books.
Phillipson, C. (2011). Developing age-friendly communities: New approaches to growing old in urban environments. In R. Settersten & J. Angel (Eds.), *Handbook of sociology of aging* (pp. 279–293). New York, NY: Springer.
Plouffe, L., & Kalache, A. (2010). Towards global age-friendly cities: Determining urban features that promote active aging. *Journal of Urban Health, 87*(5), 733–739.
Region of Waterloo. (2005). *Planning Our future: Regional growth management strategy. Kitchener: Regional municipality of waterloo*. Available: http://www.region.waterloo.on.ca/web/region.nsf/vwSiteMap/71DC3804F65AFA9485256B1A0062B956/$file/Regional%20Growth%20Management.pdf. Accessed September 8, 2011.
Seniors Canada. (2009). *Elder abuse—It's time to face the reality*. Available: http://www.seniors.gc.ca/c.4nt.2nt3c.41@.jsp?lang=eng&cid=161. Accessed September 8, 2011.
Spencer, C. (2009). *Ageism and the Law: Emerging Concepts and Practices in Housing and Health*. Toronto: The Law Commission of Ontario.
Strauss, A., & Corbin, J. (1990). *Basics of qualitative research: Grounded theory procedures and techniques*. Newbury Park, CA: Sage.
Teaff, J., Lawton, M., Nahemow, L., & Carlson, D. (1973). Impact of age segregation on the well-being of elderly tenants in public housing. *Journal of Gerontology, 33*, 130–133.
The Record. (2011). *Waterloo has high taxes, high debt, high incomes*. 11 July, page A7.
Thomas, W., & Blanchard, J. (2009). Moving beyond place: Aging in community. *Generations, 33*(2), 12–17.
Wahl, H.-W., & Gitlin, L. (2007). Environmental gerontology. In J. Birren (Ed.), *Encyclopedia of gerontology: Age, aging and the aged* (pp. 494–501). Oxford: Elsevier.
Wahl, H.-W., & Oswald, F. (2010). Environmental perspectives on aging. In D. Dannefer & C. Phillipson (Eds.), *International handbook of social gerontology* (pp. 111–124). London: Sage.
Webber, I., & Osterbind, C. (1960). Types of Retirement Communities. In E. Burgess (Ed.), *Housing the elderly in retirement communities* (pp. 157–171). Ann Arbor: University of Michigan, Division of Gerontology.
World health organization. (2002). *Active aging: A policy framework*. Geneva: WHO.
World Health Organization. (2007). *Global age-friendly cities: A guide*. Geneva: WHO.

Chapter 7
How Can a Research Program Enhance a Policy? AFC-Quebec Governance and Evaluation Opportunities

Suzanne Garon, Anne Veil, Mario Paris and Samuèle Rémillard-Boilard

7.1 Introduction

The Age-Friendly Cities model as implemented in Quebec[1] (VADA-QC) is a response to the accelerated aging of its population, due to a combination of factors, such as low birth rate, longer life expectancy linked to the expansion of medicines, health care and services, and finally, the presence of a significant cohort of "baby boomers" who has entered in the past few years in the 65 years and over age group (Choinière 2010). Quebec is home to 1,405,919 people aged 65 or over, accounting for 17.1 % of the province's population or one Quebecer out of six (ISQ 2014, p. 33). The number of older individuals is rising sharply. By 2031, a full 26 % of the Quebec population will be aged 65 or over, one Quebecer out of four. Life expectancy at birth for men and women combined now stands at 81.8 years—one of the highest in the world (ISQ 2010). This accelerated aging of the Quebec society surely explains the rapid deployment of age-friendly cities initiatives. From the

[1]Quebec represents the second largest province in Canada with 8.18 million inhabitants. It is the only Canadian province having French as official language with more than 75 % of francophone.

S. Garon (✉)
School of Social Work and Research Centre on Aging, University of Sherbrooke, Sherbrooke, QC, Canada
e-mail: Suzanne.Garon@USherbrooke.ca

A. Veil · M. Paris
Research Centre on Aging, Sherbrooke, QC, Canada
e-mail: anne.veil@usherbrooke.ca

M. Paris
e-mail: Mario.Paris@USherbrooke.ca

S. Rémillard-Boilard
School of Social Work, University of Sherbrooke, Sherbrooke, QC, Canada
e-mail: samuele.remillard-boilard@usherbrooke.ca

© Springer International Publishing Switzerland 2016

T. Moulaert and S. Garon (eds.), *Age-Friendly Cities and Communities in International Comparison*, International Perspectives on Aging 14,
DOI 10.1007/978-3-319-24031-2_7

involvement in 2006 of the AFC-QC research team in the work of the World Health Organization (WHO) to the implementation of seven pilot projects in 2008, there are now (in 2015) more than 750 age-friendly cities in Quebec out of 1133 municipalities (MAMOT 2014). More than 86 % of the Quebec population lives in a municipality that is in the process of joining this initiative.

The AFC program, which then become the governmental program *Municipalités Amies des Aînés* (MADA), is an approach that offers concrete opportunities to intervene on the eight areas[2] identified by the WHO (2007a), which combine the actions of built or social environments that can enhance the life conditions of seniors by intervening on different aspects of their social inclusion. These eight areas, which are similar to the major social determinants of health (Mikkonen and Raphael 2010), do not pose all the same challenges for a city that tries to adapt to the aging of the population. In addition, there are very large differences between urban and rural environments.

The program, as developed and implemented in Quebec, is based on a methodological approach of community development planning, which seemed to us compatible with both active aging (WHO 2002), a promising concept valued by WHO, and the bottom-up approach from the Madrid International Plan of Action on Aging (UN 2006). This positioning emphasizes on the participation of seniors in all stages of the approach, viewed as essential to ensure a relevance between the needs that they have expressed and measures to be implemented. This participation of seniors in elaborating the definition of their needs (Godbout and Charbonneau 1994) helps prevent service providers involved with seniors to exclusively determine them. This approach therefore is more oriented toward the "user led", rather than the "service led" (Carrier 2012).

In this chapter, we first describe the program approach since the beginning of the AFC-QC project undertaken in seven pilot sites between 2008 and 2013, in the line with the work of the World Health Organization (2007a), until the successive rounds of the program which benefited, as early as 2009, more than 50 cities, and then to more than 750 municipalities in 2015. In a second section, we show how this program of public action is built in parallel with a research planning structured over eight years. In a third and last section, we present one of the major contributions brought about by a combination of public programs and research on the subject, conducted in Quebec, through the concern of consistency between the research approach and the specific characteristics of AFC projects. We propose a reflection about the role of the various stakeholders in the program and the different models of evaluative research which have marked the progress of our work.

---

[2]The eight areas are: outdoor spaces and buildings, transportation, housing, social participation, respect and social inclusion, civic participation and employment, communication and information, community support and health services. Each of the areas is distributed in built environments (outdoor spaces and buildings, transportation, housing) and in social environments (the other five), as presented in Fig. 7.3.

In conclusion, we suggest the idea that issues are in the redefinition phase, whereas the Government of Quebec maintains its line of transition which modifies the parameters of governance in a meaningful way.

7.2 Principles of Action Providing the AFC-QC Model

7.2.1 The Influence of the World Health Organization

With the aging of populations, many international organizations have put forward a series of guidelines to provide a better framework for changes necessary to mitigate the impacts on different social structures, including the health and social services system (Moulaert and Biggs 2013; Moulaert and Paris 2013). Among these organizations, the WHO has particularly influenced the Quebec model by linking it to the promotion of developing adequate environments for seniors (WHO 2007a).[3] In addition, the framework for *Active Aging* (WHO 2002) was also a meaningful foundation for AFC-Quebec to build upon. 'Active ageing is {…} the process of optimizing opportunities for health, participation and security in order to enhance quality of life as people age' (WHO 2002, p. 12).

The WHO has particularly drawn the attention to the contribution of active aging in order to maintain the autonomy as shown in the following diagram (Fig. 7.1).

According to the WHO, this approach stems from the fact that aging is a process that evolves throughout life and that it remains possible to enhance the condition of human beings at any time in their life course. But still, it is the dimension of the social participation, inherent in the active aging of the WHO, which constitutes the cornerstone for understanding the development of AFCs in Quebec. Recent research underlines the importance of the participation of seniors on their mental health and their sense of well-being (Morrow-Howell et al. 2003; Van Willigen 2000), while others establish a link between participation and physical health (Bennett 2005; Bath and Gardiner 2005). In addition, the work carried out by Fougeyrollas (2009) on the disability creation process, was widely known in Quebec for several years. The concept of the disability creation process specifically raises the question of adapting environments to persons with disabilities. With that in mind, the development of physical or built environments is more respectful of various realities faced by people living with disabilities remains essential to their opportunity to participate socially.

In Quebec, active aging has been positioned within the ecological model which addresses the WHO's eight key areas of community life in an extensive and non-restrictive way, enabling it to work on a systemic basis, in synergy. We can then perform networking actions, whether between the different areas, or between

[3]The overall context of social policies which have preceded the WHO project has been described in more detail in a previously published article (Paris et al. 2013).

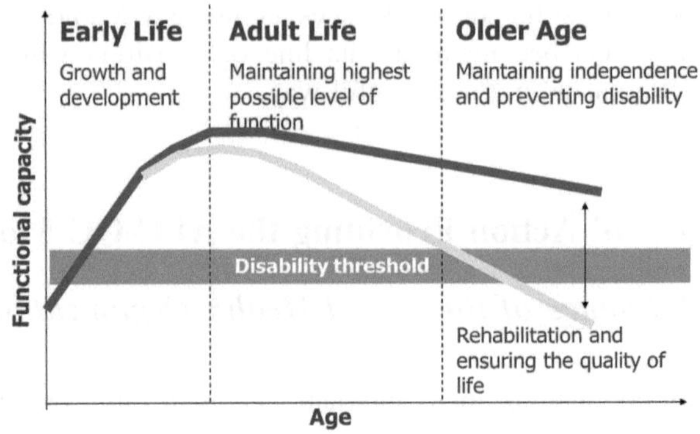

Fig. 7.1 Maintaining functional capacity over the life course. *Sources* Kalache and Kickbush (1997, p. 5) and WHO (2002, p. 14). Reproduced, with the permission of the publisher, from WHO 2002

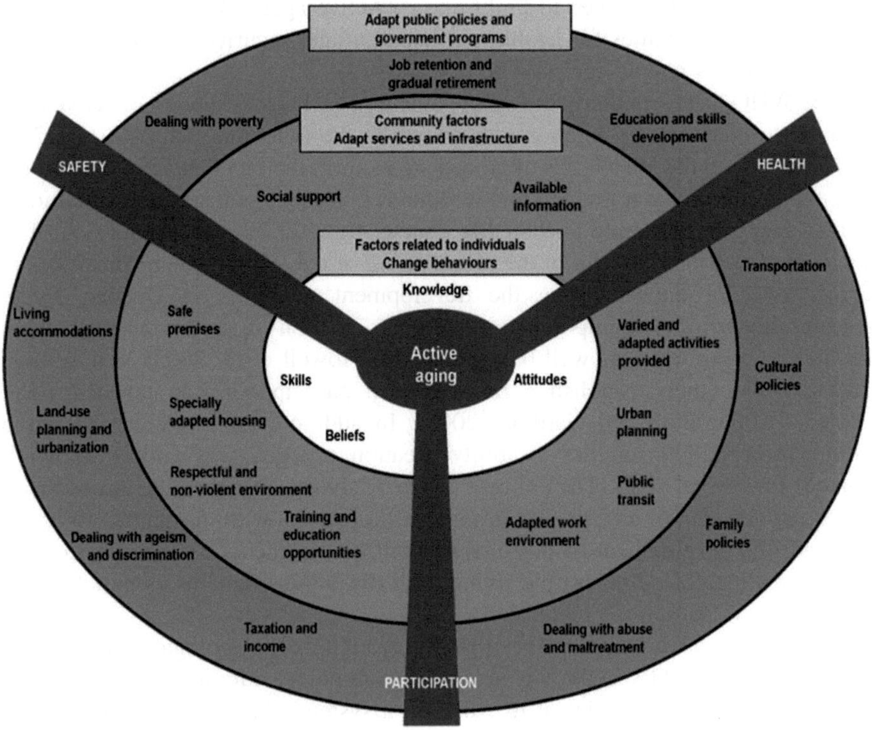

Fig. 7.2 The AFC-QC ecological model. *Source* Government of Quebec (2009). Age-Friendly Municipalities—Promoting active aging in Quebec. Quebec: Seniors' Secretariat

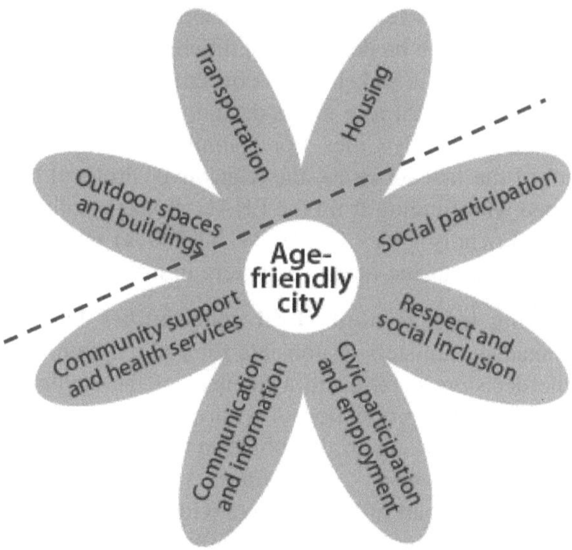

Fig. 7.3 Age-friendly city areas. *Source* WHO (2007a). Global Age-Friendly Cities: A Guide. Geneva: WHO, p. 9. Reproduced, with the permission of the publisher, from WHO (2007a)

the spheres (individual, collective, public policy). In the ecological model (Fig. 7.2), the WHO's eight areas can be points of intervention in conjunction with the three ecological spheres, sometimes echoing with one another. For example, a project supporting free public transportation during off-peak hours for seniors (services to the community), may be offered as a complement to the updating of driving skills (individual skills).

Active aging has indeed often been criticized for its potential for over responsibilisation of the individual (see the "Individualization of responsibility" section in the chapter of Moulaert, Boudiny and Paris). The ecological model avoids such criticism because it proposes an expanded vision that considers that life conditions of individuals are not only influenced by their own behaviors, knowledge, and attitudes but also by the community they live in and the existing social structures. In order to facilitate the targeting of interventions, the eight areas have been grouped according to the type of environment on which one wants to act, either the built environment, or the social environment (Fig. 7.3).

If the influence of the WHO is significant for situating the origin of the approach, then it is important to understand how the specific context of Quebec, set in the Canadian national framework, will lead to the creation of the "Quebec model."

7.2.2 The "Quebec Model" Within the Canadian Context

The Canadian Context Canada has played a critical role in the development and establishment of the AFC approach of the WHO, both through the involvement of the Public Health Agency of Canada (PHAC), and by the political commitment from ministers responsible for seniors. The way in which the AFC-QC project is carried out can be described in different ways depending on what is outlined on the spectrum of the stakeholders concerned. The Public Health Agency of Canada will play an essential role in the design, achievement, and funding of the first phase of the Age-Friendly Cities. For example, the sampling tool and data collection used by all of the first 33 cities, the Vancouver Protocol (WHO 2007b), have been developed with funding from the PHAC. It was a collaborative effort that resulted in the Age-Friendly Cities Guide (WHO 2007a) which indicates why and how we can inquire about the needs of seniors in urban areas.

Taking into account rural communities In parallel, in September 2006, the Canadian ministers responsible for seniors, both federal and provincial level, made the decision to raise the awareness of the challenges of aging in rural areas and called for the production of a specific guide for rural communities. As a result, a Canadian working group published the adaptation of the world Guide, *Age-Friendly Rural and Remote Communities: A Guide* supporting these communities in their efforts to be part of the initiative (Government of Canada 2007). The characteristics of the rural population have an influence on the lifestyle and on ways of accessing resources and services, especially when it comes to health and social services. The issues that must be taken into account for an age-friendly rural community are therefore quite different from those of the cities. In the Canadian context, the Public Health Agency of Canada continues to play a leading role with several partners and the other provinces. Because the responsibilities of the health and social services are of provincial jurisdiction, some provinces, including Quebec, have developed programs for their own territory and population.

The Quebec Context As for the Quebec case, the initial step has been done by two Quebec[4] researchers who have participated in 2006, at the invitation of the WHO, to the first modeling exercise of the worldwide age-friendly cities, in concert with researchers from 33 cities across the world, representing 22 countries.

The model adopted in Quebec has its source in the guidelines set within the WHO's initial international project,[5] but chose to refine subsequent steps in the diagnostic and differentiate its approach. Therefore, in 2008, it became necessary to include all the elements of the approach in a coherent model (Garon et al. 2011). The foundations of the model come from the development of the communities and are based on the logic model, widely distributed by the international development agencies, such as the Canadian International Development Agency (CIDA), and has been used, in the past few years, as a planning tool of various types of programs (Funnell and Rogers 2011). In the framework of the AFC experience, the main benefits of having a model are to facilitate a shared understanding of the approach to undertake

---

[4]Suzanne Garon, Ph.D. and Marie Beaulieu, Ph.D. researchers at the Research Center on Aging of the CSSS-IUGS and professors at the School of Social Work at the University of Sherbrooke.

[5]Note from the editors: for further details about the origins of the WHO program see the chapter of Plouffe, Kalache, and Voelcker.

and the adoption of a common language between stakeholders responsible for this process.

Converging with the principles derived from the WHO (active aging, the eight areas) and of the United Nations (bottom-up approach), the model is then enhanced by tools (program and service grids intended for seniors in the diagnostic, method for prioritizing projects of an action plan, self-evaluation of the approach and the effects, etc.), intended to support the achievement of each of the steps. By linking tools to the logic model, stakeholders can identify more easily their own implementation indicators. It still appears important that implementation and effect indicators remain in the hands of various actors so they can reflect on them. During the first work session with the WHO in 2007, it seemed to us that, although all the reports produced by the 33 cities contained a wealth of information, this information was not disseminated because, for the most part, contact with the municipal administration was poorly developed and no commitment was made in order to make changes or improvements. That is why, in the Quebec planning, it quickly became clear that it was to be specifically assigned to an administrative authority of the municipality. In addition, the commitment to participate in the program should be the subject of a resolution, adopted by the municipal council. Other specific features are at the heart of the Quebec model.

7.2.3 Some Distinctive Features of the "Quebec Model"

Guidelines for municipalities For cities and rural areas in Quebec, the challenge of aging remains important, in part because the acceleration in the pace of population aging has already begun. In addition, this issue reaches all social groups and affects many aspects of life in society. Municipalities usually do not consult with "specific clienteles" within their population. They generally rely on public consultations and other democratic formats which allow them to reach their citizens. So it would be well-advised to provide a protocol already validated to facilitate their approach with seniors. The consultative process allows the main interested parties to be heard about what is necessary to remain active, as well as on the solutions that they themselves perceive to reduce obstacles to their full participation. This is the first step of the "bottom up" approach.

The coordination mechanism: the steering committee With a view to supporting the aim of a rigorous and organized approach during the first five years of the AFC-QC project (2008–2013), a steering committee is set up, bringing together the major municipal stakeholders (elected officials and administration), the civil society including representatives responsible for seniors (agencies, associations), as well as health and social services. The typical steering committee is broadened by its cross-sectoral composition. Its role consists of monitoring and facilitating each of the three steps, collaborating in the implementation of actions, disseminating the information, and participating in the mobilization of stakeholders and decision-makers in the community. The steering committee plays a more extended role than the advisory committee.

Despite the existence of tags and tools, the steering committee must be notified of the importance of achieving its diagnostic as well as an action plan that relates to the municipality and to the challenges identified in the community. It is also important to respond to the needs expressed by drafting an action plan which is not the result of a single person, but from the commitment of the committee to make their community more welcoming, safer and more accessible for seniors and to work with them. Building a collaborative momentum that supports the establishment of partnerships between stakeholders is desirable, but remains in part within the power of the stakeholders themselves.

A collaborative partnership The involvement of seniors is necessary to the work of the steering committee (where they have seats allocated to them) in order to help clarify the seniors' needs in the community and to help achieving projects. With the other members of the committee, seniors become part of solutions and then fairly reflect the expression "by and for seniors." They not only influence the interpretation of needs arising from the consultation but also the strategic choices of actions included in the action plan, just as the feasibility of projects, often supported on their collective strength (associations and agencies). In this regard, the agreement signed between the city and the Seniors' Secretariat clearly states that the approach must be achieved "by and for seniors." It should not be assumed that seniors must sustain all of the projects, but rather see the creation of alliances between civil society, municipality, and service providers, in order to gather skills and resources necessary to achieve important projects for seniors. This balance in the commitment of cross-sectoral stakeholders has led to favorable results aiming at improving the life conditions for seniors in the communities (Garon et al. 2014).

Achieving the approach There are three main steps to the Quebec model: (1) the social diagnostic (community statistical portrait, profiles of services and programs for seniors, consultation on seniors' needs); (2) the action plan step; (3) implementation. The steering committee must be established earlier in the process. In its replicated model of the Quebec experience, the first step of the WHO approach will be the constitution of the steering committee.

The social diagnostic stage consists of a community statistical portrait, a profile of services and programs for seniors, and a consultation on the seniors' needs. To support the consultation, various topics of focus groups were already developed in the Vancouver Protocol, but in order to ensure its correspondence with the Quebec reality, an adaptation of it was necessary. A grid to establish the profile of services and existing programs has been designed by the AFC-QC project research team, allowing checking those who are accessible to seniors of a given community. As to the statistical profile, the purpose here is to determine the percentage of seniors living in the city, as well as other relevant socioeconomic information. The proportion of seniors is relatively easy to obtain through internet sites. This information was also a significant catalyst for certain key stakeholders, such as mayors. It helped raising awareness of the aging rate of their community, which was much higher than they had ever imagined.

The stage of the action plan led the various stakeholders to look at the record of the diagnostic, to question priorities, to translate seniors' needs into projects, and to determine the feasibility of the proposed solutions. This process allows them to get to know each other better, to build trust, and to have a more accurate perception of resources that can be mobilized. The action plan is based on a logic model approach (Funnell and Rogers 2011) resting on the principles of results-based management (CIDA 1999).

As for the implementation step, the emphasis is put on the orchestration of a whole set of means and resources to carry out the projects in the action plan. The process intends to be formative by allowing the strengthening of the group capacity, who then took advantage of this exercise to share its values, and become better acquainted with the complementary expertise of members. It is also an opportunity to see if new partnerships can be created, built on mutual confidence. The various backgrounds of members may be complex (coordination required) but can also be a source of wealth.

In the end, the approach aims at allowing seniors to evolve in their environment according to their interests and capabilities. Even if the approach is centered on seniors' needs, municipalities have actually improved some accommodations, such as crosswalks, sidewalks, bus shelters, which are actually used by the whole population, but even more so for those who, like people with disabilities or parents with strollers, often encounter obstacles in the city. The establishment of initiatives are not only used by seniors, but for people of all ages. In addition, it was also proved that the approach could be used to revitalize a community from the senior population perspective. Indeed, joint actions are possible between seniors at risk of social exclusion and other similar groups. It would be easily agreed that several actions in favor of seniors (e.g., accessible trails, automatic doors, low floor buses, intergenerational activities) are also useful for other groups of the population. This aspect appeals to municipalities which are more comfortable to meet the needs of their population in general rather than specific groups, and which are therefore on the lookout for the common denominator.

7.3 Successive Research Projects Beyond the Governmental Program

The growing importance of rapid aging among the Quebec population has convinced the Government of Quebec to act quickly. In order to be able to follow the evolution of municipalities on the ground, it became essential to support the pilot project with a reflection and an ability to objectify the results, which was done by

Focus groups (7 pilot sites)		*49 groups gathering 392 persons*	
Diagnostic and action plan 2008-2009	*N*	*Implementation (Effects*) 2010-2012*	*N*
Log books and press reviews	*12*	*Log books and press reviews **	*50*
Minutes of committee	*78*	*Minutes of committee **	*186*
Training sessions	*4*	*Focus groups**	*25*
Focus groups	*6*	*Direct observations*	*25*
Direct observations	*6*	*Wilder collaboration surveys*	*126*
Diagnostic reports	*7*	*Organizational networking surveys*	*153*
Action plans	*7*	*General survey (2013)*	*115*

Fig. 7.4 Data collection: using mixed methods to capture a complex reality

mandating the AFC-QC research team,[6] who played a similar role in the first experimentation orchestrated by the WHO. This combination "government-research" would evolve toward a project governance rooted in the reality of the municipalities, by gradually establishing mechanisms whose effects of reflection on their own action went beyond the strict mandates of each of the stakeholders. This momentum will be maintained and the mandates of the research team renewed despite three successive governments since 2007. This fact alone reflects the importance given to the phenomenon of aging and the process is well underway with the AFC.

7.3.1 First Research: Lessons Learned from the Seven Pilot Sites

The first evaluative research with the first seven AFC-QC pilot sites has led to the development of the AFC program from the Quebec Government in 2009. Results of a large data collection (Fig. 7.4) helped to understand various issues that represented the mobilization of communities toward the challenge of the aging of their population.

Upon receiving the first preliminary results, it was clear that the sites responded well to the structuring elements of the model. In fact, the AFC initiatives enable to put in place a flexible steering framework for developing projects, characterized by a cross-sectoral structure from the community, which is distinguished both by its composition and its areas of expertise (ecological perspective). Other subsequent findings have demonstrated the roles and functions of the steering committee,

[6]AFC-QC research team led by Suzanne Garon, Ph.D. through her expertise in evaluation and development of communities.

especially as a cross-sectoral coordination mechanism, which makes available a wide range of options for projects aimed at adapting built and social environments, which is at the core of the AFC. The approach encourages the adoption of an inclusive and participatory approach as soon as the committee is formed, which promotes the concertation between municipal stakeholders, regional county municipalities (RCM), health and social services centers (CSSS), socioeconomic stakeholders, in community organization, and local and rural development, not counting all the community agencies working with seniors. Community mobilization enable to associate cross-sectoral stakeholders who have the capacity, skills, and complementary resources in order to develop collectively ways to meet the seniors' needs who live in different environments (urban, rural, semi-urban).

This evaluative research spread over five years has enabled us to learn other lessons, that when this model was applied, one of the conditions of success was emerging from the process itself, the cross-sectoral collaboration (Garon et al. 2013). As for the comparison of steering committee structures, it was also demonstrated that when a city was subcontracting seniors' associations to let them manage the AFC initiatives all by themselves, it brought important limitations to the scope of the projects. The very presence of seniors and their representatives, despite their hard work, was not enough since the entire field of municipal competences was unknown to them. When the composition of the steering committee was optimal, active aging, which was linked to the social determinants of health, has encouraged various stakeholders to break the silos that organizations have created over time. The study has also demonstrated that the participation of seniors is essential; this facilitates the validation of project contents, as well as the choice of optimal approaches for seniors in the community. When municipal stakeholders are not actually sitting on the committee, inevitable limits occur, where it is impossible for senior organizations, as solid and structured as they may be, to carry out all by themselves the implementation of measures in fields as essential as transportation or housing.

Which brings us to another lesson with regard to the role of each of the stakeholders whose fundamental importance is confirmed. The elected officials, just as the administrative staff of the municipalities as well as the civil society, truly have their role to play for designing age-friendly environments (Garon et al. 2014). However, as the experience of diversified municipalities has showed us over time, there are socio-political realities, rural and isolated territories within which, even if needs are well-understood, the resources available to these communities remain restricted and pose other challenges.

The community development model, which has served as a frame to the seven pilot projects, has helped emphasize the fundamental values embedded in AFCs, particularly with respect to the inclusion of seniors at all stages as well as the involvement of cross-sectoral stakeholders in resolving issues related to built and social environments. The creation of networks and the collaboration between the stakeholders demonstrate that within such a model, practices may evolve. Evaluating the implementation and the effects has helped to raise awareness of the importance of the participatory approach and the promotion of collaborative practices, both for stakeholders in the field and for those in research, and no doubt also for public policy-makers, in particular the Seniors' Secretariat that succeeded to get its *modus operendi* through.

With these experiments from which emerged innovations at different levels (in terms of products, processes and practices), the research team wanted to better understand the conditions surrounding the emergence of social innovations as those that hinder the evolution of AFCs. At the end of the evaluative research, a new research proposal was submitted, which was accepted by the Seniors' Secretariat.

7.3.2 Second Research (2014–2017) on Social Innovations from AFCs

The first evaluative research with the seven pilot sites has led to the formulation of a hypothesis according to which, social innovations reflect the diversity of municipalities and service structures, such as the diversity of seniors themselves and that they are accountable, among other things, to community dynamics prompted by participatory governance. Each municipality and each village has distinct features that distinguish them from the others. Similarly, each of the stakeholders concerned by the AFC action exposes its culture and its practices within these dynamics which deserve to be better known, in the same way as mechanisms of emergence and sustainability of these innovations, aspects which remained poorly explored in the first research.

This area of social innovations applied to AFCs provides various related avenues, those on innovative practices (Caillouette et al. 2009), communities of practice (Wenger 1997, 2010), and on co-construction, as well as the participatory local governance, all oriented toward the development of collective forces. We also rely on Chaskin which defines the capacity building of the community as being

> the interaction of human capital, organizational resources, and social capital existing within a given community that can be leveraged to solve collective problems and improve or maintain the well-being of that community. It may operate through informal social processes and/or organized efforts by individuals, organizations and social networks that exist among them and between them and the larger systems of which the community is a part (Chaskin et al. 2001, p. 7).

We recently started the study on social innovations which is related to the stage of the implementation of an AFC approach. It is primarily the contribution of new knowledge on AFC operations with respect to social innovations that contribute to social inclusion of seniors in their community (sample A with 11 cases), as well as on ways to improve the AFC program, including from actual problems experienced in communities (sample B with 4 cases). After the seven pilot projects, we ended up with 15 AFC case studies. The sites selected (by convenience sample) are located in urban and rural areas and include RCMs, central municipalities, rural communities, as well as a large municipality. In addition to the implementation of an innovation, the selection criteria also include their geographical (regions) and economic (devitalization) situations, and other affecting the progress of their AFC approach.

In order to highlight the types of obstacles encountered by municipalities and possible solutions, the sample consists of communities that are experiencing problems to revitalize their AFC approach.

The role of the research team is modulated according to objectives targeted in the two samples. It is part of a participatory and co-constructed approach, materialized by an iterative mode of production and exchange of knowledge between researchers and communities. In all the cases, our analysis will highlight the influence of the participation of seniors on action plans and their implementation.

From this study, we sought to develop an operational model of an AFC approach for social innovations in analyzing the emergence of social innovations. The work of Fontan, Andrew, and Klein gives rise to appropriate conceptual guidelines. We primarily preferred a definition that indicates the fields covered: "{…} new social arrangements, new forms of mobilizing resources and new answers to problems to which known solutions may not be sufficient" (Klein et al. 2009, p. 3). The results are expected in 2017.

7.3.3 A Third Action-Research on Social Determinants of Health Within AFCs

In 2014, we have submitted and obtained new financing of research funded by the Canadian Institutes of Health Research (CIHR) to conduct an action-research. This new study has now taken place in four contrasting communities (i.e. with different contexts, both rural and urban areas) of the Montérégie region. Conducted in partnership with the public health department of the Montérégie and the NGO Carrefour action municipale et famille (CAMF), this research requires the collaboration of several stakeholders involved in the AFC approach, among others, seniors and the community organizations, the local practitioners, elected officials, and municipal managers.

Viewed in a health equity perspective, the action-research will contribute to the expansion of AFCs in explaining how these may have concrete benefits on the social determinants of the seniors' health. It seeks to assess the effectiveness of AFC interventions and for the purpose of developing an action model and knowledge transfer tools for communities. In response to the need and the complexity of the action on social determinants of health, it seemed appropriate to us to study the combination of missions and expertise as well as to challenge the practices of several stakeholders. In order to do so, a proposal including three strands of iterative data collection and analysis has been developed.

This research involves the production of a participatory realistic assessment (Pawsons and Tilley 1997; Kazi 2003) and the production of a logic model. This methodology allows us to put emphasis on facilitators, barriers, as well as factors that may provide benefits on the determinants of seniors' health, while considering analyzed contexts. The context, mechanisms, and benefits from AFCs are analyzed for each of the cases in the study.

Individual and group interviews are carried out with representatives of each community who agreed to participate in the research. Wishing to promote the co-construction of knowledge, the data collected in the framework of these discussions are first analyzed by the members of the team, and presented to stakeholders in the field in order to obtain a genuine dialog between research and communities. The pooling of academic knowledge and practices will help to clarify elements that work best and identify factors that make AFC initiatives fostering the reduction of health inequalities. Promote the intersection of these knowledge will allow not only to validate the findings that will emerge from these interviews, but will facilitate the transfer and dissemination of the knowledge generated with stakeholders involved in AFCs. It will enable the co-construction of knowledge by using the "teacher-learner" cycle (Nanninga and Glebbeek 2011) and rooting the project within the community. To this effect, we note that several stakeholders already wish to assess their AFC achievement from a sanitogenesis perspective (Antonovsky 1967), i.e., with greater interest for production of health than for causes of sickness. This perspective of the research will be supported to develop new tools of co-constructed evaluation.

In order to increase the dissemination of the generated knowledge, a regional day dedicated to the communities in the study, but also opened to all regional stakeholders, will be organized. This day will be designed as a discussion platform and will allow all participants to exchange with key stakeholders and to ensure that they acknowledge the results of the research and enhance them. Research results as a whole will be available through the AFC-Quebec website in 2017 (http://madaquebec.com/fr/).

7.3.4 General Considerations Over Eight Years of Research

During the massive expansion of the AFC governmental program, new collaborators were gradually added, each with its specific role. In fact, the Seniors' Secretariat was keen to see the program maintaining its internal logic, which consists of changing attitudes in order to develop a "senior reflex"[7] in the municipalities of Quebec. It is for this reason that the Seniors' Secretariat has proposed exchange and reflection spaces between these collaborators. The *Carrefour action municipale et famille* (CAMF) was appointed to provide support to the newest AFC municipalities. A national coordination committee (NCC) was formed in order to orchestrate the action at the province level. Finally, "the gathering of large cities," which was gradually developed in community of practices, brings together all of the 10 largest cities in Quebec. Their respective roles are different and in some respects

[7]Senior reflex: promoting seniors' interests in decision-making process regarding implemented programs and activities.

complementary, but especially these entities reflect an AFC governance implemented by the combined action and critical reflection that emerges from their ongoing interaction.

Le Carrefour action municipale et famille In order to support municipalities in the AFC approach, the Seniors' Secretariat has selected a non-profit organization, the *Carrefour action municipale et famille* (CAMF), with an extensive experience (25 years) in municipal policies. This organization is "{…} committed in promoting, implementing, supporting and following-up municipal family policies"[8] in Quebec. In its role, the CAMF enjoys a wide network with the municipalities, relying on a team of 26 trainers who have an expertise of the municipal environment and who travel throughout Quebec. The CAMF was quick to generate a certain enthusiasm for the AFC program. Without the contribution of this organization, it is doubtful that the number of AFC municipalities would be as high as they are today.

However, the CAMF also had an impact on the development of AFC-QC. In this regard, the research team, which provides training for CAMF trainers on an annual basis, has reacted to the fact that too often seniors are primarily considered as grandparents. Even for these experienced trainers, it is not so easy to make the transition between a familiar area, which is socially acceptable and valued as the development of municipal services for young families, to work on the development of built or social environments for seniors. It clearly could have had some consequences on the support offered to municipalities. It is in this context that the Seniors' Secretariat asked the research team to prepare a new guide which can support all stakeholders of the approach (Government of Quebec 2009). This guide has an interface with the website that allows users to find specific tools to the approach. Although primarily from Canada, more than 1000 visits are made on a monthly basis on the website. Users come from many dozens of countries and it has become, over the years, an important tool for the dissemination of the AFC approach, in Quebec and throughout the world.

Although the professional history of trainers represents an asset to assist community stakeholders in navigating in the municipal apparatus, it was found that this support can also be tainted by a preconception of what type of services[9] a municipality should offer. It was followed by a form of resistance from their part to put forward innovative scenarios that could help to solve a problem. For example, some trainers hindered the development of health and social services innovations since they consider, as did the majority of the municipal officials, that it is not the jurisdiction of the municipality to get involved in the health sector. In Quebec, as in some other Canadian provinces, health services do not fall under municipal jurisdiction, but under provincial jurisdiction. In addition, according to participants and CAMF administrators, it was found that among the training courses offered, those

---

[8][Online]. Available: http://www.carrefourmunicipal.qc.ca/index.php/camf/mission [Accessed April 20, 2015].

[9]Note from the editors: For similar tensions, see the French case.

on active aging applied to social determinants of health have helped to evolve a too narrow understanding of health, based on a curative approach. The active aging may therefore allow the different municipal stakeholders to concretely tackle the importance of health prevention and promotion, which may then be considered consistent with the role of municipalities.

The National Coordination Committee Shortly after, the Seniors' Secretariat has set up a follow-up committee for all stakeholders directly involved in the implementation of AFCs and in their continuous improvement. These stakeholders are the CAMF, the Seniors' Secretariat, and the CDRV research team. From 2010 to 2011, these committees were using occasional and targeted telephone meetings as a way of communicating with one another. It became more formal in 2012 when statutory meetings took place every six to eight weeks. In 2013, the Seniors Stakeholders Regional Committee (CTRCA for the *Conférence des tables régionales de concertation des aînés*) started as the largest group of seniors' associations in Quebec, which does not directly offer services but represents the interests of seniors in Quebec. Claiming the participatory nature of the program, CTRCA has requested to be part of this national coordination committee. One of their first claims was to get closer to the AFC steering committees in the communities, to possibly support older members of the steering committees of the participating cities.

The community of practice of Quebec's largest cities In the pilot projects, the 2008–2013 research showed that the AFC program requested from largest cities significant adjustments. The roles played by the municipal administration and the elected officials were crucial to the success or failure of its achievement. From 2012, the Seniors' Secretariat, will then invite, twice a year, all cities of more than 100,000 inhabitants to meetings for exchanging on the challenges and successes of their experience. Since then, the experience was repeated each year. The municipal representatives present alternately their successes, their difficulties, and their strategies to make their city more welcoming to seniors and, consequently, to other people living with vulnerabilities. Is this the beginning of a form of community of practice (Wenger 1997, 2010) that galvanizes knowledge sharing, learning, and change? If one considers the participation rate of these meetings and the hard work of the partners' preparation, this could be particularly promising.

Structural incentives On another level, it is appropriate to specify that the Seniors' Secretariat has a budget of approximately CDN $2 million per year. Although very limited, but in conformity with the current governmental policy on aging[10] (Government of Quebec 2012), it attracted the interest of another ministry of the Quebec's government, the Ministry of Municipal Affairs, Regions and Land

[10]Governmental policy Aging and Living Together—At home, in One's Community, in Quebec (2012). [Online]. Available: http://madaquebec.com/fr/bibliotheque/documentation/category/51-gouvernement-du-quebec [Accessed April 20, 2015].

Occupancy.[11] This ministry encourages municipalities to carry out projects on built environments favorable to seniors in an AFC approach. A budget envelope is thus set aside for municipalities that enrol in the program, to help them contribute to the funding of soft infrastructure projects related to their aging population. In this regard, as we have just seen, it is also interesting to note that many of the funded projects are related to mobility and by this very fact, provide support to other populations, people with disabilities, or young families using strollers.

Finally, the Seniors' Secretariat has also developed an annual funding program known as Quebec Age-Friendly Program (QADA). It is a form of assistance associated with calls for projects directly related to the living conditions of seniors. Initiatives were launched to break down isolation or promote social participation. The AFC governmental program has therefore a number of financial incentives to promote and support social innovations for seniors in municipalities.

7.4 The AFC Evaluation Issues

The primary mandate of this team has been to adapt and supplement the experience carried out by the WHO from 2005 to 2007 for the Quebec context. This first mandate also included the evaluation of the implantation and the first outputs of the program. We understand the importance of evaluation in the wake of challenges regarding expenditure of public funds imposed by the new public management. By contrast, it would be very naive to believe that the evaluation of such a multidimensional program as the AFC can be carried out by a few spot checks, grouped in a single research.

In the same vein, it is always difficult to define a polysemic term such as evaluation. It is generally understood that there are three ideal-types[12] of evaluative research (Desrosiers 1998; Potvin and Bisset 2008): the classical or experimental model, the model focused on users (users' led), also called the participatory evaluation, and finally the logic model, that some authors describe as "organized reflexivity" (Potvin and Bisset 2008).

The classical or experimental model is most probably well-known. It is one to which most of decision-makers such as ministries, private foundations, or international agencies tend to choose, in order to ensure that public funds are properly committed, and produce the desired outcomes. Strongly inspired by natural sciences, it assumes that the causal links between the program activities and the effects are relatively prehensile, so as to be segmented and then to be weighted in benefits.

[11]The Programme d'infrastructures Québec-Municipalités (PIQM) MADA Programme. [Online]. Available: http://www.mamrot.gouv.qc.ca/infrastructures/programme-dinfrastructures-quebec-municipalites-piqm/piqm-volet-2/ [Accessed April 20, 2015].

[12]The ideal-types (Weber 1922) are a construction allowing to highlight social phenomena in order to better understand their major components.

It works well with simple programs, whose causality is predictable (e.g., water fluoridation programs against tooth decay).

In addition, when applied to a social context, therefore complex and uncontrollable, this approach shows all its limits. Thus, even if it can determine that there has been expected change or not, which could be characterized as effects, the latter cannot be explained or even attributed to the program activities that we want to evaluate. The black box analogy, suggested by Pawson and Tilley (1997), illustrates very well the limitations of this model. This analogy consists of representing elements (inputs), which go in a box and emerge in the form of effects (outputs), without knowing if or why such effects have taken place. The results are then very hardly transferable, which does not facilitate the program implementation to other contexts. Worse still, it becomes almost impossible to adjust, along the way, activities or elements of context essential to the achievement of the program. In addition, we must add the exorbitant cost of the methodological apparatus on which is based its scientific validity and the reasoning that will follow. Despite all these constraints, this model applies to the various realities of the implementation of the AFC across the world, which is indeed interesting. And perhaps, it is in this sense that we must take greater consideration on major populational indicators, such as those used by the WHO Kobe Center or the Public Health Agency of Canada. For some indicators, the long-term outcomes must be thoughtful and anticipated as we speak today, although they cannot be tangibly perceived before 20 or 25 years.

The second mode of evaluation concerns the program logic approach. Developed among others by Chen (1990, 2005), this approach also takes other forms such as the Realistic Evaluation of Pawson and Tilley (1997) which is characterized by its proximity with the planning strategy. This approach is based on the fact that any program is evolving, migrating, or deteriorating. And since any program is a "living thing," it is essential to create spaces of reflexivity. This is precisely what may serve the evaluative research.

As stated in the description of the research sequence on the age-friendly municipalities since 2008, this model is very present in this work done by the Quebec research team. It is essential to find consistency between the selected evaluation models and the type of object to be evaluated. Thus, it remains quite conceivable to measure the impacts on the health of populations of a program or a public policy by using population data, according to the classical model. But when the content of the "black box" is essential to demonstrate the manner in which the effects are produced inside a logical chain of changes, the program logic approach, which assumes the observation or measurement of the action of the mechanisms at work or even the logic of stakeholders, will without a doubt unlock the wrong patterns and highlight the modalities, or the winning practices (what works) which guide stakeholders toward the desired outcomes in this specific program. Only then will this condition be possible to transfer programs which operate in other communities while ensuring the increase of its chances of success.

The third form grouping several variations also called "user led" (Desrosiers 1998), Negotiation Evaluation (Potvin and Bisset 2008), or even the Fourth Generation Evaluation (Guba and Lincoln 1989), are grouped under the term

Participative Evaluation. As its name suggests, it is an evaluative approach in which all stakeholders participate, both the preparation and achievement of the evaluative research. It is an evaluative model which postulates that the program and the context in which it takes place are a social construction. This is the importance of taking into consideration all the stakeholders' points of view, both users and managers. Guba and Lincoln even defend a radical constructivist standpoint, such as what the mediation work done by the evaluator, in putting into action a hermeneutic circle, becomes the reality. This position has not helped to develop this approach. Moreover, it would be a shame to see the perception of users completely excluded from an evaluative approach. It is possible to imagine several evaluation opportunities or modalities where the seniors' voice, such as the approach of Empowerment Evaluation (Fetterman et al. 2015), should be taken more into consideration in the way of considering an age-friendly city. A whole segment of this type of research remains to be developed with and by groups of seniors. This must be regarded as a promising avenue to maintain the development and relevance of AFC environments.

7.5 Conclusion

Research is an integral part of developing the AFC approach in Quebec and this, since its beginning (Moulaert and Garon 2015). The Seniors' Secretariat has certainly played an important role in making consecutive achievable research (avoidance of interruptions) but also to provide learning favorable conditions for stakeholders in grouping them, also favorable to the progress of the AFC program.

Viewed in a community development perspective, the work of the AFC-QC team gives preference to methodologies that create a genuine dialog between stakeholders involved in the approach. Fostering the sharing of common experiences is in line with the willingness of the AFC approach to break down silos and combine the expertise of a number of stakeholders and sectors. This requires however, to develop proposals which will take account of this variable and will encourage the use of a suitable methodology. In contribution to this discussion, the previous sections are intended to illustrate how the components of age-friendly municipalities in Quebec are linked in a coherent manner to the research methodology conducted by the AFC-QC research team.

In addition, according to a governmental perspective, there are other ongoing issues. If for some, the implementation of age-friendly cities has been considered as a way for the Welfare State to withdraw from areas which were under its responsibility, assistance to seniors, among others, we can also see in the Quebec experience a willingness of local governments and municipalities to re-examine their relationship with the central government, in such a way as to develop new types of governance which would give them more power and autonomy. The real challenge will be observed over time, specifically on how this governance can rightly be put in place and how the central government will be in a position to comply with these new terms (duties and responsibilities conferred to the latter) which imply the Treasury control. For example, do the accountability practices go in the same direction as this new governance? What it would be like for cities? What will happen at the end of this struggle for a new type of governance, currently very present in Quebec, in a period of turbulence around the administration of the Quebec government, with new laws abolishing regional entities? Only time will tell and the research team can be relied on to continue this analysis.

References

Antonovsky, A. (1967). Social class life expectancy and overall mortality. *Milbank Memorial Fund Quarterly, 45*, 31–73.

Bath, P. A., & Gardiner, A. (2005). Social engagement and health and social care use and medication use among older people. *European Journal of Ageing, 2*(1), 56–63.

Bennett, K. (2005). Social engagement as a longitudinal predictor of objective and subjective health. *European Journal of Ageing, 2*(1), 48–55.

Caillouette, J., Garon, S., Dallaire, N., Boyer, G., & Ellyson, A. (2009). Étude de pratiques innovantes en développement des communautés dans les sept Centre de services de santé et de services sociaux de l'Estrie. Analyse transversale de sept études de cas. *Cahiers du Centre de recherche sur les innovations sociales (CRISES)*, no. ET0903, juin.

Carrier, S., Morin, P., & Garon, S. (2012). L'approche centrée sur les résultats : l'exemple des réseaux de services intégrés pour les personnes âgées en perte d'autonomie. *Intervention, 137* (2), 47–56.

Chaskin, R. J., Brown, P., Venkatesh, S., & Vidal, A. (2001). *Building community capacity*. New York: Transaction Publishers.

Chen, H. (1990). *Theory-driven evaluations*. Newbury Park: SAGE Publications.

Chen, H. (2005). *Practical program evaluation assessing and improving planning, implementation, and effectiveness*. Thousand Oaks: SAGE Publications.

Choinière, R. (2010). *Vieillissement de la population, état fonctionnel des personnes âgées et besoins futurs en soins de longue durée au Québec*. Institut national de santé publique du Québec. Québec: Gouvernement du Québec. Available: http://www.inspq.qc.ca. Accessed July 10, 2014.

CIDA. (1999). *Results-based management in CIDA: An introductory guide to the concepts and principles*. Ottawa: Canadian International Development Agency.

Desrosiers, H. (1998). *Cadre pratique pour l'évaluation des programmes: Applications en promotion de la santé et en toxicomanie. Ministère de la santé et des services sociaux*. Québec: Gouvernement du Québec.

Fetterman, D. M., Kattarian, S. J., & Wandersman, A. (Eds.). (2015). *Empowerment evaluation: Knowledge and tools for self-assessment, evaluation capacity building, and accountability* (2nd ed.). Thousand Oaks: Sage Publications.

Fougeyrollas, P. (2009). *Mesurer la qualité de la participation sociale—l'exemple de la MHAVIE*. Présentation aux 1ères rencontres scientifiques sur l'autonomie CNSA (2009.2), Paris.

Funnell, S. C., & Rogers, P. J. (2011). *Purposeful program theory—Effective use of theories of change and logic models*. San Francisco: Jossey-Bass/Wiley.

Garon, S., Beaulieu, M., & Veil, A. (2013). *Évaluation de l'implantation et des effets—Projet Villes amies des aînés (2008–2013)*. Québec: Secrétariat aux aînés du ministère de la Santé et des Services Sociaux du Québec. Inédit.

Garon, S., Beaulieu, M., Veil, A., Paris, M., & Bigonnesse, C. (2011). L'expérience québécoise du programme "Villes-amies des aînés" de l'OMS: savoirs et pratiques tirées de l'implantation dans sept projets-pilotes. In J.-P. Viriot Durandal, & P.-M. Chapon (Eds.), *Vieillir dans la cité, Liens sociaux, intensité territoriale, et construction de l'espace public* (pp. 69–88). Paris: La Documentation française.

Garon, S., Paris, M., Beaulieu, M., Veil, A., & Laliberté, A. (2014). Collaborative partnership in age-friendly cities: Two case studies from Quebec, Canada. *Journal of Aging and Social Policy, 26*(1–2), 73–87.

Godbout, J. T., & Charbonneau, J. (1994). Le réseau familial et l'appareil de l'État. *Recherches sociographiques, 35*(1), 9–38.

Government of Canada. (2007). *Age-friendly rural and remote communities: A guide.* Federal/Provincial/Territorial (F/P/T) Committee of Officials (Seniors): Ottawa.

Government of Quebec. (2009). *Municipalité amie des aînés - Favoriser le vieillissement actif du Québec.* Québec: Secrétariat aux aînés.

Government of Quebec. (2012). *Vieillir et vivre ensemble – Chez soi, dans sa communauté, au Québec.* Québec: Secrétariat aux aînés.

Guba, E. G., & Lincoln, Y. S. (1989). *Fourth generation evaluation.* Newbury Park: SAGE Publications.

Institut de la statistique du Québec (ISQ). (2010). *Portrait social du Québec – données et analyses – édition 2010. Section sur les conditions de vie.* Gouvernement du Québec: Québec.

Institut de la statistique du Québec (ISQ). (2014). *Le bilan démographique du Québec – édition 2014. Section sur la démographie.* Gouvernement du Québec: Québec.

Kalache, A., & Kickbush, I. (1997). A global strategy for healthy ageing. *World Health, WHO Magazine, 50th Year* (4), 4–5.

Kazi, A. F. M. (2003). *Realist evaluation in practice: Health and social work.* London: SAGE Publications.

Klein, J.-L., Fontan, J.-M., Harrisson, D., & Lévesque, B. (2009). L'innovation sociale au Québec: un système d'innovation fondé sur la concertation. *Cahiers du Centre de recherche sur les innovations sociales (CRISES),* no. ET0907, Collection Études théoriques, mai.

Mikkonen, J., & Raphael, D. (2010). *Social determinants of health: The Canadian facts.* School of Health Policy and Management: York University.

Ministère des Affaires municipales et de l'Occupation du territoire (MAMOT) (2014). *L'organisation municipale et régionale au Québec en 2014.* Québec: Gouvernement du Québec.

Morrow-Howell, N., Hinterlong, J. E., Rozario, P. A., & Tang, F. (2003). Effects of volunteering on the well-being of older adults. *Journal of Gerontology, 58*(3), S137–S145.

Moulaert, T., & Biggs, S. (2013). International and European policy on work and retirement: Reinventing critical perspectives on active ageing and mature subjectivity. *Human Relations, 66*(1), 23–43.

Moulaert, T., & Garon, S. (2015). Researchers behind policy development: Comparing "age-friendly cities" models in Quebec and Wallonia. *Journal of Social Work Practice, 29* (1), 1–13.

Moulaert, T., & Paris, M. (2013). Social policy on ageing: The case of "active ageing" as a theatrical metaphor. *International Journal of Social Science Studies, 1*(2), 113–123.

Nanninga, M., & Glebbeek, A. (2011). Employing the teacher-learner cycle in realistic evaluation: A case study of the social benefits of young people's playing fields. *Evaluation, 17*(1), 73–87.

Paris, M., Beaulieu, M., Garon, S., & Bigonnesse, C. (2013). "Selon le temps, la manière" ou En quoi Villes amies des aînés se distingue des autres documents internationaux sur le vieillissement. *Canadian Review of Social Policies/Revue canadienne de politique sociale, 2* (68–69), 10–23.

Pawsons, R., & Tilley, N. (1997). *Realistic evaluation.* Thousand Oaks: SAGE Publications.

Potvin, L., & Bisset, S. (2008). There is more to methodology than method. In L. Potvin & D. V. McQueen (Eds.), *Health promotion evaluation practices in the Americas* (pp. 63–89). New York: Springer.

United Nations (UN). (2006). *Guidelines for review and appraisal of the Madrid international plan of action on ageing—Bottom-up participatory approach*. New York: UN Department of Economic and Social Affairs.

Van Willigen, M. (2000). Differential benefits of volunteering across the lifecourse. *Journal of Gerontology, 55B*(5), S308–S318.

Weber, M. (1922). *Économie et société (2 tomes)*. Paris: Pocket.

Wenger, E. (1997). *Communities of practice: Learning, meaning and identity*. Cambridge: Cambridge University Press.

Wenger, E. (2010). Communities of practice and social learning systems: The career of a concept. In C. Blackmore (Ed.), *Communities of practice and social learning systems* (pp. 179–198). London: Springer.

WHO. (2002). *Active ageing: A policy framework*. Geneva: WHO.

WHO. (2007a). *Age-friendly cities: A guide*. Geneva: WHO.

WHO. (2007b). *WHO age-friendly cities project methodology—Vancouver protocol*. Geneva: WHO.

Chapter 8
Age-Friendly Hong Kong

Grace M.Y. Chan, Vivian W.Q. Lou and Lisanne S.F. Ko

8.1 Aging Population in Hong Kong

The number of people aged 65 or above is expected to rise sharply in the next 30 years. According to the results of the 2011 Population Census, there were 941,312 older persons in Hong Kong in 2011, constituting 13.3 % of the whole population. The number of older persons increased at an average annual growth rate of 4.8 % over the 50 years (from 1961 to 2011), compared to the whole population at 1.6 % over the same period. At present, one in eight persons in Hong Kong is in this age group. By 2033, that figure rises to one in four persons (Census and Statistics Department 2012). Among older persons in 2011, 24.9 % are aged between 65 and 69 years, 46.3 % between 70 and 79 years, and 28.8 % between 80 and over. The share of the oldest people (aged 80 and over) continued to increase over the past decade. Among older persons in 2001, those aged 80 and over constituted 19.6 % only (Census and Statistics Department 2011c).

Among older persons in 2011, 91.4 % were living in domestic households. These include 12.7 % of older persons living alone, 23.6 % with spouse only, 29.7 % with spouse and children, and 21.4 % with children only. As compared with the distribution ten years ago, the proportion of older persons living with children (including those with spouse and children and those with children only) dropped from 56.8 % in 2001 to 51.2 % in 2011. On the contrary, the proportions of older

G.M.Y. Chan (✉)
The Hong Kong Council of Social Service, 12/F, Duke of Windsor Social Service Building, 15 Hennessy Road, Wan Chai, Hong Kong, China
e-mail: grace.chan@hkcss.org.hk; el@hkcss.org.hk

V.W.Q. Lou
Department of Social Work and Social Administration, Sau Po Center on Aging, The University of Hong Kong, Pokfulam, Hong Kong, China

L.S.F. Ko
Sau Po Center on Aging, The University of Hong Kong, Pokfulam, Hong Kong, China

© Springer International Publishing Switzerland 2016

T. Moulaert and S. Garon (eds.), *Age-Friendly Cities and Communities in International Comparison*, International Perspectives on Aging 14,
DOI 10.1007/978-3-319-24031-2_8

persons living with spouse only and those living alone rose by 5.2 and 1.4 percentage points, respectively over the same period.

Facing the inevitability of an aging population, the Hong Kong Special Administrative Region (HKSAR) government has never failed to include older adults in its policy addresses since 1997. In 1997, the Chief Executive of the Hong Kong Special Administrative Region made "Care for the Elderly" a strategic policy objective of the HKSAR government. In responding to the Second World Assembly on Ageing in 2002, "active and healthy ageing" was adopted as a guideline for the Elderly Commission in Hong Kong. The concept of "ageing in place" was adopted as a guiding principle by a number of bureaus with the principle of "supporting aging in the community as the core, institutional care as back-up" reiterated by the Chief Executive in his Policy Address in 2014. While 'aging in place' as a policy objective is welcomed by the public, it is important to work out a detailed framework on how the whole society can work together so as to achieve 'aging in place.' Age-friendly city framework is regarded as a promising benchmark aging policy framework internationally.

Driven by the concept of Age-Friendly City, World Health Organization (WHO) launched The Age-friendly Environments Program to address the environmental and social factors that contribute to "active ageing" in societies (WHO 2007a). The WHO developed a checklist under each of the eight themes (WHO 2007b). These checklists include details of all the core features of an age-friendly city which must also be understood and interpreted from a life course perspective since older adults do not belong to a homogeneous group when they age. To actualize the concepts of an age-friendly city according to the checklists, more than one government department needs to work together, depending on the various administrative roles of government departments. Hong Kong was not a participating member when the WHO first launched the program on age-friendly cities in 2007. Yet, being an advocate for the development of social welfare, Hong Kong Council of Social Service (HKCSS) took the lead in promoting the concept of an age-friendly city in Hong Kong by establishing the 'Age-Friendly Hong Kong Steering Committee' in 2008 with the following core tasks:

(a) To promote public understanding on age-friendly Hong Kong;
(b) With the participation of elders and stakeholders, to foster solution to improve the lives of older people; and
(c) To exchange of information and dissemination of best practices.

To achieve the above-mentioned tasks, the Committee has promoted the concept and solicited support from various nongovernment organizations (NGOs) and other local agencies at district level since 2009. Till now, age-friendly city measures have been started or implemented in 18 districts. Four task forces were also formed. They are:

(a) Task Force on Expert Group and Alliance Building;
(b) Task Force on Project Evaluation and Evidence-based Study;
(c) Task Force on Capacity Building and Good Practice Sharing;
(d) Task Force on Promotion Strategy and Appreciation Scheme.

With the groundwork of these task forces, massive programs of Age-friendly Hong Kong were commenced in 2009 in the six districts (HKCSS 2012). After three years of local participation, HKCSS wished to consolidate best practices of Age-friendly Hong Kong and identify future directions. Under the Task Force on Expert Group and Alliance Building, a mini-research project was initiated by the HKCSS aimed to consolidate the conceptualization of age-friendly city in the Hong Kong context and to further develop expertise alliance. The purposes of this exploratory study were to identify the key concerns of local experts on aging issues and older adults on each of the eight age-friendly domains, and to solicit recommendations made by different sectors to better the lives of older adults in Hong Kong. Some of the findings were shared at the First International Conference on Age-friendly Cities in Dublin in September 2011. The following sections of this chapter report the method and findings of this research project.

8.2 Research Methodology

To achieve the above purposes, a qualitative methodology with focus groups was adopted. A total of eight focus groups were conducted between September 2010 and February 2011. These groups were formed according to the eight themes advocated by the WHO Global Age-friendly City framework. In order to achieve maximize variation of focus group participants, purposeful sampling was used. For each of the focus group, participants were invited by engaging key stakeholders including representatives of older adults, relevant NGOs, relevant government Bureaus and/or Departments, relevant commercial companies, policy advocates, think tanks, and academic scholars. Each group was composed of 10–14 participants, which a total of 96 participants including 20 experts from universities and think-tanks; 37 professionals from NGOs and service providers; and 31 older adults joined these focus groups (Table 8.1).

Such heterogeneous groups, who came from different areas and were formed to provide a platform within which participants from different backgrounds could exchange their views freely on issues related to a particular theme. Interactive dynamics resulted allowing more diverse views to be heard. All focus groups were led by two of the authors and a trained research assistant. Considering potential power disparity between different stakeholders, rules of equal sharing and respect were emphasized at the very beginning of a focus group. For representatives of older adults, all of them had experiences of participating in examining age-friendly concerns in communities and their views contributed a lot to focus group discussions.

Table 8.1 Composition details of focus groups

Categories		No of persons (%)	Total
Participants	From universities	8 (40 %)	20 (100 %)
	From think tanks	9 (45 %)	
	From consultant expert	3 (15 %)	
Professionals	From NGOs	14 (38 %)	37 (100 %)
	From governmental departments	1 (3 %)	
	From architecture sector	4 (11 %)	
	From recreational sector	1 (3 %)	
	From transportation sector	2 (5 %)	
	From housing sector	3 (8 %)	
	From volunteer organization	2 (5 %)	
	From general business sector	1 (3 %)	
	From social enterprises	1 (3 %)	
	From employment service sector	2 (5 %)	
	From media	3 (8 %)	
	From medical and health sector	3 (8 %)	
Older adults	Male	26 (67 %)	39 (100 %)
	Female	13 (33 %)	
	Age 60–69	18 (46 %)	39 (100 %)
	Age 70–79	16 (41 %)	
	Age 80–89	5 (13 %)	
	Primary school level	6 (15 %)	39 (100 %)
	Secondary school level	23 (59 %)	
	Post-secondary school level	10 (26 %)	
Total		96 persons	

Each of these focus groups lasted about 2 hours. Interaction between the group members was promoted. The participants were first introduced with the concept of age-friendly city and related purposes explained. They were given detailed checklist of the designated theme and brief orientation on related concept. They were then asked to rate how much Hong Kong had achieved according to the WHO's checklist with examples on best practices. They were then invited to identify the common concerns that hinder the related development. Finally, their views were exchanged and recommendations were noted.

8.3 Research Findings

This section describes the major findings of this study based primarily on focus group discussion, with reference of previous literature, policy reviews, and census statistics.

8.3.1 Outdoor Spaces and Buildings

In Hong Kong, around 91 % of older people live in the community (Census and Statistics Department 2011c). Over half of these community-dwelling older people live in public housing, including 41.1 % of older adults in public rental housing and 16.6 % in subsidized sale flats, with 41.3 % of older adults in private housing. The living conditions in Hong Kong are generally crowded due to the high population density. The living environments of those residing in old private housings may even worse with negative impacts on older people's well-being (Phillips et al. 2005). Therefore, spacious, comfortable outdoors space, and building facilities are highly appreciated by older people.

Among the various checklist items, green spaces and walkways, outdoor safety, and public toilets did not draw much attention. Although green spaces have long been regarded as an age-friendly feature, there was not much complaint on the lack of green spaces in Hong Kong. It is not sure if the participants were content with the current green spaces or they were already well adapted to this city desert.

In regard to public toilets, Hong Kong is a small urban city in which toilets can be easily found in public amenities, such as public libraries, indoor sports centers, museums, parks, railway stations, etc. They can also be found in many private places, such as shopping malls and restaurants. Of course, public toilets may be a problem in some remote areas. It is therefore understandable that accessibility to public toilets might not be a big issue to Hong Kong older adults.

Although outdoor safety has been frequently mentioned as a problem in the overseas review, it seems to be not a big concern among our participants. The victimization rate of males aged 65 or above was 10.1 per 1000 males of same age group, and the corresponding rate of females was 14.4. Pickpocketing was reported to be the most common crime occurred among the older victims (Census and Statistics Department 2007).

8.3.1.1 Common Concerns

1. Outdoor environments
   - Height and font size of signage are inappropriate for older adults.
   - There is a lack of open spaces and appropriate public leisure facilities for older adults.
2. Outdoor seating facilities
   - There are inadequate outdoor seating areas at bus stops in particular.
   - There is a lack of instruction on the use of public sports facilities.
3. Footpaths and pedestrian roads
   - There is a lack of covered pedestrian roads.

4. Pedestrian crossings
   - The time duration for pedestrian lights is too short for older adults.
5. Drivers' attitude and traffic
   - There are too many speedy vehicles.
   - Drivers in general have bad attitude (e.g. in patient) toward driving.
6. Cycle paths and other pedestrian walkways
   - Older adults dislike using subways mainly due to that there are usually a lot of steps.
7. Service accessibility
   - There are inappropriate facilities used in information announcement on buses.
   - There are not enough rehabilitation buses.
8. Buildings
   - There is always no lift in old buildings.
   - There is a lack of public space in old buildings.
   - There are quite a number of old public rental housing built along hills.

8.3.1.2 Suggestions

1. Outdoor environments
   - More public sports facilities should be built.
   - Each District Council should actively follow-up all minor-work projects to facilitate the mobility of people with different ages and disabilities.
   - The government should work closely with law-making bodies in stipulating laws and ordinances on use of public sports facilities, and strengthen the related monitoring system.
   - District Councils should work closely with local elderly organizations to follow-up all minor-work projects in districts.
   - Safety guidelines on use of public sports facilities should be set and monitored.
2. Others
   - Older adults should be invited to express their views on development projects.
   - Controversy was found on whether or not countdown pedestrian lights should be installed. There was concern that some older adults may overestimate their ability.

In summary, all participants in the focus groups agreed that improvement had to be carried out on outdoor spaces and buildings according to the age-friendly guidelines. The suggestions made for improving the open spaces, environments and public facilities were mainly focused on two levels: policy and district involvement. The first level is policy level. It includes suggestions on how to incorporate an age-friendly city framework into policy agenda and related ordinances. The second level refers to the district level with its main concern on how to get older adults in the community involved minor-work projects.

8.3.2 Transportation

The public transport system is well-developed in Hong Kong (Szeto and Wu 2011; Loo et al. 2010). The services provided by both private franchised buses and the mass transit railway corporations cover most parts of the territory. The mass transit networks are safe, comfort, reliable, and frequent. In addition, the recent introduction of barrier-free taxis can effectively lower the stress of older adults in finding appropriate transportation means. However, there is no statutory requirement for public transport companies to design measures to accommodate the needs of older people and people with a disability. How much an older person enjoys the ride on public transportation depends on the measures taken by individual transport operators. Until now, one franchised bus company has promised to include new auxiliary facilities for the disabled, such as low-floor design and more handrails etc. yet none of these new facilities has been incorporated into the statutory policy agenda.

Of the various checklist items, only affordability, accessibility, age-friendly vehicles specialized transportation and priority seating drew the attention of the focus group participants when discussing problems related to transportation. The most highly rated item was reliability and frequency.

8.3.2.1 Common Concerns

1. Affordability
   - There is no concessionary fare on public transport for people aged between 60 and 64.
   - The fares of mass transit railway for people aged between 60 and 64 are quite expensive.
2. Accessibility
   - There is lack of comprehensive and effective feeder services in the subsidiary transport networks.

3. Age-friendly vehicles
   - There is limited number of low-floor buses.
4. Specialized transportation
   - There is limited number of taxis converted to special designed compartment which accommodated one wheelchair.
5. Priority seating
   - The priority seating measure has not been successfully implemented.

8.3.2.2 Suggestions

1. Government subsidy
   - All those aged 60 and above should be subsidized by providing them with a smartcard, with which people can enjoy concessionary fares on buses and mass transit railways. These smartcards should also be auto-credited.
   - The mass transit railway corporation should take the lead in giving concessionary fares for all those 60 and above to encourage them to go out for activities.
   - The government should supplement the private transport enterprises in providing concessionary fares for older adults and carry out the related monitoring work.
2. Accessible transport and travel
   - Accessibility
     - Comprehensive and effective feeder services in the subsidiary transport networks need to be thoroughly planned.
   - Age-friendly vehicles
     - A half step should be installed in all bus steps.
   - Transport stops and stations
     - A commonly used stored-value electronic card that can be used for most public transport, purchases in convenience, stores, fast food shops, supermarkets, which may also be used to extend traffic lights so as to facilitate older adults in road crossing.
     - Specific parking space reserved for older adults could be made mandatory in large scale parking lots.
   - Completeness and accessibility of information
     - Older adults should be informed of all those restricted areas for loading and unloading.

- Voluntary transport services and priority parking and drop-off spots
  - Barrier-free taxis should be further developed. For instance, these barrier-free taxis should be colored in a different color to be used only by older adults and disabled. Privileges should be given to encourage more taxis to be converted into barrier-free taxis, such as relaxing the restricted areas. Social enterprises can also develop the business of barrier-free taxis. These barrier-free taxi services should be provided upon reservation. Passengers of older adults and disabled and other common passengers should be charged at different rates with the use of these services.
- Accessibility and affordability of taxi services
  - The conversion scheme for barrier-free taxis should be promoted and facilitated.
  - There should be special-colored taxis for both older adults and disabled. These taxis should be exempted from applying for loading and unloading in restricted areas.

3. Public education

- Priority seating
  - Public education should be given on promoting the priority seating measure.
- Roads
  - Public awareness of road safety among the general public and drivers should be raised.
- Drivers' competence
  - Public education should be given on enhancing drivers' awareness on road safety issues and respect for older adults.
  - Bus drivers should be educated on communication skills so that they can serve older adults' needs on buses.

To conclude, all participants agreed that Hong Kong had a well-developed transportation network. Yet the situation could be better. Participants especially raised equal concerns on affordability, availability of special services, and priority seating scheme. To combat these challenges, the participants focused mainly on two issues. The first issue is the role of government. Participants argued that, besides the public transportation companies, the government should take a leading role in providing older adults with subsidy in public transport. The second issue is about how to develop a comprehensive and effective feeder services in the subsidiary transport networks and develop new special services for older people.

8.3.3 Housing

Although the government has been promoting the policies of "ageing in place as the core, institutional care as back-up," there is an immense shortage of supporting services for the frail older adults. As at February 2011, there were less than 10,000 government subsided service capacity for community support services. However, there were over 70,000 and 230,000 out of the 935,000 older adults in Hong Kong who had impairment in Activities of Daily Living (ADL) and Instrumental Activities of daily living (IADL), respectively, in 2009 (Census and Statistics Department 2005, 2009).

Most of the participants were concerned about housing issues, including the problems that they were currently facing and solutions they thought appropriate. Most of their views were centered on housing stock, in line with literature (Yung and Lee 2012; Cheng et al. 2014). There was little comment on the software like community integration.

8.3.3.1 Common Concerns

1. Affordability
   - Older adults have to use up their savings on regular housing maintenance.
2. Housing facilities
   - There are inappropriate lift services (including the escalators and moving walkways are too fast) for older with mobility difficulties.
3. Housing design
   - It seemed less possible to have designs which are suitable for both young and old.
   - The existing Senior Citizen Residences Home designed by the Hong Kong Housing Society is marketed to middle-class older adults in general.
   - There is a lack of tailor-made housing for older adults due to market driven orientations of land developers.
   - The toilet of interim houses is too small for older, in particular who use walking aids.
4. Housing maintenance
   - There is a lack of money for housing maintenance among those living on Comprehensive Social Security Assistance (CSSA).
5. "Ageing in place"
   - The funding criteria of The Home Environment Improvement Scheme for Elderly are too stringent.
   - There is lack of policy on how to achieve the ideal of "ageing in place" for those frail older adults.

6. Housing options

   - Under the current public rental housing policy, owner-occupiers are not eligible to apply for public rental housing.
   - There are too many frail adults who think residential home care services is their only choice.
   - Older adults face a dilemma on their choice of housing. On one hand, older adults are not willing to live in residential homes. On the other hand, poor intergenerational relationship and stress caused by the need of taking care of frail elderly family member may push older adults to opt for residential care home services as their only available choice.
   - There is a lack of tailor-made facilities in public housing for older adults.

7. Living environment

   - The social networks of older adults living in new towns may be affected by inadequate community facilities in these new towns.

8.3.3.2 Suggestions

1. Housing policies

   - Affordability

     – More public housing should be built.

   - Barrier-free designs

     – Standardized labels should be used and promoted to identify those aged-friendly housings for education purpose.
     – Evaluative mechanisms for assessing whether or not housing and facilities are age-friendly should be established.
     – Barrier-free design should be included in the approval regulations of all redevelopment projects in old towns.
     – Open design and modular design should be introduced to meet the needs of people of different ages.
     – Special lifts should be made for older with special needs.

   - Other housing designs

     – New forms of Senior Citizen Residences Home should be built to meet the needs of older adults with different socio-economic backgrounds.
     – Hong Kong Housing Authority and Hong Kong Housing Society should take the lead in developing special housing for older adults.

  - Two household types should be considered in housing designs. The first is households with highly independent living older person, and the second is family unit with elderly households.
  - Measures should be taken to allow older adults to live in the same building or same housing estate of their family members.

- "Ageing in place"

  - Measures of how older adults living in old buildings should be supported for housing maintenance should be further examined.
  - More appropriate and adequate facilities for older adults, such as facilities for the blind and universal design with double swing door, should be built into existing buildings.
  - Community-dwelling older adults should be automatically linked up to community health care system.

- Housing options

  - The eligibility for public rental housing should be unrestricted to elderly owner-occupiers.
  - Public education should be given to older adults to let them know that residential care homes are not their only option.

2. Housing finance

   - Older adults receiving CSSA should be supported with social capital for housing maintenance.
   - Reverse mortgage should be promoted to older adults to ensure they have regular protection and income.

3. Long-term care services

   - Adequate community care support services and residential care services are needed.

4. Residential care home services

   - A scoring system and public monitoring system may be introduced to enhance better quality services and provide older adults more choices (Cheng et al. 2013).

5. Public education

   - Public education should be given to the general public on how to take care of their older family members. For instance, home modification may be needed if an older family member becomes frail.
   - Housing Society Elderly Resources Center can play a key role in this regard. For example, the Center could raise public awareness on home accidents and provide the public with advice on home modification accordingly.

A consensus was reached on the topic of housing. In brief, Hong Kong must implement measures to improve housing-related issues. For instance, with regard to housing options, the improvement in service quality and facilities of existing residential care homes should be a priority in the short term, while better designs of community environments and housing should be the foci in the long term. In regard to education, more public education is needed on the concerns of housing arrangements of older adults and their caregivers including the need for community care.

8.3.4 Social Participation

It is well recognized that The Hong Kong government advocated for the social participation of older adults by setting up a total number of 211 District Community Centers for the Elderly (DECC), Neighborhood Centers for the Elderly (NEC), and Social Centers (SE) for the Elderly according to population distribution at district level (Law et al. 2006). Engagement in social activities and peer functions enable older adults to maintain their energy and activity level and hence their physical and psychological wellbeing. Among the various social participation items, the finding of increased networking was noted. The least discussed item was community integration.

8.3.4.1 Common Concerns

1. Types and accessibility to events and activities
   - Socially disengaged older adults do not have much information on which activities they could join and whom they could approach for help.
2. Affordability
   - Social lives of older adults are usually affected by their differential financial affordability. For instance, the affordability of the paid opera or movie ticket, or even the paid TV channel for World Cup Football Match.
3. Range of events and activities
   - There are older adults who live too far away from any of the elderly centers so they seldom participate in any activities organized by these elderly centers.
4. Facilities and settings
   - There is a lack of policies which support and facilitate older adults to be connected among them without the need to participate in activities organized by elderly centers.

5. Promotion of activities

   - The use of promotional flyers, pamphlets, website, and press releases may not be effective in transmitting information to older adults because of the high illiteracy rate of older adults.

6. Activities for individuals and groups

   - The neighborhood relations are generally poor in the urban community.

8.3.4.2 Suggestions

1. Research

   - The age distribution of older adults in the coming twenty years should be taken into consideration when designing the types and number of activities for older adults.
   - More analyses should be conducted to estimate the types of activity needed by people of different ages, and more resources should be allocated accordingly.

2. Flow of information

   - More diversified audio-visual communication modes, such as devices, radio or television broadcasts, should be used to increase the chance of contact with older people.
   - Neighborhood relations programs on strengthening the intergenerational neighborhood networks, such as "Pilot Neighborhood Active Ageing Project" or "Intergenerational Membership Program" for local district should be done to build harmonious and inclusive community atmosphere and promote social respect for the older people.

In brief, a total of eight participants opined that measures should be taken to enhance the social participation of older adults. Only one participant, who had volunteering experience, expressed a strong view by saying that the situations must be improved. The three main concerns of the participants were the types and accessibility to events and activities, the range of events and activities, and facilities and settings. For the future, the participants believed two things could be done: first, information and broadcast transmitted to older people should be easily managed by older people. Second, to eliminate the labeling effects on older adults and strengthen the mutual support network between older adults and younger generations, public education on cultivating harmonious neighborhoods and caring for older adults should be strengthened (Wang et al. 2013).

8.3.5 Respect and Social Inclusion

Although Hong Kong is still a Chinese dominant society, regretfully, there was an observation, shared by the focus group participants, that older adults are neglected and disrespected. This may be due to the negative stereotype toward older adults. They are usually perceived as slow, less able, and poor. What makes this situation even worse are the changing family structures and the weakening of family support in a globalized Hong Kong. With the increase in both nuclear and more mobile families due to changes of economic structure in Hong Kong since late 1980s, older adults find no role to play in this contemporary family structure, and this may directly influence the declining concept of 'respecting the old.' Family and societal resources appear also to be more inclined toward young people instead of older people. This public attitude is supported by the findings of a study on perceived importance of health services (Mak et al. 2011, 2012; Woo 2012). The study shows that only 3.7 and 4.6 % of respondents, respectively stated that priority resources should be given to those aged 75 and those living in residential care homes. Yet the corresponding percentage for sick children was 30.2 %.

Similar findings were also observed in this study. Respect and social inclusion are problems for older adults in Hong Kong. The focus group participants' concerns covered all the items enlisted in the WHO checklist.

8.3.5.1 Common Concerns

1. Social inclusion and social respect
   - The social respect for older people is generally poor.
   - Transportation cost is still regarded as a burden to older people even with the concessionary tickets. Sometime frail elders on wheelchair who are in need of taxi or soon to be old adult without benefit from concessionary tickets will have less motivation to go out. This directly reduces the social participation of older adults.
2. Social awareness of respect for older persons
   - The youth do not always respect older persons.
   - Elder abuse occurs from time to time, and family members are usually found to be the abusers. In addition, there are still quite a number of unreported cases of elder abuse.
3. Intergenerational family interactions
   - The young generations do not like living with older adults, leading to the occurrence of elderly only households.
   - Children and parental–child relationships are the common foci of most family activities nowadays, and so older persons are seldom considered in family issues or are even overlooked.

4. Civic education

   - Respect for older persons and how to get along with them are not included in the moral education programs for primary and secondary students.

5. Receptivity of older adults in community

   - There are too many unfavorable conditions for organizing public activities for older adults. For instance, some shopping malls do not welcome older adults. Their estate managements may turn down the application of organizations serving older adults if these organizations want to organize some activities for older adults. These estate managements consider the image of older adults is incompatible with that of their shopping malls.
   - There are too limited aged-friendly facilities, such as seats, in most shopping malls.

6. Financial factors

   - Although there is huge number of laborers who are now approaching the retirement age, the financial protection given under the current mandatory provident fund (MPF) is not enough to give them a decent financial protection for their retirement lives.
   - There are still many older adults who earn very little and who do not have MPF protection. They even refuse to join any social activities to save money for other needs, such as medical consultations.
   - The government has put too much resource on capital and estate markets but very little on welfare and elderly services.

8.3.5.2 Suggestions

1. Civic education

   - Respect for older persons and how to get along with them should be included in the moral education programs for primary and secondary students.
   - Committee on the Promotion of Civic Education should support more civic education to arouse the social awareness of older adults.

2. Social corporate responsibility

   - The government should take measures to recognize and encourage private enterprises which fulfill social corporate responsibility in implementing age-friendly policies.

3. Mass media

   - The mass media should produce more television and radio programs on how to get along with older adults, and they should be viewed in programs which are popular among young people.

To conclude, seven participants expressed the view that the current situations "need to be improved" and four participants revealed that the situations "must be improved." All the participants agreed that social inclusion, social respect, and civic education are the keys to improving the wellbeing of older adults. In view of these, the current educational policy has to be reviewed on how 'filial piety' and 'respecting the old' can be effectively integrated into the school curricula. Committee on the Promotion of Civic Education, Education Bureau, and Employees Retraining Board should also take the lead in this regard. Laws with which older adults' rights are protected should also be established.

8.3.6 Civic Participation and Employment

In general, participating in existing elderly volunteer services is not welcomed by the current older population. There were only 3.6 % of people aged 65 and above joining the older volunteer services, as compared to the corresponding 16.9 % among those aged 15–24 (Census and Statistics Department 2003). For employment, other than those high-tech or high-level managerial positions, the work opportunities for older adults or even the middle aged are slim. The labor force participation of those aged 65 and over greatly dropped from 13 % in 1996 to 5.6 % in 2010 (Census and Statistics Department 2011a). Findings of a survey on Hong Kong employers' views on the working elderly revealed the key problem: a common phenomenon of age discrimination at work (Cheung et al. 2011). It was found that the employers commonly perceive that older adults have difficulties in adapting to new work, learning new tasks, and cooperation with younger colleagues.

8.3.6.1 Common Concerns on Civic Participation

1. Options for volunteer work
   - The current older volunteer services are very limited in scope. As such, retirees with professional knowledge feel discontented with the limited opportunities with which they can utilize their expertise. Similarly, only a few choices are available for the oldest old if they want to participate in older volunteer services.
2. Accessibility
   - There is inadequate information of volunteer services.
   - There is a lack of financial incentives, such as transport allowance, to motivate older adults to participate in social activities.

3. Civic participation

   - There is no comprehensive and well-defined strategy on elderly voluntary service development.

4. Others

   - Although the government has continuously supported older volunteer services, its support is usually granted to NGOs for related service operation. The direct support to the older volunteers is relatively slim.

8.3.6.2 Suggestions on Civic Participation

1. The government should conduct thorough studies on service indicators and other supporting measures before it develops a long-term plan on older volunteer services.
2. The government can also strengthen publicity on positive images of older adults so that older adults may more easily find appropriate volunteer work.
3. NGOs can consider forming uniformed groups for older adults to increase their service diversification.

8.3.6.3 Common Concerns on Employment

1. Employment opportunities

   - Workers aged 60 and above are not covered under the current employees' compensation insurance scheme, and this directly decreases the employability of older adults.
   - Older adults usually do not have many work opportunities due to the negative images portrayed of them.
   - It is not easy for older adults with low or no skills to find a job during economic downturn yet they are the neediest.

2. Training

   - There is neither tailor-made retraining program nor detailed statistics on the effectiveness of current re-employment counseling for older adults.

3. Pay

   - There are more working opportunities for female older adults yet their salaries are generally lower than those of male older adults.

4. Others

   - There is no detailed statistics on employment and unemployment of older adults, and so no appropriate supporting and counseling service are available.

8.3.6.4 Suggestions on Employment

1. Employability
   - The government should increase the employability of older adults by providing more training opportunities and subsidies.
2. Incentives for employers
   - The government should establish policies, such as relaxation of tax incentives and employees' compensation insurance etc., to lengthen the working age of older adults.
   - The government should facilitate and encourage social enterprises to recruit older adults. Social enterprises can also adopt different modes in recruiting older adults, such as flexible working modes and reasonable remuneration.

To encourage flexibility in work, enterprises should extend the retirement age and adopt a flexible retirement system. For instance, older adults could be given different employment modes (such as full-time, part-time, or contract-based capacity) to continue their services with companies after their formal retirement. With such practices, older adults could smoothly transit from full-time workers to full-time retirees.

To enhance the civic participation of older adults, older volunteers should be publicly recognized. Most of the participants agreed to strengthen the support for older volunteers. These works include how to portray the positive images of older adults through the use of promotional campaigns; how to increase the scopes and nature of older volunteer services; how to strengthen the information flow between NGOs and older volunteers; and how to expand the insurance coverage so that older volunteers can also be covered.

8.3.7 Communication and Information

Good communication and information dissemination channels are necessary so that no older person is isolated simply because of lack of information. However, there are still 35.8 % of older adults who have no schooling or only preprimary education, while only one-fourth had completed secondary and higher education (Census and Statistics Department 2008). Given this background, the majority of current older adults may not be able to master the information disseminated if information is in text and numbers. A recent survey showed that only 9.4 % of older adults had used a personal computer in the past 12 months (Census and Statistics Department 2011b). The focus group participants had many concerns regarding each of the WHO checklist items.

8.3.7.1 Common Concerns

1. Platform for information flow and broadcasts
   - Currently, the most basic information on elderly services is available only on the internet. Older adults may therefore encounter difficulties in accessing this information since most of them do not have basic computer knowledge. Also, the number of computers for use by staff providing support services to older adults is limited in NGOs.
2. Communication mode and feasibility
   - Use of internet for information dissemination is not effective for older adults with some specific needs, such as the blind, illiterate, and hidden lonely older adults.
   - Information dissemination through family members is also not reliable.
3. Accessibility of oral communication
   - One's social network is most important for older adults to disseminate and receive information.
4. Printed communication
   - Printed materials are not a popular communication mode among the older cohorts. Instead, the interpretation by a third-person is comparatively common.
5. Automatic communication devices and equipment
   - Use of automatic communication devices without hotline services makes older adults puzzled and related services inaccessible.
6. Accessibility to computers, internet, television, and radio broadcasts
   - There is a lack of broadcast channel tailor-made for older adults.
   - There is no television program through which younger generations can learn more about older adults to enhance harmonious communication.
7. Language for communication
   - English is usually the common language on internet. Yet classes teaching older adults on the use of internet and English are limited.
   - The expressions, vocabularies, techniques, and pace used in broadcasts are not easily understood by older adults.
8. Quality of communication
   - Older adults are usually portrayed as vulnerable and resource seekers on internet.

8.3.7.2 Suggestions

1. Program production
    - More programs targeted older adults should be produced when Hong Kong starts to use the digital broadcasting system in the near future. These programs should be used as platforms for information dissemination on one hand and for older adults to express their views on public policies on the other hand.
    - Contents and formats of broadcast channels and programs tailor-made for older adults should be further examined.
    - Information disseminated in these programs should emphasize respectful and harmonious communication with older adults rather than portray negative images.
    - More broadcast programs should be produced to educate the younger generations on how to get along with older adults.
    - There should be district broadcast channels with which District Councils can produce programs to disseminate information for older adults at district level.
2. Roles of NGOs
    - Live broadcasts and replays should be made available to older adults in District Elderly Community Centers.
    - NGOs should develop programs to train some volunteers to be 'age-friendly ambassadors' at district level. These ambassadors can assist NGOs to disseminate information to older adults. How to effectively communicate with older adults should, therefore, be one of the main focuses in the related training.
3. Silver market
    - More education and information are needed to enable and facilitate young designers to create new computer networks, software, and other related services to meet the needs of older adults in the market.
    - A 'silver hair market' or 'grey market' for older adults should be developed.
4. Public education
    - Older adults should be encouraged to voice their views in the form of a parliament.

A majority of the participants opined that the current situations of communication and information had to be improved. Their main concerns were around 'platform for information flow and broadcasts,' 'communication mode and feasibility,' and 'quality of communication'. In brief, to connect older adults with society, the priorities are to strengthen the communication flow by adopting approaches that older adults can easily manage, to eliminate the labeling effects on older adults by doing more public education, and to enable older adults to get more community support by mobilizing neighborhood resources at district level.

8.3.8 Community Support and Health Services

Older adults in Hong Kong are generally in good health. More than 75 % of community-dwelling older adults expressed that they did not need any assistance from others in their daily living. However, among the remaining 25 % who needed assistance, slightly more than a half of them were unable to find any help (Census and Statistics Department 2009; Mak et al. 2012; Woo 2012). Moreover, under the current division of labor in the government, there is more than one bureau and department responsible for policy and implementation on community support and health services. Among all the various bureaus and departments, the key ones are the Food and Health Bureau and the Labor and Welfare Bureau. For instance, older adults usually go to clinics run by either the Hospital Authority or Department of Health for health services. But they will go to Social Welfare Department for community support services and community mental health services. There is room for better coordination between health services and community support services for older people. Nearly all focus group participants had many concerns regarding each of the WHO checklist items.

8.3.8.1 Common Concerns

1. Service accessibility

   - Medical consultations and dispensaries are usually located at different places. Such arrangement brings difficulties for older adults, in particular, those with more than one chronic disease.
   - Older adults need to go to other districts for service since the current elderly health centers provide uniform quota services in each district but have not increased the quota in accordance with the number of older population in districts.

2. Offer of services

   - Because of communication problems among health care professionals, many older patients may need to repeat examinations in hospitals. This directly increases the wastage in health workforce and related resources. It is also time-consuming to older patients.
   - No seamless service is available in the current health care system. The specific details of support services have not been well planned whenever there is a launch of new service. Similar inconsistencies and segregation are

especially common in provision of 'integrated home care services' and 'enhanced home care services.'

3. Support services for caregivers

   - Needs of caregivers of older adults are not met by the current services. Long waiting times for residential care home services for older adults is an example.

4. Service scope

   - Some services are good but are not commonly known by older people while some services are so limited that older adults need to wait for a long time for the services.

5. Quality of services

   - There is no standard arrangement in the service provision in elderly health centers.
   - Doctor-shopping is prevalent among older adults and this directly affects their drug management and hence the effectiveness of drugs.
   - Changes of dosage and color are quite common in drugs provided by government clinics. Yet older adults quite easily fail to adapt to these changes if they do not consult their doctors or pharmacists.
   - It is not easy for older patients and their caregivers to follow the instructions printed on the medicine containers if information is printed in English.
   - There are still advertisements on Chinese patent medicines in which exaggerated efficacy have been found.

6. Voluntary support

   - The key problem of the current health care system is the incomprehensive and not thorough service provision but not if volunteers are available to accompany older patients for medical consultation.

7. Emergency planning and care

   - The current telephone booking system does not allow older adults to make appointment one day in advance. Besides, the system provides older adults with only one choice of appointment.

8.3.8.2 Suggestions

1. Preventive care

   - Department of Health should adopt some innovative methods to strongly promote preventive care. For example, older adults should be encouraged to

attend workshops on different types of diseases. Exercise sessions in parks could also be continuously organized and supported.
- Oral hygiene should be considered in service scope of elderly health centers.

2. Service scope

   - The law should be timely reviewed and revised to cater for the needs of older adults and their caregivers. For example, laws and ordinances should be set up on advanced nursing practices so as to release the pressure of older adults and their families in facing death. In addition, the private niches for cremated remains should be regulated. More public niches should be built.
   - Older adults should be given choices after they have completed their contribution to the medical insurance premium.

3. Data management

   - Electronic medical records should be fully developed and widely used in different departments, such as family doctors. Individual patients should also be allowed to read their own record for reference.

4. Public–private interface

   - The government should take the lead in establishing cooperative relationships between public and private health care sectors so as to provide patients with seamless services.

There were seven out of eleven participants expressed the view that the current situations "need to be improved" or "must be improved." Of the different items, service accessibility was the prime concern of participants, followed by service scope and support services for caregivers. To solve the current problems, participants commonly agreed that different official departments should work closely with various professional bodies to provide diversified preventive health services for older people. Also, the government should take a leading role to facilitate the cooperation among various departments in the public sector and between public and private sectors to provide older people with seamless services.

8.4 Discussion and Conclusion

In this section, we describe briefly the concept of age-friendly cities in local context, followed by an overview of the development and discussion on the future directions of age-friendly strategies in Hong Kong.

8.4.1 Critique to the Concept of Age-Friendly Cities in Chinese Culture

The WHO age-friendly cities guidelines and checklist are comprehensive and detailed. However, the concept of age-friendly city has not been concisely defined in Chinese context. Until now, the connotation is around a city within which the design is suitable for people to grow old, and older adults are usually the focal point. It is understandable that an age-friendly city should be a friendly city to not only older people but also people of different ages and with different backgrounds. A city can thence be named as an age-friendly city only when it can accommodate even the needs of the most vulnerable, the most deprived, or the frailest groups of people in a city. Yet, the age-friendly city guidelines and checklists are "urban life in developed countries" oriented. Most of the eight themes are about how to provide suitable, accessible and affordable physical and social environments for older people to engage in an urban community life. There is little discussion on how to provide older people with their basic social and income protection. Given the limited resources in any society, such orientation is understandable. How to strike a balance among people of different ages and with different backgrounds for an age-friendly city is a question to be answered.

Second, it is not clear which specific measures can achieve which particular theme since some measures can be used to achieve more than one theme. For instance, measures to increase the employability of older adults can enhance both "respect and social inclusion" as well as "civic participation and employment." There are no measurable policy outcomes under the age-friendly framework that have been identified. Due to the comprehensiveness of the framework, the framework does require that different government bureaus and departments may implement policies that serve different purposes across different themes.

Last but not the least, the current WHO age-friendly city guidelines and checklists still have room to work out clear directions for a government to establish policy priorities in achieving an age-friendly city. The key concern is which target group should be given priority in policy planning. For instance, should the young–old, the old or the oldest-old be listed first in a policy agenda? Should healthy older adults or the frail be given prior attention in policy planning? Should the most social and financial deprived older adults be targeted? Should the focus be on older adults living in newly developed communities or those residing in old buildings of old districts? A recent study that examined age-friendly indicators and happiness among healthy older adults revealed that communication and information, transportation, outdoor spaces and buildings, respect and social inclusion were significantly associated with sense of happiness (Hong Kong Young Women Christian Association 2013).

Under normal circumstances, a society's development usually considers the young rather than old populations in its policy making. As such, many social, economic, and physical developments are centered on both hardware and software that suit the needs of the young populations. These can be easily observed from the short duration of traffic lights for pedestrians, the stagnant retirement policies, the programs broadcasted in both televisions and radios and the like. This is reasonable if a society is still at its early stage of development when the population is young. However, when a society has been developed with urbanization and modernization, its population will become aging. This aging population is now a global phenomenon. Policies over-emphasizing the young populations may thus not be favorable to further development of the society.

Regardless of the developmental stage of a city, government has been playing a key role in making an age-friendly city. Its significance does not lie on its dominant position in guiding and monitoring the implementation of all age-friendly related policies, but on its leading role in providing platforms and environments with which the public and NGOs can go hand in hand to advise the government on related policy making and cultivate age-friendly environments.

8.4.2 International Good Practices in Developing Age-Friendly Strategies

Four observations can be made that might have implications to Hong Kong.

8.4.2.1 Leading Role of Government

One of the key observations on reviewed cities and countries are the leading roles of the government in showing their commitment to the age-friendly framework. Even though different mechanisms have been developed under different names, the roles of government are essential (The New York Academy of Medicine 2012). Currently, the local NGOs have been playing a crucial bottom-up role in adopting the idea to develop Hong Kong as an age-friendly city. However, government's top down role cannot be excluded. The commitment and the role of the Hong Kong government in Age-friendly City are not yet well discussed. However, at least the agenda was put into the consultation of Population Policy in Hong Kong in 2013. To materialize the ideal other cities have achieved, the leading role of government needs to be thoroughly deliberated in term of achievable target sets.

8.4.2.2 Bottom-Up Approach with Top-Down Backup

With positive governmental support and backup, the bottom-up approach has been proved to be a successful method in actualizing an age-friendly city concept in many overseas cities and in Hong Kong too. In other words, if a government works closely with NGOs and even private sectors to participate and to discuss how the age-friendly city checklists can be implemented, such bottom-up approach can effectively fulfill its functions.

8.4.2.3 District-Based Initiatives

Since older adults are the ones who know very well about their own district environments and their own needs, district-based initiatives should be highly encouraged according to international good practices. These initiatives can include planning, policy implementing, and monitoring at district level. Hong Kong has started to implement district planning since early 2000. District Council has more resources to allocate to fulfill local needs, which can be regarded as an advantage in promoting age-friendly Hong Kong.

8.4.2.4 Comprehensive and Regular Data Bases

To enable better planning of any policy, comprehensive and regular data collections are very crucial. It is thus suggested that once a clear framework for the development of an age-friendly city has been formulated, related informations to serve as age-friendly indicators should be gathered on a regular basis. Related statistics should then be documented and analyzed by both those interested departments as well as the concerned NGOs to formulate population policy. Optimally, data transparency can be achieved so that the public can participate in evaluating whether certain policy objectives have been achieved or not. Unfortunately, during the process of conducting the mini-research and report writing, difficulties to allocating updated data have been encountered.

8.4.3 Age-Friendly Hong Kong—Where We Are? Where Are We Going?

Based on historical development in Hong Kong, some of its hardware infrastructures have been developed as a universal service. "Transportation" and "health services" are some of these examples. It is therefore understandable that the majority of focus group participants, including expertise and older adults' participants, were content with these aspects. However, this does not mean that such facilities are tailor-made for older adults. It only implies that the transportation and health services in Hong Kong are regarded as fulfilling needs in this aging population area. On the contrary, "Outdoor spaces and building" and "housing" were the two disappointing aspects frequently mentioned by focus group participants, in particular older adult participants. One of the possible reasons is that older adults, as compared to other age group people, are more in need of outdoor spaces and housing needs to sustain their quality of life and aging in place. What is worse is the 'age-unfriendly' social element or what we called 'the software.' As reflected in our research findings, "respect and social inclusion" were the most unsatisfactory aspect among focus group participants. In other words, age discrimination still prevails, with older adults not finding it easy to be respected or to be socially included by the society.

The leading role of Hong Kong Council of Social Services (HKCSS) in making Hong Kong an age-friendly city is remarkable. Being a council that coordinates social service agencies, HKCSS has been a key representative of many local NGOs to advise the government on many major social policies. Although its more than 440 agency members is relatively small in number when compared with the total of 3000 registered NGOs in Hong Kong, the majority of these agency members are well-established NGOs with a long history of servicing the community. Moreover, HKCSS also has established regular platforms to exchange views with key government Bureaus such as Welfare and Labor Bureau and Health and Food Bureau. The concerted efforts made by these NGOs under the leadership of HKCSS are definitely beneficial in actualizing the concept of age-friendly city in Hong Kong.

The unique role of HKSAR government is another feature compared to international cities and countries. To materialize the concept of age-friendly city, collaborations at district level are the most common form found in Hong Kong. It is believed that people in district know what the most urgent needs are for older people living in the same district; therefore, district partnerships can better prioritize resource allocation according to the needs of the local older people. This alliance also has its drawbacks. Age-friendly elderly platforms have been established in 18 districts, in which older adults have been empowered to raise their concerns, advocate changes, negotiate with government departments or business companies, and even report to media to attract public concerns. Changes have been observed in some districts in regard with prolonged time for road crossing, building lifts for passenger's bridges, setting priority seats for seniors at Mass Transit Railway, etc. However, In case an age-friendly measure involves a change in structural or policy level, such as universal retirement protection, building design, and outdoor spacing, the involvement of government rather than District Council should work more efficiently and effectively in formulating the related policy change.

In order to make Hong Kong an age-friendly city, we would like to make the following recommendations.

First, the concept of age-friendly city is recommended to be thoroughly deliberated among Hong Kong society. Not only among the policy makers, government officials, and NGOs, but also commercial markets and older people themselves. One of the key questions is whether age-friendly Hong Kong is an appropriate policy framework for Hong Kong to face challenges of an aging society. If not, what are the better alternatives? While international good practices show that the leading role of the government in promoting age-friendly city is essential, further studies are recommended to examine what are the barriers, politically and administratively, for Hong Kong government to show its commitment. Would it be possible that an age-friendly Hong Kong can be achieved without a leading government? Would a district strategy be a good practice in Hong Kong?

Second, identify and develop age-friendly Hong Kong measures to evaluate age-friendly Hong Kong initiatives is a key message. After years of development at the district level, it is recommended that measurable outcomes can be identified and developed so that policy initiatives can be evaluated and made transparent to the public. It is important to notice that an age-friendly city framework advocated by WHO promotes an mixture of top-down and bottom-up approaches, full participation of all stakeholders including the government, the NGOs, the markets, and older adults and their family members are one of the key measures to achieve such integration. It would be better if the WHO framework could provide measurable policy outcomes. Moreover, no international benchmarking data are available, which makes the implementation of the framework challenging. Hong Kong can take the initiative to establish measurable outcomes for Asian communities, to make contribution to the international communities. In 2011, the Framework of Age-friendly Hong Kong Index (AFHKI) has been developed by HKCSS. With the solid experiences, we hope to work with related countries and communities that share similar socio-economic and cultural context to contribute to the age-friendly movement.

Third, at each district level, it is recommended to review age-friendly initiatives and consolidate good practices. On one hand, good practices can be shared with those newly participated districts. On the other hand, it would be important to review whether a selective or comprehensive approach should be adopted. The international review showed that all reviewed cities and countries adopted a comprehensive approach, which implies that they commit to work on all eight domains and set policy objectives correspondingly. However, in Hong Kong, with understandable limitations, district initiatives are focusing on some domains as a start and gradually extend to all eight domains. As at 2013, although appreciations are awarded to more than 450 initiatives, product and services on both district and territory level (Hong Kong Council of Social Service 2013), whether such a focus gears innovative age-friendly strategy or implies development limitations due to other recognized reasons, need further examination.

Last but not the least, being a Chinese society with a very unique historical pathway, Hong Kong has a mixture of influx of East and West cultures. We are very proud of our self-reliant work ethic and also a well-established legal and administrative system. The government emphasized again and again that core values such as "rule of law, freedom, equality, openness, and a corruption-free society" would be safeguarded (Hong Kong SAR Government 2004). However, it is timely for Hong Kong to reflect whether we only strive for being a "leading position in the areas of finance, trade and economy, and communications." For a sustainable city, social development and economic development shall go hand in hand. It is really disappointing to observe that our Chinese tradition on 'respecting older people' is less being recognized and treasured, which definitely creates an opposite force for achieving social and economic development in Hong Kong in a long-run. It is a critical time to question people: for children and youth; how do you wish Hong Kong to treat your grandparents and grand grandparents? For adults; how do you wish Hong Kong to treat your parents? For older adults; how do you wish Hong Kong to treat you and your peers? Would it be more desirable that we can be a leading city in both economic and social development?

References

Census and Statistics Department. (2003). *Thematic Household Survey Report no. 14.* Hong Kong: Census and Statistics Department (Online). Available from http://www.statistics.gov.hk/publication/stat_report/social_data/B11302142003XXXXB0100.pdf. Accessed September 8, 2012.

Census and Statistics Department. (2005). *Thematic Household Survey Report No. 21.* Hong Kong: Census and Statistics Department (Online). Available from http://www.statistics.gov.hk/publication/stat_report/social_data/B11302212005XXXXB0100.pdf. Accessed September 8, 2012.

Census and Statistics Department. (2007). *Demographic trends in Hong Kong 1981–2006.* Hong Kong: Census and Statistics Department (Online). Available from http://www.statistics.gov.hk/publication/stat_report/population/B1120017022007XXXXB0200.pdf. Accessed September 8, 2012.

Census and Statistics Department. (2008). *Hong Kong 2006 Population By-census Thematic Report: Older Persons.* Hong Kong: Census and Statistics Department (Online). Available from http://www.statistics.gov.hk/publication/stat_report/population/B11200532006XXXXB0100.pdf. Accessed September 8, 2012.

Census and Statistics Department. (2009). *Thematic Household Survey Report No. 40.* Hong Kong: Census and Statistics Department (Online). Available from http://www.statistics.gov.hk/publication/stat_report/social_data/B11302402009XXXXB0100.pdf. Accessed September 8, 2012.

Census and Statistics Department. (2011a). *Table: Labor force and labor force participation rate —by age group.* Hong Kong Statistics Table—Census and Statistics Department (Online). Available from http://www.censtatd.gov.hk/hong_kong_statistics/statistical_tables/index.jsp?subjectID=2&tableID=008. Accessed September 8, 2012.

Census and Statistics Department. (2011b). *Thematic Household Survey Report no. 48.* Hong Kong: Census and Statistics Department (Online). Available from http://www.smokefree.hk/UserFiles/resources/Statistics/Thematic_Household_Survey_No48.pdf. Accessed September 8, 2012.

Census and Statistics Department. (2011c). *2011 Population By-Census.* Thematic Report: Older persons. Hong Kong: census and Statistics Department, Hong Kong Special Region Government.

Census and Statistics Department. (2012). *Hong Kong Population Projections 2012–2041.* Statistics Department, Hong Kong Special Administrative Region.

Cheng, S.-T., Lum, T., Lam, L. C., & Fung, H. H. (2013). Hong Kong: Embracing a fast aging society with limited welfare. *The Gerontologist, 53*(4), 527.

Cheng, K. H., Phillips, D. R., Siu, O.-L., & Yeh, A. G. (2014). Patterns of residential adjustment for older person: Who will recover and how do they recover? A study in different residential environments in Hong Kong. *Social Indicators Research, 119*, 295–319.

Cheung, C., Kam, P. K., & Ngan, R. M. (2011). Age discrimination in the labor market from the perspectives of employers and older workers. *International Social Work, 54*(1), 118–136.

Hong Kong SAR Government (2004). *HK's core values remain intact* (Online). Available from http://www.info.gov.hk/gia/general/200406/07/0607240.htm. Accessed September 8, 2012.

Hong Kong Young Women Christian Association. (2013). *Happiness and age-friendly indicators among community-dwelling healthy older adults.* Hong Kong: Hong Kong Young Women Christian Association.

Law, C. K., Chui, E. W. T., Lee, K. M., Ko, L. S. F., & Wan, C. W. (2006). *Evaluation study of the re-engineered neighborhood elderly centers and district elderly community centers.* Hong Kong: The University of Hong Kong.

Loo, B. P., Chen, C., & Chan, E. T. (2010). Rail-based transit-oriented development: Lessons from New York City and Hong Kong. *Landscape and Urban Planning, 97*(3), 202–212.

Mak, B., Chau, P., & Woo, J. (2012). Prioritization in health care and ageism. In J. Woo (Ed.), *Aging in Hong Kong: A comparative perspective* (pp. 257–268). New York, NY: Springer. International Perspectives on Aging 5.

Mak, B., Woo, J., Bowling, A., Wong, F., & Chau, P. H. (2011). Health care prioritization in ageing societies: Influence of age, education, health literacy and culture. *Health Policy, 100*, 219–233.

Phillips, D. R., Siu, O.-L., Yeh, A. G., & Cheng, K. H. (2005). The impacts of dwelling conditions on older persons' psychological well-being in Hong Kong: The mediating role of residential satisfaction. *Social Science and Medicine, 60*(12), 2785–2797.

Szeto, W., & Wu, Y. (2011). A simultaneous bus route design and frequency setting problem for Tin Shui Wai, Hong Kong. *European Journal of Operational Research, 209*(2), 141–155.

The Hong Kong Council of Social Service. (2012). 起動全城香港長者友善社區.

The Hong Kong Council of Social Service. (2013). *Age- Friendly City and Communities—What is involved?* The Hong Kong Experience.

The New York Academy of Medicine. (2012). *Age-friendly NYC* (Online). Available from http://www.nyam.org/agefriendlynyc/. Accessed September 8, 2012.

Wang, M. P., Viswanath, K., Lam, T. H., Wang, X., & Chan, S. S. (2013). Social determinants of health information seeking among Chinese adults in Hong Kong. *PLoS ONE, 8*(8), e73049.

WHO. (2007a). *Global Age-friendly Cities: A Guide* (Online). Available from http://www.who.int/ageing/age_friendly_cities_guide/en/index.html. Accessed September 8, 2012.

WHO. (2007b). *Checklist of Essential Features of Age-friendly Cities* (Online). Available from http://www.who.int/ageing/publications/Age_friendly_cities_checklist.pdf. Accessed September 8, 2012.

Woo, J. (Ed.). (2012). *Aging in Hong Kong: A comparative perspective*. New York, NY: Springer. International Perspectives on Aging 5.

Yung, B., & Lee, F.-P. (2012). 'Right to housing' in Hong Kong: Perspectives from the Hong Kong community. *Housing, Theory and Society, 29*(4), 401–419.

Chapter 9
The Implementation of Age-Friendly Cities in Three Districts of Argentina

Nélida Redondo and Silvia Gascón

9.1 Introduction

Argentina is currently in the advanced stages of its demographic and epidemiologic transition. This has brought a process of demographic aging which will intensify and extend progressively—although in different degrees—in all jurisdictions of the country. Additionally, the progressive postponement of death to extreme ages in life is pushing up the number and proportion of people aged over 80, including those who reached 100 years and more, to unprecedented levels in the country's history.

The Argentine demographic transition came about in a context of early and rapid urbanization. By the beginning of the twentieth century more than half the country's inhabitants lived in urban areas. According to data from the last population census conducted in 2010, the current urban population has risen to 91 % of the total inhabitants in the country. People aged 65 and over show a higher tendency to live in large urban centers than the total population: in 2010, 92.3 % of the people aged over 65 lived in urban areas. It should be noted that the Argentine National Institute for Statistics and Census (INDEC) defines urban population as those who

N. Redondo (✉)
University of Buenos Aires, Buenos Aires, Argentina
e-mail: Redondo.nelida@gmail.com

N. Redondo
Sociology of the Aging, ISALUD University, 3 de Febrero 969—5° Piso, 1426 Ciudad de Buenos Aires, Argentina

S. Gascón
Social Welfare and Social Services Management, Alcalá University, Alcalá, Spain
e-mail: sgascon@isalud.edu.ar

S. Gascón
Iberoamericana Organization for Social Security (OISS), Management of Gerontology Services, ISALUD University, 7th St. No. 1369, 1901 Ringuelet, La Plata Buenos Aires, Argentina

© Springer International Publishing Switzerland 2016

T. Moulaert and S. Garon (eds.), *Age-Friendly Cities and Communities in International Comparison*, International Perspectives on Aging 14,
DOI 10.1007/978-3-319-24031-2_9

reside in towns of 2000 people and over, while those living in smaller villages or in scattered rural areas belong to the rural population. The latter displays two major population groups according to their place of residence: (a) grouped rural population, which comprises residential districts with less than 2000 inhabitants; (b) dispersed rural population, made up by people who reside in the open country without forming part of population centers.

As in most of Latin American countries, Argentine economic, social, and human developments have followed different trajectories in the various geographical regions of its national territory and across social strata. As a result, demographic aging is more pronounced in the big cities of the provinces in the central-littoral and Cuyo regions, the wealthiest in the country.

The Argentine epidemiologic transition followed the same distinct regionalized trends, and by the year 2001, the difference between the jurisdiction with the highest level of human development—the Autonomous City of Buenos Aires (CABA)—and the jurisdiction with polarized behaviour, i.e., the lowest life expectancy at birth—the northeastern province of Chaco—showed a lower average life expectancy by 5 years for males and 6 years for females in the latter province. In the CABA, the difference by gender has intensified in recent years (Caviezel 2008).

At the same time, the drop in mortality intensified throughout the national territory, but did not result in similar health-illness patterns. While healthy life expectancy or life expectancy free from permanent limitations rose in the CABA, life expectancy with permanent limitations grew in the province of Chaco, including life expectancy with higher degrees of disability due to the coexistence of two or more permanent limitations (Belliard et al. 2013).

In this chapter, the results of the implementation of the Age-Friendly Cities (AFC) project in three Argentine cities are analyzed in the light of regional inequalities.

Section 9.2 contains a brief reference to AFC project in Argentina. In Sect. 9.3, the social and demographic profile of each of the three cities is described. Section 9.4 includes a comparative analysis of the strengths and weaknesses voiced by the elderly inhabitants, as well as providers and informal caregivers in specific areas researched by the Vancouver Protocol. Section 9.5 examines the progress achieved under the plan for improvements in each city in the context of the specific characteristics relating to governance of each region. Lastly, the conclusions highlight the particular characteristics of the Argentine case and provide a discussion of the lessons learned from an experience that may be useful to understand the situation of countries with intermediate economic growth such as Argentina.

9.2 Age-Friendly Cities in Argentina

The Argentine project was carried out in different geographical locations, with differing sizes and varied social and economic environments:

(a) the city of La Plata, the capital of the wealthy province of Buenos Aires;
(b) the city of Resistencia, capital of the province of Chaco in the country's northeast, one of its poorest regions;
(c) the city of Lezama, a small town with 5000 inhabitants, also located in the prosperous province of Buenos Aires.

The level of progress achieved under the project to make these cities age-friendly is different. Nevertheless, diagnostic studies have been completed in the three cities by applying the Vancouver Protocol, making the results of the three research studies comparable. This facilitates the identification of the different urban and social situations in which the lives of elderly persons unfold and the problems that need to be addressed to guarantee their social inclusion.

The research was conducted in these three cities by *Universidad ISALUD*, a private nonprofit academic organization devoted to the study and training of human resources in different socio-sanitary areas, including the field of gerontology. The gerontology area carries out research, community extension programs, and undergraduate and graduate education, in addition to adult training courses. The Director of the Master in Gerontology Services Management at *Universidad ISALUD,* was part of the World Health Organization (WHO) designed AFC Project task force that met in Vancouver in April 2006. As a result, the city of La Plata was included in the initial world round of 32 member cities. In September 2006, the diagnostic research stage commenced in La Plata in strict compliance with the approach and guidelines of the protocol designed at the Vancouver meeting.

To carry out the study in Argentina, *Universidad ISALUD* formed a work team tasked with preparing the field work, selecting neighborhoods, recruiting participants, and translating the material produced by WHO into Spanish. The project was implemented by a project leader and a lead researcher, both members of the *Universidad ISALUD* (Gascón et al. 2008).

In 2012, the team also conducted the diagnostic study in the city of Resistencia, Chaco. This took place through a cooperation agreement between *Universidad ISALUD* and the School of Government of the Chaco Province. Lastly, the project was implemented in the city of Lezama in 2013. The research was led by a postgraduate student as part of her thesis project for the Master in Gerontology Management Services.

In order to carry out the AFC Program, in the city of La Plata, *Universidad ISALUD* and the *Red Mayor* have been working together since 2006. In recent years the municipality has adopted a more proactive role in the implementation of the project. In Resistencia, *Universidad ISALUD* and the School of Governance of the Province of Chaco, to which the former renders consulting and technical assistance services in several areas, agreed to work together to implement the Age-Friendly

Table 9.1 People interviewed in Argentina AFC program by city and group, 2006–2013

Focus group	La Plata	Lezama	Resistencia	Total
People 60–74 years old lower class	23	7	11	41
People 60–74 years old upper middle class	15	13	10	38
People 75 years old and over lower class	16	6	7	29
People 75 years old and over upper middle class	14	10	6	30
Informal caregivers	7	7	9	23
State providers	6	9	1	16
Private provider	7	7	8	22
NGO providers	7	0	1	8
Total of people	95	59	53	207

Cities project. As a result, in the year 2012, a diagnostic study was conducted in the city. Lastly, the diagnostic study was carried out in the city of Lezama in 2013 under the framework of the Master in Gerontology Services Management of *Universidad ISALUD*.

The final samples including all the people interviewed in the three cities are presented below. The participants' opinions about the eight domains included in the Vancouver Protocol were investigated in each group (Table 9.1).

9.3 Socio-Demographic Profiles of Cities

9.3.1 La Plata City

La Plata is the capital city of the province of Buenos Aires, the country´s main province. It is situated in the wet pampas, 56 km southeast of the Autonomous City of Buenos Aires (CABA), capital of the Republic of Argentina. Its climate is temperate, with a mean annual rainfall of 1023 mm. It spans an approximate surface area of 926 km^2. In 1998, the city was included in the list of places to be recognized as Cultural and Historical Heritage of Humanity by the General Conference of the UNESCO. The layout of the city was designed before its foundation, and it was built as a well-planned model metropolis, with magnificent public buildings and numerous parks and green areas.

The city is recognized for the quality and diversity of its cultural and educational activities. It has three universities that are renowned in Latin America: the National University of La Plata (UNLP), the Technological University (*Universidad Tecnológica*), and the Catholic University (*Universidad Católica*). La Plata has a vastly rich history in terms of culture, ideological debate and scientific research. The provincial institutions that represent the three branches of Government (Executive, Legislative and Judicial) are located in this city. In terms of healthcare, the city has private and public health establishments (run by the provincial or municipal government), with a combined capacity of approximately six thousand hospital beds, of which 70 % are run by the provincial government.

In the year 2010, the district of **La Plata** had a population of 654,324 inhabitants (INDEC 2014). As a result of the sharp demographic aging in the district, the population is predominantly female: women account for 52.2 % of the total population. According to data from the latest population census (2010), the population aged over 65 accounted for 11.3 % of the total population, showing a pronounced aging process in the city. In 2010, the inhabitants aged over 80 represented 3 % of the total population of the district, which compares with 2.7 % in the previous census conducted in 2001.

La Plata has several universities. This causes fluctuations in the number of youths entering the city that can reduce the proportion of older persons in the total population. It also explains the broadening of the age pyramid in the segments comprising people aged 20–29 years. Nevertheless, the growth of the 80-and-over age group has remained steady and has even increased, indicating aging at the peak of the age pyramid due to the drop in mortality among the elderly population (Fig. 9.1).

9.3.2 *Lezama City*

Lezama is the capital city of the district of Lezama, which formerly belonged to the neighboring district of Chascomús. It was granted independence in 2009, based on

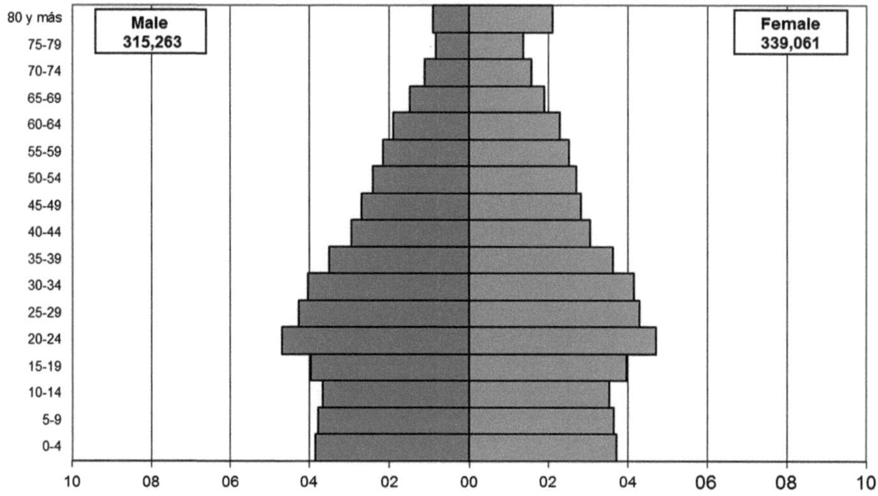

Fig. 9.1 Gender and age structure of the population in the district of La Plata, capital of the province of Buenos Aires, 2010. *Source* INDEC. National Population, Households, and Dwellings Census, 2010

the former township and Colony named after Manuel J. Cobo. The district is situated 157 km to the south of the Autonomous City of Buenos Aires, with which it is connected via a highway and a major railroad. The city of Lezama lies in a rural area which is dedicated mainly to livestock production. Like all urban settlements in the Argentine wet pampas, the city was organized around the train station and the route that connects it with the Federal Capital. As a result, the city is divided into the West neighborhood (*Barrio Oeste*) and the East neighborhood (*Barrio Este*), also known as *El Tero*. The civic center and main square are in the West neighborhood and are divided by the national highway no. 2 (or *Autovía 2*), which, as described by the neighbors' testimonies, is a major barrier for the inhabitants communication and access to services.

The district of **Lezama** is also located in the province of Buenos Aires. According to the 2010 census, the population in the district reached 5,629 inhabitants. The demographic aging of the district is very marked and exceeds the national average. The population aged over 65 represented 13.9 % of the total, while those over 80 accounted for 3.3 % of the total, demonstrating an aging process of the older population in the district. Its farming economy and the reduced size of its urban population center are the main factors behind the emigration of youth from both sexes aged 20–30 in search of higher education in the provincial capital (City of La Plata) or the Federal Capital. The population structure by age and gender is presented in the following graph (Fig. 9.2).

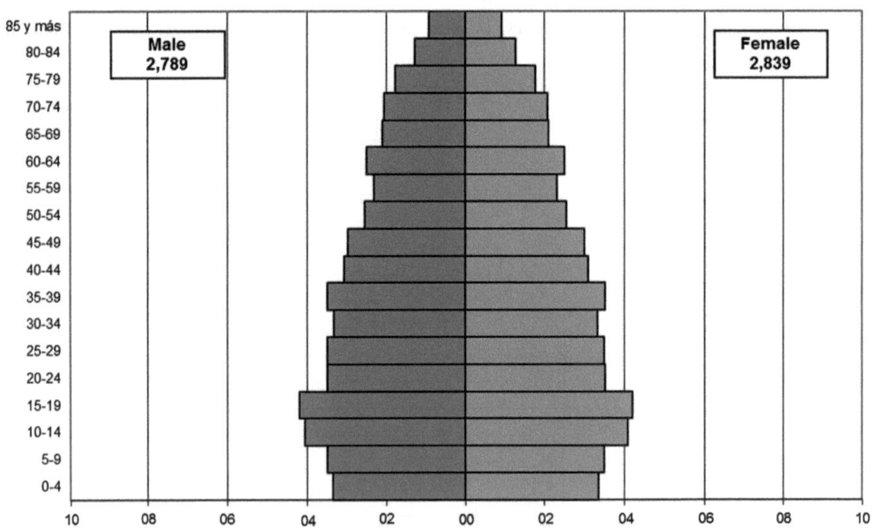

Fig. 9.2 Population pyramid by gender and age for the district of Lezama, province of Buenos Aires, 2010. *Source* INDEC. National Population, Households, and Dwellings Census, 2010

9.3.3 Resistencia City

The **city of Resistencia** is in the north of the country, on the border with Paraguay. It was established as a colony toward the end of the nineteenth century with the arrival of the first settlers—a group of Friulian immigrants from Italy. The establishment of settlers and of authorities was met by resistance and fighting from the native population. Once the town had been founded, Italian, Spanish, and French immigrants flowed in continuously, occupying the settlements encouraged by the possibility of owning their own plots in fiscal land.

Unlike the typical grid plan used in the layout of most old Argentine towns, Resistencia features spacious streets and avenues, wide footpaths, and ample two-lane streets. The avenues that make up the original main area have large parks or six lanes for traffic.

The center of the city is a park that spans more than four hectares, *Plaza 25 de Mayo*. Four main avenues start out from the center of this park at right angles, and all other streets spread out from these avenues, which mark the beginning of street numbering. Other four parks make up the original green area in the center of the city. The largest park, *Parque 2 de Febrero*, spreads across close to five hectares and surrounds the city center. A riverside park or *Paseo Costanero* was inaugurated in the year 2006. It is a natural extension of the previous park that runs alongside the Río Negro River, one of the main attractions in the city.

Given the importance of the city, some of the first high schools in the region were established in Resistencia. At present, there are two state-run universities: the oldest is the *Universidad Nacional del Nordeste*, or National Northeast University, whose main premises are in the cities of Resistencia and Corrientes. This university was created in 1956 as an academic offshoot of the University of Tucumán and the University of the Littoral (*Universidad de Tucumán* and *Universidad Nacional del Litoral*). The second, Resistencia Regional School or *Facultad Regional Resistencia*, was created as an academic branch of the National Technological University or *Universidad Tecnológica Nacional*, in 1959. The two universities attract a significant number of students from the area, from the rest of the province of Chaco, from the provinces of Corrientes, Formosa, North of Santa Fe, and from Misiones to a lesser extent. There are also representations of private universities, although their number and career offering is in constant decline.

In terms of economic, social, and human development, the province of Chaco is at the opposite pole from the province of Buenos Aires: while the latter is the richest province in the country and generates more than 50 % of the national Gross Domestic Product, the province of Chaco is one of the poorest. As a result of this regional disparity, the demographic and epidemiological transition in the northeast city lags far behind its wealthier counterpart. According to the 2010 census, the population of the city of Resistencia reached 290,723 inhabitants. Although the city's population is already aged, the percentage of inhabitants aged over 65 is below the national average, at 7.9 % of the total population, while those aged over 80 accounted for 1.7 % of the total population in Resistencia. The demographic

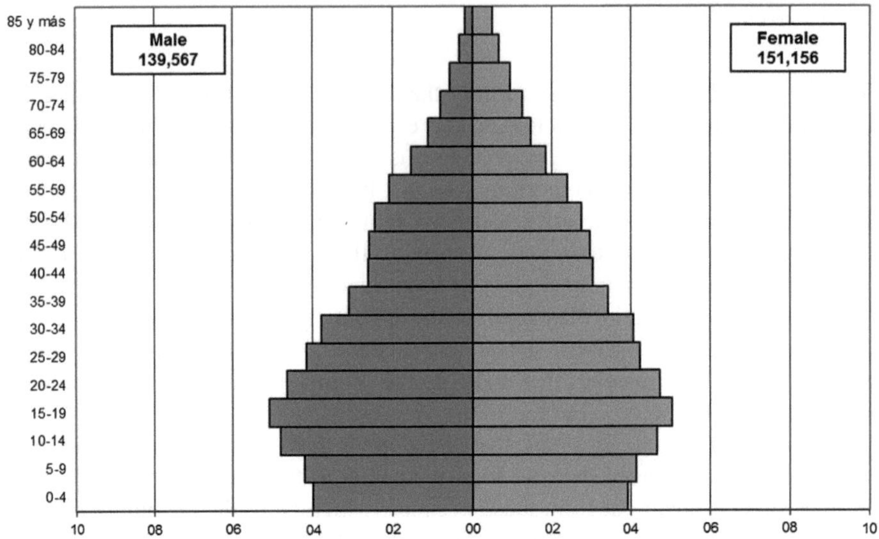

Fig. 9.3 Population pyramid by gender and age of the city of Resistencia, province of Chaco, 2010. *Source* INDEC. National Population, Households, and Dwellings Census, 2010

aging of the city is attributed to the progressive drop in fertility and mortality rates. Similarly to the city of La Plata, as the provincial capital and seat of various national state and private universities, Resistencia attracts a student population between the ages of 15 and 29. Following graph presents the structure of the population of Resistencia by gender and age (Fig. 9.3).

9.3.4 Comparison of Social and Demographic Characteristics of the Three Cities

The next table presents the main social and demographic characteristics of the three cities in comparative form. Of the three cities included in this study, Lezama features the sharpest demographic aging, and aging of the older population. Since it is located in the wealthiest rural region of the country, it also presents the lowest proportion of elderly persons living in homes with unmet basic needs, i.e., in structural poverty due to a lack of minimum infrastructure, housing, or other elements associated with structural poverty. The elderly population in La Plata shows the lowest percentage of illiteracy, in line with its fame as a prestigious university-city. Lastly, of the three cities, Resistencia presents the lowest levels of demographic aging and of aging of the older population, and the highest percentage of illiterate older population and inhabitants with unmet basic needs (Table 9.2).

Table 9.2 Selected social and demographic characteristics: La Plata, Lezama, Resistencia, 2010

Characteristics of the population	La Plata	Lezama	Resistencia
% people aged 65 and over relative to total	11.4	13.9	7.9
% people aged 80 and over relative to total	3.1	3.3	1.7
Number of people aged 60	101,345	1061	32,702
% people aged 60 and over relative to total	15.6	18.8	11.2
% of illiterate people aged 65 and over	1.9	2.2	6.8
% people aged 65 and over with at least one UBN[a]	2.7	0.9	4.4
Number of inhabitants	654,324	5629	290,723

[a]*UBN* Unmet Basic Need. This indicates the percentage of the population aged 65 and more that lives in structural poverty
Source INDEC. National Population, Households, and Dwellings Census, 2010

This regional disparity typical of Latin American countries, including Argentina, led to a particular demographic and epidemiological process which has been called "polarized epidemiological transition" (Frenk et al. 1989), and is characterized by a sharp difference in population health-illness patterns according to their geographic location or social strata. As a result of these differences, the prolongation of the average life in the more prosperous regions was associated with a longer healthy life expectancy. In the more backward regions, however, life expectancy has increased but features more years with permanent limitations (or dependencies) stemming from physical or cognitive problems (Belliard et al. 2013).

The following table shows the differences between the interior of the province of Buenos Aires, the area the cities of La Plata and Lezama belong to, and the province of Chaco. The older adult population of Chaco shows the highest prevalence of at least one permanent limitation, and a rate of permanent cognitive limitations that is above the national average. On the opposite pole, the interior of the province of Buenos Aires presents a lower prevalence of disabilities compared with Chaco and with the national average under the two categories analyzed in Table 9.3.

Table 9.3 Prevalence of population aged 65 and over with at least one permanent limitation and with cognitive limitations

Jurisdictions	Prevalence of population aged 65 and over, with at least one permanent limitation, %	Prevalence of population aged 65 and over, with cognitive limitations, %
Interior of prov. Bs. As (La Plata and Lezama)	35.8	4
Chaco (Resistencia)	52.3	8.7
Country total	41	5

Interior of the province of Buenos Aires and Chaco, 2010
Prevalence of permanent limitations: % of people with permanent limitations relative to the total population in that age group
Source INDEC. National Population, Households, and Dwellings Census, 2010

The prevalence of permanent limitations associated with physical or cognitive problems is an indicator that is highly relevant for assessing the needs for adjustment of existing social, health and urban services in order to meet the accessibility requirements of elderly persons with disabilities.

9.4 Results

Notwithstanding the socio-demographic and socioeconomic differences referred above, the opinions presented by the elderly persons, the informal caregivers of elderly dependent persons, and the service providers from the three cities, agreed on the following as positive aspects for the everyday life of the older population: (a) the solidarity displayed by neighbors and relatives; (b) the possibility of social participation in senior citizens centers and retiree clubs; and (c) the availability of health services. They also agreed in pointing out the following negative aspects of the three cities: (a) public transport; (b) public spaces, in particular the poor state of footpaths; and (c) the lack of political participation in defense of their rights.

In the three cities analyzed, reference to solidarity received from community members, relatives and neighbors in the provision of support for everyday living needs of elderly persons stood out. Additionally, the network of retiree centers and senior citizen clubs exists throughout the national territory, with independent representations in each neighborhood. These centers carry out recreational and cultural activities, "mini" tourism programs, and provide a meeting point that is highly valued by the elderly population. However, respondents recognized that these are single-generation segregated meeting points, and that opportunities for participation in intergenerational activities are infrequent, and there are no forums for political participation in which they may defend the interests of their age group. The health coverage and medical attention available to the elderly people from the three cities was also highly valued. However, opinions regarding accessibility to services were dissimilar.

Lastly, another issue on which the interviewees from the three cities coincided was the insufficient provision of two public services that are vital for the social integration of the elderly: safe footpaths and public spaces on the one hand, and public transport on the other hand. The poor upkeep of urban infrastructure, exemplified by broken footpaths, parks in need of maintenance, and absence of public restrooms and suitable benches are the issues mentioned most frequently by the groups of respondents from the three cities, coupled with complaints relating to public transport. In Lezama, a small town, there is no public transport that connects it with the neighboring towns. In Resistencia and La Plata, both cities with more than 200,000 inhabitants, public transport frequently changes its regular route, and does not provide easy access for elderly persons with disabilities or with mobility problems. Taxis or hire cars constitute alternate forms of transport that are too costly for the low income of retired or pensioned persons.

In view of this convergence of opinions from highly differentiated spheres, it is to be assumed that the aspects they refer to relate to a particular form of social reproduction typical of societies with intermediate economic development: informal networks established by kinship or vicinity that cover the gaps in the public service offering.

Conversely, opinions regarding the age-friendly status of the cities differed noticeably according to three variables: (a) the size of the urban population centers; (b) the urban design; and (c) the regional location.

9.4.1 The Size of the Cities

The three cases studied in Argentina are related to two large urban population centers with 300,000 and 600,000 inhabitants, and one small town with a population of little more than 5000. In other words, we have two polarized situations in terms of the size of the cities. This constituted a significant variable in terms of the differences in living conditions and quality of life of the elderly population.

Safety is a typical problem of larger cities. Both in La Plata and in Resistencia, the elderly population is particularly vulnerable to delinquent or criminal activity, which is frequent in Latin American cities and in particular in large Argentine cities. Elderly people are exposed to safety problems (robbery) when they go to collect their retirement pay or pension, at bus stops, in the street, or in their own homes. This prevents the suitable social integration of older adults since they must be home before dark, avoid certain areas of the city, and restrict home assistance or services to preserve their safety.

Another typical problem of the two large cities is the lack of easy access to health services. The excessive bureaucratic processes required to gain access to services, and the obligation to request appointments in person, force elderly people to travel across districts. Inadequate public transport forces them to resort to the more expensive hire cars or taxis, or to pay a messenger service to carry out the procedure on their behalf.

Problems relating to traffic, which older persons rate as chaotic, are also typical of the two larger cities, and restrict the mobility of the elderly population for fear of accidents.

In Lezama, however, no reference was made to safety problems. On the contrary, opinions coincided in that the elderly population is familiar with the environment and with the neighbors, they feel safe moving around within their community, and are socially integrated. In their own words, "*we all know each other and help each other.*"

They move around in a small environment of primary community relations. However, given its small size, the city lacks public transport that connects it with the neighboring districts. Additionally, only primary health services are available in the district, making it necessary to move to other larger districts to see specialists or undergo complex studies. The absence of public transport forces the elderly people to use hire cars. Another consideration is that the driver must wait while the passenger attends an appointment, increasing the cost considerably. Retirement pay is not sufficient to cover many trips of this type.

9.4.2 Urban Design of the Cities

Another variable that is closely related with the quality of life of the elderly population is the urban layout. Both the city of La Plata and Resistencia were designed at the end of the nineteenth century as the provincial capitals of Buenos Aires and Chaco, respectively. The two cities feature a modern shape that differs visibly from the traditional design of the colonial cities of other Argentine or Latin American provinces. They have wide footpaths and many parks and green areas for recreation. The city of La Plata was designed by a renowned urban architect who made it especially beautiful. As the capital of the country′s richest province, La Plata grew in an orderly fashion, although the outlying neighborhoods feature constructions of lesser architectural value, and the original centralized layout that includes a park every seven streets was not respected.

In the city of Resistencia, on the other hand, the city center presents a modern urban design, featuring wide avenues interspersed with parks. Its growth, however, was disorderly, and the newer neighborhoods that developed in the periphery do not have paved streets, footpaths, or green areas. The lower economic development of this province was determinant of the lower urban development of its capital city.

Lastly, Lezama is a town situated in a rural area that surrounds the train station and spreads alongside the highway that connects it with the Autonomous City of Buenos Aires, the federal capital city. It has a single main park and is divided into two neighborhoods by the highway. The older neighbors mention that a bridge was requested to allow circulation between the two neighborhoods. The bridge was built but the neighbors do not use it. Consequently, the two sectors of the city remain disconnected. The absence of public green areas limits open air activities and outings. The absence of footpaths is another characteristic of this small district which poses a problem for pedestrian circulation.

Despite complaints relating to lack of proper urban maintenance, the elderly people living in La Plata consider that the city design is one of its strengths, contributing to the good quality of life in the city. On the opposite pole, the neighbors of Lezama refer to the lack of an age-friendly urban layout as the main weakness posed by this city. The elderly people of the city of Resistencia are pleased with the design of their city center; however those living in the peripheral neighborhoods view the problems relating to the poor urban design as an unfriendly aspect to the detriment of their quality of life.

9.4.3 Regional Inequalities

The regional context to which each of the cities analyzed belongs is another differentiating factor that is determinant of the quality of life perceived by the neighbors. The prevalence of disabilities is higher in the city of Resistencia. It must be taken into consideration that the province suffered dengue epidemics between 2004 and 2006; in other words, its health-illness pattern unfolds in a scenario of counter epidemiologic transition (Frenk et al. 1991). As a result, opinions from neighbors refer predominantly to the disadvantages for circulating in the city for people in wheelchairs and with permanent physical disabilities due to the poor urban upkeep and flawed public transport system. These complaints increase in the outlying neighborhoods.

In the city of Lezama, which is situated in a rural area, the lack of sufficient beds in the sole geriatric home in the district is a matter of concern for the neighbors. This is attributed to the fact that the elderly rural population moves to the city when they begin to suffer chronic disabilities or health problems that require medical attention.

In La Plata, the provincial capital and university-city par excellence, opinions are predominantly favorable regarding the availability of a wide educational and cultural offering for elderly persons. In summary, the regional differences are a determining factor of the positive or negative opinions relating to the age-friendly nature of each city.

9.5 Plans for Improvements Based on Assessment Findings

A particular feature that differentiates the execution of the AFC project from other cases where local government authorities undertake implementation is that project responsibility resides with *Universidad ISALUD*, a nongovernment organization. To date, the plan for improvements is only executed in the city of La Plata. In Resistencia the diagnostic study has been conducted and the results have been presented to participants, nongovernment organizations, and government authorities. The strategic plan for improvements was drawn up and proposed to the provincial government, but to date has not been implemented. Lastly, at the time of writing this chapter, the analysis of the diagnostic research in Lezama was in its final stage.

9.5.1 Planning and Implementation in La Plata

In 2007 and once the diagnostic stage had concluded, the plan for improvements was designed and set in motion. At the date of this chapter writing, its execution was ongoing. The plan was drawn up using a participative planning approach. As mentioned earlier, *Universidad ISALUD* and the *Red Mayor de La Plata*, both NGOs, were responsible for preparing, executing and assessing the work carried out under this plan. One of the main results obtained over the seven years since the commencement of the participative planning and execution has been the institutional consolidation of the *Red Mayor* in the city. As a result of the systematic and continuous action of the members of this Network of Elderly People, the Age-Friendly Cities project gained visibility and became viewed as reliable.

It must be highlighted that the *Red Mayor* is an organization that brings together older adult volunteers who are committed to defend the specific rights and interests of the elderly population. These years of practical work have consolidated a steering group made up by a reduced number of members of the Network who took on project leadership and promoted collective mobilization actions, lobbying and social activism. The objective of the group is to generate accessible or 'friendly' places in the city along the guidelines of the plan for improvements drawn up, on the one hand, and introduce variations where these are politically timely, feasible in terms of the current context or of the emergence of new contexts that require urgent solutions, on the other hand.

One of the greatest weaknesses throughout these seven years has been the lack of insertion of the project in the municipal government scheme, and the lack of external financial support to enable *Universidad ISALUD* and the *Red Mayor* to assign paid staff to the execution of the plan for improvements.

The absence of a sustainable articulation mechanism with the government authorities and planning areas of local government, and the low degree of support from other social players, constituted an obstacle during the initial project implementation phase. These shortcomings are attributed in part to the initial lack of interest of the elderly people from the Network responsible for project leadership and of other NGOs who were supporting it, in establishing contact with the city government. Citizen distrust of people in government stems from the existing rupture between civilian society and the local State.

This lack of interest in seeking the involvement of other social players and of the local government changed as progress was made in the identification of priorities and in deciding on the best solutions for the problems identified. The plan for improvements in the city gathered momentum as from 2009 in the framework of a change in the local administration. Additionally, *Fundación ISALUD* implemented a training and education project for volunteer leaders which strengthened the management skills of the *Red Mayor*. As a result of the favorable changes in the sphere of government and in the civilian steering group, successive improvements were addressed.

During 2013, the municipal government of the city of La Plata conducted an assessment of the actions carried out. The resulting report listed the following improvements achieved in the city:

Public spaces

- Aerobic circuits and age-friendly parks with activities free of charge.
- Pruning and felling of trees and reforestation activities were increased.
- Public toilets.
- Improvements in public transport.
- Improvements in traffic planning, with the incorporation of routes requested by the neighbors.
- Road signaling and identification of bus stops was improved significantly.
- A road education and driver awareness campaign was carried out.
- Traffic lights with visual countdown for pedestrian crossings were installed.
- New facilities and repair of existing ones.
- Implementation of magnetic cards for paying bus fares, with discount for retired persons.

Safety

- Lighting improvements.
- Installation of security cameras.
- Incorporation of Neighborhood Watch that accompanies the police in crime prevention.

However, two of the issues most frequently requested by the elderly population and other age groups have not been addressed to date:

- Repair of footpaths, construction of access ramps, and bicycle paths to avoid their unrestricted circulation which poses a threat to pedestrian safety.
- Concerning safety, crime rates that target the elderly population continue to increase in the city. Coordinated actions are currently being implemented by the province and the municipality to stop crime.

9.5.2 Institutional Factors and the Plan for Improvements

The normal execution of the plan for improvements in the city has been negatively affected by institutional factors such as the Argentine style of government and the emergence of natural disasters or social crises. It must be taken into account that, from a technical viewpoint, the original reform plan proposed to follow defined stages: (a) validation of the diagnosis made by a majority of citizens from La Plata; (b) participative design of a strategic plan of action; (c) design of trial urban intervention actions to assess suitableness of improvements for solving the problems identified; and (d) programming of the next step of making the actions extensive to the entire urban layout.

The interruption of public policy due to changes in government was the first institutional aspect that arose as an obstacle for the completion of the original program.

A detailed description of events provides an understanding of the impact of this aspect. Immediately after the diagnostic stage, the summary results of the study structured under core actions were presented at an open plenary meeting attended—by special invitation—by the participants of the focus groups with the aim of implementing a plan for improvements to address the main issues identified by the neighbors at group meetings. The diagnosis and core actions proposed were approved unanimously by all the participants. Subsequently, they were submitted formally to the government authorities for approval. This happened in 2007 at a time when the municipal government was completing its term in office, and the AFC project was not among its priorities.

Consequently, in 2009 the AFC project had to be resubmitted to the newly elected mayor, together with its fundamentals, the results of the diagnostic study carried out, and the core actions envisaged for implementation. A private meeting was held at which the mayor was informed of the components of the AFC project, and it was agreed to call it the "Elderly Person Friendly City" project in La Plata. This change in name had also been proposed by the focus groups during the diagnostic study and by the mass media, who considered the new name a clearer way of presenting the project. It should be noted that for a recently elected official with a five-year term ahead, the programming of improvements requested by elderly persons and other groups of neighbors represented an attractive challenge.

The city's NGOs, including the *Red Mayor*, also limited the scope of the actions proposed under the plan for improvements. The main problem is the lack of funding to support the tasks carried out by the volunteer base and to undertake actions for improvement independently, regardless how simple they may be. The availability of this type of volunteer is limited given that volunteers participating in the project must pay their own travel expenses and cover any other expense required in connection with the project. Additionally, these organizations have slim chances of obtaining financial support for their actions because of the broad demand and scarce supply of funding for NGOs. Due to these limitations, and those relating to the political sphere, *Universidad ISALUD* did not obtain the necessary funding to prepare and implement the Plan for Strategic Improvements originally envisaged.

The experience carried out in the city of La Plata is the only one existing to date in the country. Consequently, it is not possible to affirm that this pattern is associated with Argentine institutions or with the particular dynamic of the city. However, the interruption in the consideration of the plan for improvements submitted to the government authorities of the province of Chaco due to a change in administration shows similarity with the institutional dynamic seen in La Plata. We hope that the recording of events carried out and results obtained, and of the techniques that favored or hindered the execution of the actions, will enable a comparative analysis of the successive implementation of plans for improvement to make cities more age-friendly, and subsequent identification of common factors that outline a national pattern in the subject matter.

9.6 Conclusion

Argentina is a country with an aging population, where the number of people aged 80 and over is rising progressively. The vast majority of the older adult Argentine population lives in urban areas. As in other middle-income countries, the supply of public services and urban maintenance presents notorious flaws that have a particular impact on the quality of life of the elderly population residing in the country's urban centers. The AFC project is very apt to encourage the materialization of adaptations that will make Argentine cities more age-friendly.

Since its origin in 2006, the AFC project has been supported by *Universidad ISALUD* with the aim of incorporating in the agenda of local governments concern for the execution of improvements in urban areas and in public services to make them accessible to the growing numbers and rising proportion of elderly people residing in their sphere.

From 2006 to date, the project has spread to three cities that differ in size and regional location, however, the plan for improvements to the city was only implemented in La Plata. The experience gained so far shows that the institutional weaknesses constitute an obstacle for the advancement of the project. Nevertheless, the involvement of NGOs proved effective in granting visibility to the impact of demographic aging on everyday life in cities. The results of these studies also show that the size of the cities, their level of human, economic, and social developments, and the urban design of each city are variables that determine the differing quality of life of elderly people and their access to full social inclusion.

The bottom-up methodology proposed by the Vancouver Protocol, its research through focus groups, and the guidelines for presenting results and reaching a consensus on the plan for improvements are in themselves privileged and effective tools for incorporating in the agenda of local, provincial, and national governments the need to adapt to the existing environments to make them accessible for the elderly population (or age-friendly). Additionally, the group meetings specifically directed at the discussion of topics that concern the elderly population, without any form of political limitation and accepting the differences of opinion, proved an innovative methodology for the people summoned, who were accustomed to participating only through opinion surveys conducted during pre-electoral periods, market research, or television ratings. The relaxed–heated discussion of everyday issues in the groups, and the presentation of results in an auditorium with the presence of government officers offered a forum that offers greater importance to the elderly neighbors of the city. Attendants at every meeting showed explicit gestures of emotion and satisfaction regarding the full list of proceedings carried out during the diagnostic research stage and the uninterrupted implementation of actions for improvement promoted by the *Red Mayor* and *Universidad Isalud* throughout these years.

The institutional capacity to promote citizen participation in decisions is low in the three local governments in which the project is developed. During implementation of the project, we observed that the diagnosis was made on the basis of the views of different groups of citizens and that it responded partly to their demands.

In the city of La Plata, formal structures of "agreement" are scheduled but have not been finalized and are so far to end them. In contrast, the situation in the three cities is too far to the "decision making" shared among citizens and government officials.

The experience carried out to date in Argentina also demonstrates that the AFC project can be implemented effectively by private nonprofit organizations in countries with intermediate economic development, despite the funding restrictions and greater institutional weaknesses. Private management facilitates project continuity through changes in local government administration, and promotes project extension between subnational governments run by different political parties.

Nevertheless, there is also a pressing need to strengthen processes from the bottom up to obtain the commitment of government authorities to include the requests of the elderly population in their government programs.

Argentine practice shows the suitableness of using flexible planning techniques and adopting a pragmatic approach in the use of management tools in order to increase the capacity to adapt to unexpected changes. Persistence in keeping the goal of making cities age-friendly as a priority is key in contexts where institutional uncertainty and financial restrictions prevail.

Finally, it is worth highlighting the major role of volunteer organizations run by elderly people and the importance of strengthening their management capacity to defend their particular interests and rights. The involvement of organizations of elderly people independent of the government sphere creates a virtuous cycle since it enhances their methods for building power in terms of their needs being met, at the same time guaranteeing project continuity unhindered by unfavorable institutional or financial circumstances.

References

Belliard, M., Massa, C. & Redondo, N. (2013). Análisis comparado de la esperanza de vida con salud en la Ciudad Autónoma de Buenos Aires. DGYC, Government of the City of Buenos Aires. *Población de Buenos Aires, 10*(18), 7–29.

Caviezel, P. (2008). La mortalidad en la Ciudad de Buenos Aires entre 1947 y 2001. Dirección General de Estadística y Censos. *Población de Buenos Aires, 5*(7), 75–100.

Frenk, J., Bobadilla, J. L., Sepúlveda, J., & López-Cervantes, M. (1989). Health transition in middle-income countries: new challenges for health care. *Health Policy and Planning, 4*(1), 29–39.

Frenk, J., Frejka, T., Bobadilla, J. L., Stern, C., Lazano, R., Sepúlveda, J., & José, M. V. (1991). La transición epidemiológica en América Latina. *Boletín Oficina Sanitaria Panamericana, 111* (6), 485–496.

Gascón, S., Redondo, N., Aguilar, M., Tordó, M., & Elicabe, I. (2008). *Ciudades amigables con la edad: Aplicación del Protocolo de Vancouver (OMS) a la ciudad de La Plata.* Serie de Estudios ISALUD, 5. Buenos Aires: Universidad ISALUD.

National Institute for Statistics and Census (INDEC). (2014). Available: www.indec.mecon.gov.ar. Accessed August 18, 2014.

Chapter 10
Age-Friendly New York City: A Case Study

Lindsay Goldman, Shauneequa Owusu, Caitlyn Smith, Dana Martens and Meghan Lynch

10.1 Background

New York City, home to 8.4 million people, is a highly attractive place for older adults to live, featuring sophisticated public transportation, vast cultural resources and historical traditions, celebrated landmarks and open spaces, varied housing options, excellent medical centers and health-related support structures, large networks of economic activity, and a wide array of social services. Over 1 million New York City residents, or 12 % of the total population, are age 65 or older, a number projected to increase by 40.7 % over the next 20 years (New York City Department of City Planning 2013). Nearly 97 % of these people reside in the community rather than in group quarters (U.S. Census Bureau 2013a). Often among the most long-term and civically engaged residents, older New Yorkers are a critical resource to the City.

The views presented in this chapter are those of the authors, and not necessarily those of The New York Academy of Medicine, or its Trustees, Officers, or staff.

L. Goldman (✉) · S. Owusu · C. Smith · D. Martens · M. Lynch
The New York Academy of Medicine, 1216 5th Avenue, New York, NY 10029, USA
e-mail: goldman@nyam.org

S. Owusu
e-mail: sowusu@nyam.org

C. Smith
e-mail: csmith@nyam.org

M. Lynch
e-mail: mlynch@nyam.org

© Springer International Publishing Switzerland 2016

T. Moulaert and S. Garon (eds.), *Age-Friendly Cities and Communities in International Comparison*, International Perspectives on Aging 14,
DOI 10.1007/978-3-319-24031-2_10

Age-friendly NYC is an unprecedented public–private partnership between the New York Academy of Medicine (NYAM), the Office of the Mayor, and the New York City Council dedicated to making New York City a better place to grow old by promoting active engagement in all aspects of city life. Using the World Health Organization's framework for age-friendly cities, NYAM and the City of New York conducted a comprehensive assessment of the age-friendliness of New York City in 2007 and developed an action plan to address identified issues (World Health Organization 2007).

Through community consultations, focus groups, interviews, and surveys, over 2000 older adults told us about their daily lives. The data gathering was conducted in 14 neighborhoods throughout the City in six languages across a wide range of settings, including religious institutions, senior centers, community-based organizations, assisted living facilities, and local businesses. Roundtable discussions were held with business, housing, social services, health, transportation, and higher education experts. As part of our initial assessment, we also looked at the existing literature within each of the eight domains of an age-friendly city and mapped data describing older New Yorkers and their environments. Our findings are synthesized in *Toward an Age-friendly City: A Findings Report* which has served as a touchstone for all of our subsequent efforts (Finkelstein et al. 2008).

Each of Age-friendly NYC's partners has played a unique yet complementary role in responding to the findings of the assessment phase.

10.1.1 The New York Academy of Medicine

Founded in 1847, the New York Academy of Medicine (NYAM) is an independent, nonpartisan, nonprofit institution whose mission is to improve the health of people living in cities, with a focus on disadvantaged and vulnerable populations. Through research, evaluation, education, community engagement, and evidence-based advocacy, NYAM works to create environments that support healthy aging; to strengthen systems that prevent disease and promote the public's health; to eliminate health disparities, and to preserve the heritage of medicine and public health. NYAM draws on the expertise of its professional staff, as well as 2,000 elected fellows from across the professions.

NYAM assisted the World Health Organization (WHO) in developing the age-friendly cities framework and serves as the WHO's Collaborating Center on aging, globalization, and urbanization, providing technical assistance to age-friendly initiatives around the world. With funding from the Office of the Mayor, NYAM serves as the secretariat for Age-friendly NYC. As such, NYAM cultivates strategic partnerships with the nonprofit and for-profit sectors, staffs the Age-friendly NYC Commission (detailed below), and develops the evidence base to support citywide and sector-specific interventions and policy change.

10.1.2 Office of the Mayor

The Office of the Mayor is the executive branch of City government that oversees over 40 City agencies, as well as numerous issue-oriented boards and commissions. The Mayor is responsible for preparing and administering the City's annual expense and capital budgets and financial plan. While NYAM conducted the initial community assessments and literature reviews, the Mayor's Office led an assessment of the age-friendliness of programs and services provided by City government agencies, held meetings of key agency leaders, and conducted a series of seminars on topics such as promoting active aging and universal design. This review yielded a blueprint consisting of "59 initiatives" to improve the quality of life of older New Yorkers (New York City Office of the Mayor and New York City Council 2009). Beginning in 2010 and continuing today, the Mayor's Office has been working to coordinate across City agencies to implement these recommendations which are focused on four areas: community and civic participation, housing, public spaces and transportation, and health and social services. The Mayor's Office maintains accountability through consistent documentation of progress and metrics in written reports (New York City Office of the Mayor 2013a), and Age-friendly NYC has been integrated into the City's overall performance management structure (New York City Office of the Mayor 2013b).

10.1.3 City Council

The City Council is the legislative branch of City government. Led by the Speaker, 51 Council Members are elected every 4 years and represent districts of approximately 160,000 people. In addition to its legislative role and oversight powers over City agencies, the Council approves the City's budget and has decision-making powers over land use issues. Through the budget, the Council establishes priorities, allocates resources and sets the policy agenda for the year. The Council makes direct investments into Age-friendly NYC through the allocation of discretionary funding for initiatives at the neighborhood level.

10.1.4 Age-Friendly NYC Commission

Appointed by the Mayor and approved by the City Council Speaker, the Age-friendly NYC Commission[1] is comprised of leaders from government (administrative and legislative) and from fields including business, education, health care, law,

[1][Online]. Available: http://www.nyam.org/agefriendlynyc/about-us/commission-for-afnyc.html [Accessed December 12, 2014].

architecture, housing, and social work. The Commission is charged with monitoring the progress of the "59 Initiatives" and providing dynamic and innovative leadership to engage the private sector in changing the culture of New York City to become more inclusive of older people. Seated in 2010, the Age-friendly NYC Commission is guided by the following underlying principles that are aligned with the WHO framework:

- An aging population is an opportunity, not a crisis.
- Older people must be involved in all phases of problem identification and resolution within a community.
- Initiatives should address the full diversity of a community's older population with respect to functional capacity, economic resources, and demographics.
- Initiatives should be grounded in evidence.
- All sectors (public and private) must be engaged in developing solutions to eliminate barriers to older adults' full participation in society.

The expertise and energy of older New Yorkers informs the Commission's agenda. From 2010 to 2013, the Commission convened working groups focused on four priority areas identified during community consultations: higher education, commerce, professions, and age-friendly neighborhoods.[2] The Commission's efforts served to reinforce and complement the Mayor's 59 Initiatives.

10.2 Approach and Strategies

The New York Academy of Medicine employs a three-pronged approach to address complex issues of urban health. We begin by reviewing existing evidence and data, we then consult the people affected by a given problem, and in collaboration with those people, convene all relevant sectors to develop creative solutions collectively. The social determinants of health provide an underlying framework for all of our work:

> The social determinants of health are the circumstances in which people are born, grow up, live, work and age, and the systems put in place to deal with illness. These circumstances are in turn shaped by a wider set of forces: economics, social policies, and politics (World Health Organization n.d.).

Age-friendly NYC has followed this model from its inception in 2007. However, in contrast to a focus on large-scale capital investments to address social determinants of health, Age-Friendly NYC promotes policies and practices that enhance access to existing resources and advocates for low-cost, simple improvements to

[2][Online]. Available: http://www.nyam.org/agefriendlynyc/docs/AF-Accomplishments_Report-8.pdf [Accessed December 12, 2014].

infrastructure. Age-Friendly NYC leverages its partners in the public, nonprofit, and for-profit sectors to maximize the social, physical, and economic participation of older people and thereby improve population health.

Age-friendly NYC was implemented during the great recession in the United States (which lasted from December 2007–June 2009) when overall government spending was constricted. Over the years, Age-friendly NYC has received a blend of public and private support for core administrative functions as well as project-specific expenses. However, Age-friendly NYC's primary approach is to advocate for more efficient and effective spending rather than for additive dollars. The Age-friendly NYC initiative is an example of how to make significant impact at little to no marginal cost by incorporating the perspectives of older adults into ongoing planning processes. The primary mechanisms through which Age-friendly NYC effects change include:

- Increase utilization of existing assets.
- Develop local solutions to local problems.
- Promote economic security and spending.
- Leverage private sector resources.

10.2.1 Increase Utilization of Existing Assets

Age-friendly NYC works to ensure the City's assets are appropriate for and useable by older people. During consultations, older people reported that while they love New York City's wealth of educational, cultural, and commercial opportunities, lack of information and transportation, safety concerns, and cost often contribute to diminished access to these resources and the social and intellectual stimulation they afford (Finkelstein et al. 2008). This is especially problematic for older people who are already at high risk of social isolation, including those who live alone (31.3 %); live in poverty (18.5 %); live with disabilities (36.6 %); do not speak English (33.6 %); and identify as lesbian, gay, bisexual or transgender (LGBT) (U.S. Census Bureau 2013b). Because social isolation is associated with re-hospitalization (Mistry et al. 2001), premature mortality (Steptoe et al. 2013), and depression (Cacioppo et al. 2006), it is in the City's best interest to ensure older people remain involved in city life (Sederer 2006).

10.2.1.1 Raise Awareness of Resources

To increase older people's awareness and utilization of existing assets and thereby promote social connectivity associated with better health and resilience to stress (Klinenberg 2002), Age-friendly NYC has curated and disseminated information about arts and culture, higher education, and local businesses. Age-friendly College

Link, an online searchable database,[3] connects older people with courses, programs, events, and activities available at New York City colleges and universities. Profiles in this database are intended to serve as a gateway by providing basic information about school offerings, special opportunities for older people, and links to individual school websites for more information. Age-friendly NYC also worked with the NYC Department of Cultural Affairs and the Alliance for the Arts in creating city- and borough-wide guides to programming, discounts, and facility accessibility at every cultural institution in the City. In collaboration with two particularly committed City Council Members, NYAM surveyed grocery stores to highlight best practices in serving older consumers and created Age-friendly Grocery Guides within the Members' respective districts. The guides include amenities like the ability to shop by phone, shop online, delivery options, prepared goods available in single portions, handicap accessible aisles and restrooms, seating provided or available on request, the presence of a water fountain, multiple floors with an escalator/elevator, and availability of a senior discount.

By curating content, Age-friendly NYC was able to scan the City to see where opportunities for older people were abundant or in short supply. However, compiling this information was labor and time intensive, and the utility of the content and various formats (i.e., print versus online) has not been analyzed. In addition, whether these compendiums have inspired new age-friendly programming or policies remains largely unknown. The information is now outdated, and this strategy must be re-evaluated.

10.2.1.2 Reappropriate Resources

Where older people are aware of local resources but find them unwelcoming or inaccessible, Age-friendly NYC has looked to modify and reappropriate these resources. One of the most pressing and common concerns to emerge from discussions with older adults is the need for new and improved transportation options. Transportation is essential to older adults' quality of life, health, independence, social interaction, and ability to contribute to their communities—including their ability to participate in the local economy, both as consumers and workers (Rittner and Kirk 1995).

New York City has a very robust system of public transportation; every day, 7.6 million people ride the subways or the buses (New York City Office of the Mayor: Special Initiative for Rebuilding and Resiliency (SIRR) 2013). Yet the accessibility of public transportation is varied, with many subway stations inaccessible to those with mobility impairment and some areas of the City devoid of adequate transportation infrastructure (New York City Department of City Planning 2012). A paratransit program serves those with disabilities, but rides must be scheduled in advance and wait-times can be long. As the older population increases,

[3] [Online]. Available: www.agefriendlycolleges.org [Accessed December 12, 2014].

demand for the service and associated costs are also projected to increase (Lave and Mathias 1998). In response to these challenges, Age-friendly NYC has expanded public transportation options for older people using school buses and taxis.

Through a partnership between the Department of Education and the Department for the Aging, school buses can be used by senior centers to transport older people on shopping trips to obtain healthy and locally grown food when buses are not needed for children during school hours. From 2013 to 2014, over 1,500 older people, primarily from underserved, low-income communities, accessed healthy and fresh food through this program, known as MarketRide (New York City Office of the Mayor 2013a). For those with mobility impairment, a taxi voucher program is being piloted as an alternative to paratransit. The pilot is a partnership between the Department for the Aging and the Mayor's Office for People with Disabilities. For a small enrollment fee, older people are given prepaid debit cards to use in taxis. A preliminary analysis of 400 customers found the average per-trip cost of a taxi to be 70 % less than the cost of an accessible van. In addition to reducing costs, this program increases flexibility, mobility, and independence by allowing customers to hail a cab to take their trip, rather than having to schedule a trip in advance and experience long wait-times for van pick up. An "Accessible Dispatch Pilot Program" enables people to use a phone or app to request a wheelchair-accessible taxi, and half of all New York City taxis will be accessible by 2020. We hope to demonstrate the cost-effectiveness of these programs and scale them up accordingly in the future.

10.2.1.3 Make Improvements to Resources

One of the major benefits for older people living in New York City is the proximity of stores, services, and cultural institutions (Roux et al. 2007). A 2006 survey found that 52 % of New York City-dwelling respondents were likely to walk to a destination rather than use another form of transportation (Stowell-Ritter et al. 2006). There is considerable evidence that walking is associated with better physical and mental health (Simonsick et al. 2005). However, during consultations, older people reported significant barriers to pedestrian safety including inadequate street crossing times, poorly maintained sidewalks, and lack of seating. As one person told us, "I age every time I have to cross the street."

The NYC Department of Transportation (DOT) established the Safe Streets for Seniors Program in 2008. DOT evaluates conditions in areas with high rates of senior pedestrian fatalities or injuries and then develops and implements mitigation measures, such as extending pedestrian crossing times at crosswalks to accommodate slower walking speeds, constructing pedestrian safety islands, widening curbs and medians, narrowing roadways, and installing new stop controls and signals. DOT solicits ideas and feedback through presentations and workshops at senior centers and community boards and also partners with the Department for the Aging to coordinate outreach and share resources. Through this process, over 600 dangerous intersections have been redesigned, and senior pedestrian fatalities have decreased 21 % citywide, from 58 senior fatalities in 2008 to 46 in 2012. These changes have made streets safer for all New Yorkers.

The CityBench program is another initiative to increase walkability. Through a federal grant, DOT is installing 1,500 attractive and durable benches around the city, particularly near senior centers and housing; hospitals and community health centers; commercial zones and shopping districts; and municipal facilities (e.g., public libraries, schools). Individuals and communities can request a bench in a specific location, and older people report having made new social ties with people who frequent the same benches at the same times.

Finally, in response to feedback from older people that bus shelters often lacked seating and felt unsafe, 4,000 new bus shelters have been installed. The new shelters have seating and the walls are transparent, addressing concerns about the old shelters which hid their interiors from view. These shelters are paid for by advertisements projected on their sides. DOT has replaced almost every pre-existing bus shelter and has installed additional bus shelters at locations throughout the five boroughs identified by older people and community leaders.

Beyond impressive numbers of lives saved and injuries averted, there is tremendous value for local community development efforts in the changed perception among older adults that they can feel safe and independent walking their streets, shopping, and going about their daily lives. Anecdotal evidence suggests that these interventions are also serving to decrease social isolation and increase social cohesion within communities as people become more comfortable leaving their homes, utilizing public spaces, and interacting with neighbors (Sugiyama and Thompson 2007).

10.2.2 Develop Local Solutions to Local Problems: Age-Friendly Neighborhoods Initiative

Neighborhood identity is strong among all New Yorkers but is especially important for older people, who have often seen their communities through decades of change. Understanding that each of the City's neighborhoods has both assets and deficits, older New Yorkers emphasized the importance of developing targeted local solutions to complement and inform citywide policy changes.

First introduced by the Age-Friendly NYC Commission in 2010 under the name "Aging Improvement Districts," the "Age-friendly Neighborhoods" model leverages the City's organizational infrastructure, specifically the 51 City Council Districts and 70 Business Improvement Districts (BID),[4] to advance

[4]BIDs are a key public/private partnership in New York City and have helped revitalize neighborhoods and catalyze economic development throughout the City. The City's 70 BIDs annually invest more than $100 million worth of programs and services in neighborhoods across the five boroughs. [Online]. Available: http://www.nyc.gov/html/sbs/html/neighborhood/bid.shtml [Accessed December 12, 2014].

neighborhood-level, age-friendly priorities through community and economic development efforts.

The model employs a four-phased approach to mobilize older people and other local stakeholders to build on strengths and address obstacles. The first phase involves a comprehensive assessment of a neighborhood's age-friendliness across the eight domains of an age-friendly city. This is done through town hall meetings and focus groups with several hundred older people, as well as asset mapping and demographic analysis. During the second phase, an advisory committee, co-chaired by a representative of City Council and either a representative of a Business Improvement District or a community-based organization, and consisting of at least one older person, and six to eight additional community leaders, is convened to review the findings of the assessment phase. In the third phase, an action plan is devised with short and long-term goals to optimize the social, physical, and economic participation of older people through enhanced connections to services, entitlements, and opportunities, new programming, and improvements to the physical environment. Finally, during the fourth phase, the advisory group tracks implementation progress and outcomes and publicizes successes. The neighborhood action plan is revisited annually to promote a cycle of continuous improvement (The New York Academy of Medicine 2012).

Some of the improvements catalyzed by the three Age-friendly Neighborhood Initiative pilot communities from 2010 to 2013 are as follows.

10.2.2.1 East Harlem

The City's first Age-friendly Neighborhood was launched in East Harlem in collaboration with then City Council Member Melissa Mark-Viverito. When discussing local recreational opportunities, older people revealed that they had not used public pools in decades because they felt uncomfortable and unsafe among all of the children and teenagers. This finding was relayed to the Department of Parks and Recreation which piloted senior-only swim hours at one public pool, known as "Senior Splash." The program was so popular (often over 100 people in attendance) in East Harlem that it was expanded to 16 pools throughout the City and augmented with water aerobics instructors. A preliminary evaluation of this program indicated that older people who participated in regular water aerobics demonstrated greater lower body strength and flexibility than those who did not (Asphalt Green 2013). A video was produced to demonstrate challenges faced by older residents of public housing when accessing laundry facilities, and in response, the New York City Housing Authority began investigating the issue and opened a new laundry room in one East Harlem development. To address a perceived dearth of cultural opportunities, El Museo del Barrio, a small neighborhood museum, started an advisory committee of older adults and expanded its programming, which had been primarily focused on youth.

10.2.2.2 Upper West Side

The second Age-friendly Neighborhood, Manhattan's Upper West Side (UWS), was initiated by former Council Member Gale Brewer (now Manhattan Borough President). One of the highlights of the work in this community was "Grow Green, Age Well" to help older people access healthy foods by matching greenmarkets with senior centers and creating a West Side Senior Supported Agriculture (WSSA) Food Box Program to provide older adults with bimonthly affordable, locally grown produce. Based on the success of these initiatives, the Department for the Aging changed its policy to allow senior centers to buy food directly from farmers. The Lincoln Square Business Improvement District produced a "Community for All Ages" brochure and webpage, which identifies places to stop, sit, and relax, free public art and monuments, public restrooms, and accessibility features.[5]

10.2.2.3 Bedford-Stuyvesant

The New York Academy of Medicine partnered with the Local Initiatives Support Corporation NY (LISC-NY) and the Coalition for the Improvement of Bed-Stuyvesant (CIBS) to launch an Age-friendly Neighborhood in the Bedford-Stuyvesant (Bed-Stuy) section of Brooklyn. This successful collaboration resulted in a thorough community planning process that yielded eight priority areas, including increasing financial literacy and improving access to banking facilities, housing seniors in safe and affordable dwellings, revitalizing Bed-Stuy to enhance the use of public streets and facilities by seniors, and supporting local businesses to become age-friendly. Bed-Stuy has effectively incorporated age-friendly principles into their economic revitalization plan, and other communities have followed suit. The Myrtle Avenue Revitalization Project (MARP) now funds and oversees an Age-friendly Neighborhood initiative in Brooklyn's Fort Greene and Clinton Hill neighborhoods.

10.2.2.4 Every District by 2018

A highly effective strategy to engage and support older people, Age-friendly Neighborhoods were endorsed by the Speaker of the City Council, Melissa Mark-Viverito, and funded by the Council to expand into 10 new communities in 2015. The Speaker has committed to bringing Age-friendly Neighborhood efforts to all 51 Council Districts by 2018 (Pazmino 2014).

---

[5][Online]. Available: http://lincolnbid.org/documents/LSAgeFirendlyBrochure.pdf [Accessed December 12, 2014].

10.2.3 Promote Economic Security and Spending

Financially secure older people help to decrease reliance on social insurance programs and drive economic growth through increased consumer spending that results in job creation. Older people represent nearly 20 % of New York City's workforce, spend $70.1 billion annually, tend to shop locally, and provide countless unpaid caregiving hours that enable others to work (AARP Public Policy Institute 2014).

10.2.3.1 Workers

Working longer leads to increased financial security among older people (Johnson 2005). In New York City, 26.3 % of the age 60 and over population and 15.2 % of those 65 and over are employed—some by choice, others by necessity (U.S. Census Bureau 2013a). Reflecting larger national trends, older adults play an increasing and significant role in New York City's workforce but face significant barriers to ongoing participation. Challenges include lack of adequate computer skills and/or access to training, employers' perceived reluctance to hire older workers, the pressures of maintaining a work-life balance, and the scarcity of "bridge" jobs that support a gradual transition to retirement (Finkelstein et al. 2013).

The majority of older workers in the City (60.2 %) are employed in the City's top industries, including: health care, professional services, entertainment, retail sales, finance, and education (Fiscal Policy Institute 2011). Older adults hold a disproportionate percentage of jobs (greater than 20 %) in the following industries: manufacturing, transportation, education, medical, social services, and public administration. In those industries with high proportions of older workers that are projected to grow (medical and social services), the concentration of older workers suggests the need to extend the careers of older workers or risk impending skill shortages (Finkelstein et al. 2013).

Age-Friendly NYC promotes job retention and employment opportunities for older adults through dissemination and public recognition of exemplary employer practices. In the fall of 2012, the New York Academy of Medicine researched current workplace practices and policies to create an evidence-based compendium of strategies that characterize age smart workplaces. NYAM established and staffed an Age Smart Employer Research Advisory Panel comprised multidisciplinary experts in the fields of aging and work. The research entailed a systematic review of recent academic literature from relevant fields (aging, business, organizational theory, psychology, occupational health, and health economics) as well as business and media reports that highlight effective employer strategies for integrating and managing an aging workforce. The New York Academy of Medicine held consultations with key stakeholders regarding diversity, organizational management, and employee benefits.

Findings and recommendations are detailed in *The Age Smart Employer Compendium of Strategies and Practices* (Finkelstein et al. 2013). The *Compendium* presents practices shown to achieve employers' goals and benefit employees. Practices are organized into five areas: Recruiting Older Workers; Optimizing the Performance and Productivity of Older Workers; Reducing Unwanted Turnover by Boosting Engagement of Older Workers; Retaining Older Workers While Controlling Health Cost; and Promoting Meaningful and Streamlined Transitions from Work to Retirement. Key policies and practices with demonstrated results are presented in each section. For example, practices identified to effectively recruit older workers include: partner with community-based educational and nonprofit organizations, use staffing services that attract older workers, advertise at "senior job fairs," and write job advertisements that include explicitly age-friendly language. The *Compendium* is a functional guide that features effective models currently in use and presents compelling value propositions for employers, employees, and society.

The *Compendium* served as the foundation for the application and selection criteria for the first Age Smart Employer Awards, a partnership between the Robert N. Butler Columbia Aging Center and the New York Academy of Medicine, funded with support from the Alfred P. Sloan Foundation. The Age Smart Employer Awards honor employers who value workers of all ages and aims to support businesses looking to maximize the potential of their workers as they age by providing research and best practices. Upon extensive review, an external Age Smart Employer Selection Committee chose nine finalists, who were then required to allow their employees to complete an anonymous survey about the age-smart workplace practices and organizational culture outlined in the employer's initial application. Four winners were selected—Montefiore Medical Center, Pfizer Inc., Settepani Restaurant and Bakery, and Renewal Care Partners—representing industry leaders in important economic sectors for New York City.

However, small employers (defined by New York State as those with fewer than 100 employees) were not adequately represented in the applicant pool because the application eligibility criteria and language were biased toward larger employers. Evidence was required to demonstrate the efficacy of the applicants' policies and practices, and most small businesses lack the capacity and resources to conduct an internal or external evaluation to generate such evidence. Yet small businesses do implement strategies, often on an ad hoc basis, to recruit, retain, transition, and promote the health of workers of all ages. These strategies capitalize on their small size and ability to be responsive to the needs of individual workers. For example, one small business applicant confers enormous flexibility with work process and location to retain sales staffs ranging in age from 50 to 90 who have invaluable relationships with clients. Another described a series of modifications to an employee's responsibilities, work location, and benefits to accommodate caregiving responsibilities (though the employer offers no formal "caregiver benefit"). One employer described a trusted bookkeeper's ten-year transition to retirement (though the employer has no "retirement planning program"). Several small businesses described informal mentorships across generations, often with younger workers learning customer service skills and older workers learning to use computer-based technology.

10.2.3.2 Consumers

From a consumer standpoint, age-friendly businesses ensure goods and services are equally accessible for older adults as for other age groups and that goods and services are offered in a respectful, safe and welcoming environment. The Age-friendly Local Business Initiative (AFLBI) assists small businesses in implementing low-cost or no-cost changes to products, services, store layout, and customer service that enhance the older customer experience. With these changes, businesses are in position to market their age-friendly features. From the business perspective, the objective is to attract, engage and retain older adults as customers to increase the bottom line. AFLBI creates a virtuous circle that reinforces itself and expands through a positive feedback loop. A few businesses implement and promote age-friendly features and as a result obtain a competitive advantage; more establishments recognize the business benefits of becoming age-friendly and must similarly adapt to compete effectively in the changed marketplace (Li 2014).

The first Age-friendly Local Business Pilot Project was launched in East Harlem in 2011. Businesses were approached individually, provided with a resource guide, and encouraged to carry out simple enhancements, such as offering chairs for older adults to sit and rest. Businesses were also given window decals so that older adults would know which businesses were age-friendly in the neighborhood. The East Harlem Pilot was well received, and a similar pilot was launched on the Upper West Side. A small study was conducted to evaluate the impact of these efforts on businesses. The results, though not definitive (because of a small sample surveyed over a short-time period in a single neighborhood), nevertheless contributed to the business case by showing that the participating businesses had higher average cash receipts than nonparticipating, similar businesses within the four-month study period (baseline two-month period followed by a two-month intervention period). The New York Academy of Medicine scaled AFLBI citywide using this model, and in the following months, canvassed small target areas across all five boroughs of New York City to engage 1,000 new small businesses.

Those businesses received an "Age-Friendly Business" decal, a resource guide on how to become age-friendly, and a chair to put in their business—which became a symbol of the initiative. They were also listed in borough-specific business directories on the Age-friendly NYC website. Due to the number and geographic distribution of the businesses, in-person follow-up visits to provide monitoring and technical assistance were not practical. Focus shifted from working directly with small businesses to engaging business-serving organizations such as Business Improvement Districts, chambers of commerce, and merchants associations. This strategic shift allowed AFLBI to leverage, rather than attempt to replicate, the established relationships, knowledge, and trust that *local* organizations had built over time with *local* businesses to improve *local* engagement and sustainability.

Business-serving organizations were educated on how to use community assets under their control (e.g., safety, sanitation, events) to maximize the social and economic participation of older residents. New tools, trainings, and promotion channels have been developed to support this effort, as well as to help business owners identify and implement age-friendly business features (The New York Academy of Medicine 2014). These resources were created in consultation with our business-serving organization partners and small businesses in their catchment areas to ensure that the messaging resonated. A training was also created to empower older adults to have meaningful—and non-confrontational—discourse with local businesses in their neighborhood on improvements they would like to see.

Surveys and interview tools were created to help capture why small businesses thought becoming age-friendly was important, how they did so, and why they think other businesses should do so. Responses from these surveys and interviews are being used to create business profiles showcasing age-friendly features of participating businesses and have been promoted online by Age-friendly NYC and by the corresponding business-serving organizations.

10.2.4 Leverage the Private Sector

Age-friendly NYC has leveraged New York City's many professional associations as gateways to reach large groups of people open to thinking about their work in new ways and to seeing the aging of the population as an opportunity for professional growth. We focused on professions that frequently affect and are affected by older adults, including architecture and design, law, pharmacy, library sciences, and urban planning. Age-friendly NYC Commission members encouraged their own networks to consider their profession from an "age-friendly lens" and champion the cause. Because the message was delivered by a colleague rather than an outsider representing aging services or public health, it strongly resonated within professional associations. This resulted in increased awareness of the value of older people, significant improvements to practice, and the emergence of new areas of focus and activity.

Co-branded tools such as "Top Ten Lists" and "Best Practices" brochures, were created in collaboration with professional associations and disseminated to members. These short, easily digestible pieces provide concrete suggestions and resources for those who interact with and should consider older people throughout the course of doing business. In some cases, these lists catalyzed more substantive conversations within a profession. For example, following Age-friendly NYC and the New York Public Library's publication of "Top Ten Ways to Make Your Library Age-friendly," the library system provided a series of trainings for staff, brought together outreach coordinators to share best practices around older adults, and partnered with the Department for the Aging to host borough-wide meetings of library branches and senior centers to share information. The library system also added special senior sections to its website and hosted technology classes and panel discussions around issues important to older adults.

The American Institute for Architects New York Chapter (AIA-NY) embraced Age-friendly NYC wholeheartedly. In addition to generating a "Top Ten Ways for Architects to Become Age-friendly," and "Urban Design and Architectural Guidelines for an Age-Friendly New York City," they started a Design for Aging Committee with over 60 members that has now been meeting monthly for 3 years. The Committee has incorporated age-friendly design into its policy platform and testified at a City Council hearing on senior housing. They also provided feedback to the Department of City Planning on how zoning regulations might be changed to build more units of affordable senior housing.

In May 2013, AIA-NY hosted an all-day charrette called "Booming Boroughs," bringing together over 100 people of different disciplines to design solutions for four different types of housing typical to New York. A subgroup of the Design for Aging Committee is now fundraising to conduct a feasibility study to implement some of the ideas that emerged from the charrette within public housing. Spearheading this initiative is a nonagenarian architect who designed some of the City's public housing developments and was inspired to adapt them to engage older tenants through improvements to the buildings' communal spaces. An upcoming competition to identify and catalyze innovative approaches to senior housing in New York City will continue to elevate design for older people as an architectural priority.

Other notable examples of professions that have adopted age-friendly practices include the American Planning Association (APA) which created a framework for planners to promote aging in community (American Planning Association 2014); and the New York City and New York State Bar Associations which hosted an Aging and Life Planning Forum and performed a review of all of their age-friendly activities.

10.3 Insights and Challenges

10.3.1 "Top Down and Bottom Up" as an Effective Strategy

Age-friendly NYC employs a "top-down and bottom-up approach" which targets both macro-level systems and structures and the numerous subsystems and structures that constitute the whole. "Top-down and bottom up" allows for complementary methods of data collection (quantitative and qualitative) and for local voices and innovations to shape City policy (McKinlay 1992). The approach requires the buy-in and sustained involvement of elected officials, community and business leaders. While NYAM serves as an advisor, information clearinghouse, and convener, its government, nonprofit, and for-profit partners must implement, model, and publicize policy and programmatic changes to make NYC age-friendly. Simultaneously, older adults must also mobilize to demand ongoing respect for their rights, responsiveness to their needs, and recognition and application of their strengths.

Our "top-down and bottom-up" approach allows for both citywide and neighborhood-specific interventions, as well as scaling up or down, and in some cases, retooling of initiatives as appropriate. For example, as previously mentioned, the local business initiative began by working with individual businesses and evolved to attain greater reach and impact by collaborating with business-serving organizations and better integrating local businesses into age-friendly neighborhood-level efforts. Similarly, an Age-friendly Ambassadors program created by NYAM using a train-the-trainer model to educate older people across the city on basic tenets of the active aging framework must now be better connected to local advocacy channels, such as City Council district offices and the Age-friendly Neighborhoods initiative.

The flexibility of "top down and bottom up" enables Age-friendly NYC to be responsive to emerging issues and changing landscapes. Following Hurricane Sandy, tens of thousands of older people were trapped in high-rise buildings and private homes for extended periods without access to food, water, medical attention or medication, and older people made up 25 of the 44 deaths in New York City (Zimmerman et al. 2014). Using Age-friendly NYC as a platform to leverage relationships with multiple City agencies, the private sector, and philanthropy, the New York Academy of Medicine embarked on a one-year study of older people in disasters with a special focus on Hurricane Sandy.[6] Guided by a policy advisory committee comprised of high-level leadership from across sectors and disciplines, NYAM conducted focus groups in seriously affected communities with older people and those who assisted them and analyzed population-level and storm-related data. The findings and recommendations of this study, reported in *Resilient Communities: Empowering Older Adults in Disasters and Daily Life*, are serving to inform new areas of focus for Age-friendly NYC, including the public safety of older adults under both routine conditions and disasters, and a redoubling of efforts to increase affordability, accessibility, and security of senior housing (Goldman et al. 2014).

10.3.2 Challenges Ahead

10.3.2.1 Addressing Population Diversity

With a new City administration in place as of January 2014, Age-friendly NYC undertook an analysis to determine the prevalence and geographic spread of age-friendly activities throughout the City to develop a strategy for our second phase of implementation. We learned that though 58 % of older people reside in Brooklyn or Queens, much of our work has been concentrated in resource-rich Manhattan and that we have not been sufficiently responsive to the diversity of the

[6]This project was funded by The New York Community Trust and the Altman Foundation.

older population (U.S. Census Bureau 2013b). Nearly 50 % of older people were born outside of the United States, and the proportion of older adults age 65 and over who speak English "less than very well" is 33.6 %.[7] While our consultations were done in five different languages, our materials and toolkits have not been translated, and our team has limited language capacity. Through our Age-friendly Neighborhoods Initiative, we hope to rectify this imbalance by building relationships with culturally and linguistically competent providers from within each neighborhood.

The number of chronic health conditions often increases with age, and older New Yorkers over age 65 experience various disabilities at four to ten times the rate of people under age 65 (U.S. Census Bureau 2013b). However, disability does not have to mean dependence. To facilitate ongoing independence, Age-friendly NYC has largely worked at the community level, focusing on improvements to physical infrastructure, programming, and commerce. At the level of the individual, an indispensable network of aging and disability service providers and advocates provides a range of supportive services, such as case management, home-delivered meals, eviction prevention, senior centers, and elder abuse services. These services are a lifeline for those who use them and have enabled thousands of people to remain more comfortably at home. To effectively address the needs and channel the strengths of older people regardless of functional capacity, Age-friendly NYC must work in collaboration with rather than in parallel to this network to assure goals and strategies are complementary rather than competitive. Working more closely with these organizations will help to assure the sustainability of age-friendly change and to achieve a greater impact in exceedingly complex domains such as senior housing which has been beset by years of federal funding cuts.

10.3.2.2 Measuring Health Outcomes

While significant work has been done to assist cities and communities in assessing their age-friendliness across a range of indicators using existing data which can be tracked over time, less work has been done to measure the health outcomes associated with age-friendly cities (World Health Organization 2014). While some interventions yield tangible results, such as a reduction in senior pedestrian fatalities or increased strength resulting from an exercise program, others are not so easily quantified, such as the overall impact of a bench. The fundamental goal of age-friendly cities is to enable older people to remain in the community by maximizing their social, economic, and physical participation to reduce or reverse disability, yet this is also very difficult to measure at a population level given the myriad systemic and environmental variables at play. Nonetheless, future research

---

[7]The most prevalent non-English languages spoken include Spanish, Chinese, and Russian (New York City Department for the Aging 2013).

is needed to demonstrate the improved quality of life, increased social capital and cohesion, and better health anecdotally associated with Age-friendly NYC and other age-friendly cities around the world.

Acknowledgements First and foremost, the authors would like to thank older New Yorkers who have generously shared their stories, ideas, and wisdom to make New York City a better place to grow old. We appreciate the efforts of all of the organizations, government agencies, foundations and businesses who have partnered with us to make Age-friendly NYC so successful.

We would like to give special thanks to the following individuals and institutions: Mayor Michael Bloomberg and Speaker Christine Quinn who were the first investors in Age-friendly NYC; The Age-friendly NYC Commission; Mayor Bill de Blasio, Speaker Melissa Mark-Viverito and members of the New York City Council who have committed to continue to make New York City age-friendly and have generously funded the initiative; Manhattan Borough President Gale Brewer who spearheaded the Upper West Side Aging Improvement District and continues to support Age-friendly NYC; Deputy Mayor Lilliam Barrios-Paoli and Kristin Misner, Mayor's Office; Diana Ayala, Office of Speaker Melissa Mark-Viverito; Commissioner Donna Corrado and Deputy Commissioner Caryn Resnick of the Department for the Aging for their critical contributions and collaboration on program implementation; Dr. Ruth Finkelstein and Dorian Block, Robert N. Butler Columbia Aging Center, for their initiation and leadership of Age-friendly NYC; and John Beard and Alex Kalache for their vision and support. We gratefully acknowledge The New York Community Trust, The Fan Fox and Leslie R. Samuels Foundation, the Altman Foundation, the Rudin Family Foundation, The Alfred P. Sloan Foundation, the Isaac H. Tuttle Fund, the Bennack Polan Foundation, the Tides Foundation, Atlantic Philanthropies, and Grantmakers in Aging.

Finally, we thank the dedicated Age-friendly NYC team under the steadfast leadership of Dr. Jo Ivey Boufford and Dr. Anthony Shih.

References

AARP Public Policy Institute. (2014). *State of the 50+ in New York City*. Available: http://www.aarp.org/content/dam/aarp/research/surveys_statistics/general/2014/State-of-the-50-Plus-in-NYC-2014-AARP-res-gen.pdf. Accessed December 12, 2014.

American Planning Association. (2014). *Aging in community policy guide*. Available: https://www.planning.org/policy/guides/pdf/agingincommunity.pdf. Accessed December 12, 2014.

Asphalt Green. (2013). *Senior summer swim 2012 at Thomas Jefferson Park a collaboration between asphalt green and union settlement*. New York, unpublished document.

Cacioppo, J., Hughes, M., Waite, L., Hawkley, L., & Thisted, R. (2006). Loneliness as a specific risk factor for depressive symptoms: Cross-sectional and longitudinal analyses. *Psychology and Aging, 21*(1), 140–151.

Finkelstein, R., Garcia, A., Netherland, J., & Walker, J. (2008). *Toward an age-friendly New York City: A findings report*. NYC: The New York Academy of Medicine. Available: http://www.nyam.org/news/docs/AgeFriendly.pdf. Accessed December 12, 2014.

Finkelstein, R., Roher, S., & Owusu, S. (2013). *Age-smart employer NYC: A compendium of strategies and practices*. NYC: The New York Academy of Medicine. Available: http://www.nyam.org/age-smart-employer/documents/ASE_Compendium.pdf. Accessed December 12, 2014.

Fiscal Policy Institute. (2011). *Analysis of 2011 American Community survey micro data*. New York, unpublished document.

Goldman, L., Finkelstein, R., Schafer, P., & Pugh, T. (2014). *Resilient communities: Empowering older adults in disasters and daily life*. NYC: The New York Academy of Medicine. Available: http://www.nyam.org/news/docs/pdf/Resilient_Communities_Report_Final.pdf. Accessed December 12, 2014.

Johnson, R. W. (2005). *Older Americans' economic security*. Washington D.C.: The Urban Institute. Available: http://www.urban.org/uploadedpdf/900887_working_longer.pdf. Accessed December 12, 2014.

Klinenberg, E. (2002). *Heat wave : A social autopsy of disaster in Chicago*. Chicago: University of Chicago Press. Available: http://www.worldcat.org/title/heat-wave-a-social-autopsy-of-disaster-in-chicago/oclc/47971411&referer=brief_results. Accessed December 12, 2014.

Lave, R., & Mathias, R. (1998). *State of the art of paratransit*. Washington D. C.: Transportation Research Board. Available: http://onlinepubs.trb.org/Onlinepubs/millennium/00107.pdf. Accessed December 12, 2014.

Li, X. (2014). *Engaging small businesses to foster age-friendly communities: A case study of an age-friendly local business initiative in New York City*. Cornell University Masters Exit Project in the Department of City and Regional Planning.

McKinlay, J. (1992). Health promotion through healthy public policy: The contributions of complementary research methods. *Canadian Journal of Public Health, 83*(Suppl 1), 54–67. Available: http://www.bvsde.paho.org/bvsacd/cd68/paho557/s8.pdf. Accessed December 12, 2014.

Mistry, R., Rosansky, J., McGuire, J., McDermott, C., & Jarvik, L. (2001). Social isolation predicts re-hospitalization in a group of older American veterans enrolled in the UPBEAT Program. *International Journal of Geriatric Psychiatry, 16*(10), 950–959.

New York City Department for the Aging. (2013). *Profile of older New Yorkers*. NYC: New York City. Available: http://www.nyc.gov/html/dfta/downloads/pdf/demographic/profile_olderNYRS_0911_newsummaries_update.pdf. Accessed December 12, 2014.

New York City Department of City Planning. (2012). *Community district needs Staten Island fiscal year 2013*. NYC: New York City. Available: http://www.nyc.gov/html/dcp/pdf/pub/sineeds_2013.pdf. Accessed December 12, 2014.

New York City Department of City Planning. (2013). *New York City population projections by age/sex & borough 2010–2040*. NYC: New York City. Available: http://www.nyc.gov/html/dcp/pdf/census/projections_report_2010_2040.pdf. Accessed December 12, 2014.

New York City Office of the Mayor. (2013a). 59 initiatives age-friendly NYC: A progress report. NYC: New York City. Available: http://www.capitalnewyork.com/article/city-hall/2014/10/8555075/mark-viveritos-senior-citizen-agenda. Accessed December 12, 2014.

New York City Office of the Mayor. (2013b). *Mayor's management report*. NYC: New York City. Available: http://www.nyc.gov/html/ops/downloads/pdf/mmr2013/2013_mmr.pdf. Accessed December 12, 2014.

New York City Office of the Mayor, & New York City Council. (2009). *Age-friendly NYC: Enhancing our city's livability for older New Yorkers*. NYC: New York City. Available: http://www.nyam.org/agefriendlynyc/docs/NYC_Age_Friendly_reportEnhancing-Livability.pdf. Accessed December 12, 2014.

New York City Office of the Mayor: Special Initiative for Rebuilding and Resiliency (SIRR). (2013). *A stronger, more resilient New York*. NYC: New York City. Available: http://www.nyc.gov/html/sirr/html/report/report.shtml. Accessed December 12, 2014.

Pazmino, G. (2014). Mark-Viverito's senior citizen agenda|Capital New York. *Capital*. Available: http://www.capitalnewyork.com/article/city-hall/2014/10/8555075/mark-viveritos-senior-citizen-agenda. Accessed December 12, 2014.

Rittner, B., & Kirk, A. (1995). Health care and public transportation use by poor and frail elderly people. *Social Work, 40*(3), 365–373.

Roux, A., Evenson, K., McGinn, A., Brown, D., Moore, L., Brines, S., & Jacobs, D. (2007). Availability of recreational resources and physical activity in adults. *American Journal of Public Health, 97*(3), 493–499.

Sederer, L. (2006). Depression, social isolation and the urban elderly. In *Conference on Geriatric Mental Health*. New York. Available: http://www.powershow.com/view/290d0-OGZkN/Depression_Social_Isolation_and_the_Urban_Elderly_Conference_on_Geriatric_Mental_Health_powerpoint_ppt_presentation. Accessed December 12, 2014.

Simonsick, E., Guralnik, J., Volpato, S., Balfour, J., & Fried, L. (2005). Just get out the door! Importance of walking outside the home for maintaining mobility: Findings from the women's health and aging study. *Journal of the American Geriatrics Society, 53*(2), 198–203.

Steptoe, A., Shankar, A., Demakakos, P., & Wardle, J. (2013). Social isolation, loneliness, and all-cause mortality in older men and women. *Proceedings of the National Academy of Sciences of the United States of America, 110*(15), 5797–5801.

Stowell-Ritter, A., Bridges, K., & Sims, R. (2006). *Good to go: Assessing the transit needs of new york metro aarp members*. Available: http://assets.aarp.org/rgcenter/il/ny_transit_06.pdf. Accessed December 12, 2014.

Sugiyama, T., & Thompson, C. (2007). Outdoor environments, activity and the well-being of older people: conceptualising environmental support. *Environment and Planning A, 39*(8), 1943–1960.

The New York Academy of Medicine. (2012). *Creating an age-friendly NYC one neighborhood at a time*. NYC: NYAM. Available: http://www.nyam.org/agefriendlynyc/docs/Toolkit_Report_0321-VA-new.pdf. Accessed December 12, 2014.

The New York Academy of Medicine. (2014). *Age-friendly business resource guide*. NYC: NYAM. Available: http://www.nyam.org/agefriendlynyc/docs/AgeFriendlyBusinessGuide.pdf. Accessed December 12, 2014.

U.S. Census Bureau. (2013a). *2013 American Community survey 1-year estimates*. Available: http://factfinder2.census.gov/faces/tableservices/jsf/pages/productview.xhtml?pid=ACS_13_1YR_S0103&prodType=table. Accessed December 12, 2014.

U.S. Census Bureau. (2013b). *American community survey, 2008–2012 American community survey 5-year estimates*. American FactFinder. Available: http://factfinder2.census.gov. Accessed December 12, 2014.

World Health Organization. (2007). *Global age-friendly cities: A guide*. Geneva: WHO. Available: http://www.nyam.org/agefriendlynyc/docs/WHO_Global_Age_friendly_Cities_A_Guide.pdf. Accessed December 12, 2014.

World Health Organization. (2014). *Measuring the age-friendliness of cities: A guide to using core indicators*. Kobe: WHO. Available: http://www.who.int/kobe_centre/ageing/age_friendly_cities/AFC_Indicator_Guide_Pilot_English.pdf?ua=1. Accessed December 12, 2014.

World Health Organization. (n.d.). *Social determinants of health. Key concepts.* Geneva: WHO. Available: http://www.who.int/social_determinants/thecommission/finalreport/key_concepts/en/. Accessed December 12, 2014.

Zimmerman, R., Li, W., Gambatese, M., A, M., Lasner-Frater, L., Van Wye, G., Sun, Y. (2014). *Summary of vital statistics, 2012: Executive summary*. New York. Available: http://www.nyc.gov/html/doh/downloads/pdf/vs/vs-executive-summary-2012.pdf. Accessed December 12, 2014.

Chapter 11
"Age-Friendly Cities" in France: Between Local Dynamics and Segmented Old Age Policy

Simone Pennec and Françoise Le Borgne-Uguen

11.1 Introduction[1]

In France, the "Age-Friendly Cities" approach (AFC) of the World Health Organization (WHO) took place within the framework of a national program "Aging Well." The ideas proposed in this chapter are based on our researches on public policies and their local applications which, as AFCs or the "Aging Well" plan, relate predominantly on aging populations. They concern the adaptation of housing, mobility, styles of neighborhood, professional and family services, social and health vulnerabilities. Several research concerted actions were carried out with local communities in order to codevelop or evaluate innovations and partnership projects (technologies, sociability, etc.). In 2011, the *Laboratoire de sociologie de l'Université de Bretagne Occidentale* (Laboratory for Research in Sociology of the University of Western Brittany-UBO) was asked by a medium-sized city to conduct the diagnosis phase of the AFC program. This research was supported by two Quebec university-based researchers.[2] In addition to this action research, this text presents two other cities: the largest city (by population) of the same department although not committed to the AFC approach, as well as the regional metropolis which was the first to present its AFC action plan. These three case studies make it possible to identify old age policies which differ according to their objectives and

[1]Our warmest thanks to the editors who translated this chapter, including original quotations.
[2]Suzanne Garon and Marie Beaulieu, Research Centre on Aging, University of Sherbrooke, who were originally responsible for the research program on AFC in Quebec.

S. Pennec (✉) · F. Le Borgne-Uguen
Laboratory for Research in Sociology (Labers), University of Western Brittany (UBO), EA 3149 Brest, France
e-mail: simone.pennec@univ-brest.fr

F. Le Borgne-Uguen
e-mail: Francoise.leborgne-uguen@univ-brest.fr

© Springer International Publishing Switzerland 2016

T. Moulaert and S. Garon (eds.), *Age-Friendly Cities and Communities in International Comparison*, International Perspectives on Aging 14,
DOI 10.1007/978-3-319-24031-2_11

the methods used in each urban unit. Finally, we also took into consideration data disseminated by the 19 French cities recognized as "AFC cities" by the WHO in 2014 as well as by the AFC francophone network created in January 2012.[3]

Two main observations are identified in these studies: the first shows that the segmentation of local policies constitutes a serious drawback to efforts of cross-cutting AFC approach. The second stresses on the significant importance of logics which structure the field of aging in France in terms of environments and forms of citizenship and powers.

11.2 Incentives or Strengthening Effects of Local Policies?

The WHO's Age-Friendly Cities program (2007) has been defined and disseminated at the same time as the French plan *Bien Vieillir-Vivre Ensemble* (2007) (Aging Well-Living Together). If the disseminated models are similar, they may differ as for the objectives: as far as the WHO is concerned, the emphasis must be put on the participation of seniors in adapting environments, while the plan "Aging Well" highlights more on prevention when it comes to "successful aging." The local application of these programs is experiencing a diversity of objectives, forms of procedure and sectoral consultations. It is reflected by the heterogeneous results according to the areas promoted as analyzed in these three urban contexts.

11.2.1 A Comparison Between the WHO Program and the French Program

When promoting their programs, the WHO and the French State share the same definition of quality of life. However, there are several distinctions to be made regarding fields of actions and environments as well as procedures to be implemented.

11.2.1.1 At the WHO: Environments for "Active Aging" Through the Participation of Seniors

The WHO's approach seeks to build welcoming environments for older adults, in the framework of "active aging" of the Madrid International Plan of Action on Ageing in 2002, which was resumed in 2005 at the 18th World Congress of Gerontology in Rio de Janeiro. The objective is to: "Create urban environments that enable older people to age actively and continue to participate in social life [for]

[3]The network was launched during the REIACTIS symposium in Dijon, 2012 (REIACTIS is the "Réseau d'Étude International sur l'Age, la Citoyenneté et l'Intégration").

optimizing opportunities for health, participation and security in order to enhance quality of life as people age" (WHO 2002, 2007). The fundamental focus underlines: the promotion of active and healthy aging and of the support of social participation of older people in developing their environment. For the WHO's "Ageing and Life Course" team:

> The project was based on the proposal that an age-friendly city encourages active ageing; such a city *maximizes opportunities for health, participation and security in order to enhance quality of life as people age*. Also, in the *WHO's Active Ageing* concept, the word 'active' refers to continuing participation in social, economic, cultural, spiritual and civic affairs, and not just the ability to be physically active and/or to participate in the labour force (Petitot et al. 2010, p. 233).

The work carried out by the research team of Quebec helped identify forms of social participation and tools for assessing seniors' needs (community building), within the built and social dimensions (Garon and Veil 2011). The community building is seen as a "global process by which the driving forces of a local community banded together to work as a unit in order to achieve a common goal" (Garon et al. 2012, p. 76).

11.2.1.2 In France: "Aging Well-Living Together," A Labellization of Local Initiatives

In opposition to the WHO, the French government does not propose a standard method nor provides for significant funding to support its approach. It is limited to labelling cities on the basis of initiatives and projects that they submit. If few awards have been distributed to certain cities, no overall funding was allocated to this program.

The French policy for seniors also includes a section entitled "Supporting Aging Well Policies at Local Level" (no. 7) (*Développer le Bien Vieillir au niveau local*). The plan submitted in 2007 "aims at proposing ways for a 'successful aging,' both from the individual and social relations standpoints, highlighting the organization of relevant preventive measures" (Petitot et al. 2010). This objective takes shape through a call for tenders in June 2009.

> The label is created to help French towns to meet the challenges of aging, on how best to support local dynamics through methodological assistance, and fostering interactions; it is also important that people should know about cities' achievements in the field of aging well, promoting the most innovative and reproducible projects and contributing to the emergence of a national network where experiences can be shared (Association des Maires de France 2009, p. 1).

In February 2010, the label "Aging Well" will thus be awarded to 34 cities. But faced with the overlapping of designations and the variation of local policies, France will negotiate with the WHO for the use of the 2007 "Age-friendly Cities" World Guide. In 2009, France requested the recognition of cities labelled "Aging Well" for the label "Age-friendly Cities." In addition, sometime previously, other French cities had undertaken actions with the WHO to implement the AFC approach without soliciting the French national labelling; they will then become pioneers, following the example of Dijon, in initiating the francophone network of "Age-friendly Cities" with their Quebec counterparts.

11.2.2 *Contrasted Local Usage: Procedures, Mobilized Actors, and Outcomes*

In the absence of national funding, municipalities have seized the labelling "Aging Well" in a variable way and often as an opportunity to enhance the local social action, sometimes extended to environmental and cultural actions, as well as greater participation from seniors. Similarly, the AFC labelling does not include direct financial support from the National Government which suggests in part the strong dissimilarity of implemented protocols and targets according to territories. In this regard, the three cities present diversified situations concerning procedures being used, the partners and the expected results. As for the other French regions, the recent nature of the "Aging Well" labelling, and even more with the AFC approach, prompts us to remain cautious in analyzing data.

11.2.2.1 Case 1: Mobilizing AFCs to Strengthen Existing Partnerships Without Applying the Diagnosis

In order to gain the AFC-WHO recognition, the regional metropolis, identified as the first case, introduced its "action plan" already initiated on the basis of the WHO's eight areas, without going through the diagnosis stage. This way of doing things is experienced as a gain of time and energy allowing to focus on ongoing operations which intersect with various lines of the protocol (housing, roads, technology, consultation, sociability) and to enhance the visibility of municipal social actions. This city benefits from actions previously undertaken and visibility in this area that it is to extend, particularly in redefining activities and consultations according to the new allocation of functions between the various decision-making levels in France (Health Regional Agency, Department, or City). In fact, the distance with the WHO approach is assumed in the name of the legitimacy of "social action" against the long-standing perception of assistance, but also in order to cope with the loss of a part of its responsibilities by the transfer to the department of a large part of these policies (in particular, the allocation of the Personal Autonomy Allowance: PAA).

The search for local synergies leads to bring together elected officials and services as well as professional organizations beyond health and social sectors, by integrating issues such as housing, environment, sociocultural activities, etc. The issue of aging is set out in relation with actions from the "urban policy" such as neighborhood or community actions. The regular collaborations with various partners (retirees, neighborhood, tenants, and institutional associations from different areas) give rise to workshops and debates around mutual proposals. Consultations with retirees go beyond structured organizations in order to better integrate representatives of neighborhood and tenant, in presence of existing and established associations having more members.

Mixed urban development leads to the launching of actions such as the construction of an intergenerational residence in 2009 and also to introduce a 20 % quota of accessible housing for seniors in programs promoting and regenerating certain neighborhoods. Moreover, the implementation of the accessibility plan may include walking tours of urban settings. This helps locate suitable facilities for mobility and sociability. The close collaboration with the Local Information and Coordination Center (CLIC) strengthens the consistency of activities and experiences; it would allow seniors, their relatives and professionals to be provided with updated and extensive information. National and international workshops are locally organized with the participation of retirees, professionals and they include interventions from national, regional, local elected officials as well as experts. These workshops can be organized into highly symbolic places such as official buildings from the Department or the Region and may benefit from an increased visibility.

Obtaining the title "Age-Friendly City" just confirms the visibility of the municipal social action and also reinforces the synergy between municipal services. This dynamic as well as the objectives will allow the city to publicly present its "AFC action plan" in October 2012, thus becoming the first city of the western region to be part of this AFC initiative. This policy comprises three major areas: "Advocating an age-friendly housing; Taking actions to prevent isolation; Improving information and communication intended for seniors." In this first case study, the AFC approach operates as *ex-post* legitimation. It is part of a coordination of a local aging policy built on multi-partnership type in order to better address various aspects of the aging process.

11.2.2.2 Case 2: Performing AFC Diagnosis to Build Synergies Over Time

For other cities, such as our second case study, the formulation of a diagnosis has given rise to a consistent working time which has brought together diversified partnerships depending on the municipalities. As a result of this step, the development of action plans has been the subject of more restrictive consultations, sometimes limited to municipal services and elected officials. In this point, we consider the initial implementation of the AFC program from the diagnosis step initiated by an average-sized city (70,000 inhabitants) in which are located the department's administrative bodies (the prefecture and the General Council).

A diagnosis stage to redefine the old age policy and federate the services

In case 2, the commitment to the AFC approach was initiated by the deputy mayor responsible for social actions. This commitment was an attempt to improve the position of this area of municipal action into hierarchies, both within internal hierarchies of elected officials and municipal services, and toward targeted population.

Obtaining the French label "Aging Well" at the time where negotiations between the French government and the WHO led to the merging of "Aging Well" with the AFC approach, the city decided to establish a thorough and concerted diagnosis in requesting university expertise. The AFC approach is introduced by French sociologists in link with the Quebec researchers from Sherbrooke (S. Garon and M. Beaulieu). Then, the Communal Social Welfare Center (CCSA) will brief the city on consultations with the senior population and with services (shops, health), while the university will be responsible for group interviews and the presentation of results. The topics covered are aligned with the AFC protocol whose implementation is exposed to elected officials and representatives as well as partners mobilized by the CCSA, in consultation with the municipality's management team. Performing the diagnosis is a way of redefining objectives, partnerships, and giving a new impulse to local dynamics related to the senior population (Table 11.1).

The high number of encountered persons will be subsequently put forward to highlight the diagnosis, with retirees' associations as to professionals or elected officials who questioned the findings of the diagnosis. This numerical argument supports the credibility of the diagnosis, at both local and international levels. It shows the significant importance of demonstrated figures even if the approach has a qualitative background. Numbers represent unquestioned evidence, which leads to clarify the situation with partners, elected officials, and professional services about the limits of accounting approaches and the specific interests of focus groups, while distinguishing them from simple telephone surveys or the ones conducted in the street.

Table 11.1 People interviewed in case study

Focus Group (n = 13)	
Older people	**97**
By age groups	
People 60 to 69 years old	35
People 70 to 79 years old	38
People 80 to 89 years old	21
People 90 to 99 years old	3
By sex	
Female	63
Male	34
By living situation	
People living alone (*1 living in institutional setting, 1 in his child's home*)	38
People living in couple (*2 living with their children*)	59
Carers	**18**
Informal caregivers	2
Professional providers	16
Total of people	**115**

People have been selected in electoral files by respecting 2 groups of age by neighborhood (4 neighborhoods in total), respecting the sex ratio in these local populations and respecting two different groups in terms of level of resources

Another example, despite the sampling, representativeness is not always assured, which was the case for the "family caregivers." Considering older persons we met, if some of them participate in the local retirees' association, the absence of other associations is noticeable. For example, none comes from the University of the Third Age and volunteer associations related to health, neighborhood or culture. As for the professionals, they were predominantly from health and social services, from professions or institutions, which has strengthened the convergence of their statements and propositions which are closer from the very older persons. This characteristic overlaps with earlier work on this city (Douguet and Pennec 2002). Initially deemed as a facilitator, thanks to sectoral knowledge, this "path dependency effect" has, on the contrary, resulted in difficulties to redefine, from the different areas suggested by the WHO, the boundaries of action in the field of aging. Of this fact, the repositioning of the types of actions already implemented has remained weak as was the formulation of new issues. If right from the beginning of the approach, the reflections had integrated more of the municipal services, they probably would have fostered more extended collaborations between sectors such as housing, transportation, social cohesion, culture, and health. These factual elements have led to carry out a 'parallel' analysis of proposals of absent players, issued on other scenes and which are therefore little integrated in the analysis because formulated outside the protocol.

Steering committee, follow-up and development of the action plan

The composition of the steering committee, the implementation, the monitoring, and the development of the action plan locate the partnership between the areas of action of professionals and of associations. Outside the health and social services sector, actors primarily come from municipal departments such as roads and housing, under the incentive effect of validating the approach by the city council and by the steady participation of the mayor (launch, steering committee, presentation of results, etc.). The social and cultural sectors as well as educational services are poorly represented. As for seniors, apart from the ones from institutionalized associations, they are also little represented; only Neighborhood Advisory Councils responded to requests. For retirees' associations, the ongoing participation of the city's retirees association can be seen at different stages. By contrast, the absence of associations in the field of health and of those of older people and their families limits the questioning, as demonstrated by the difficulties to meet "caregivers." Similarly, retirees from the University of the Third Age (U3A) are not associated with the approach. These features emphasize the partitioning of all these organizations in France and also how the implementation of the cross-sectoral approach is proven to be complex.

The results of the diagnosis (Le Borgne-Uguen and Pennec 2012) will be the subject of two presentations: first to the steering committee, and then at a meeting where the UBO shares the presentation time with the deputy mayor responsible for social actions and with CCSA (they present future actions). During the first presentation, two points will concentrate a significant part of the issues and debates: one relates to urban environments, especially transportation, and the second one to social participation (recreation and civic life). Transportation has given rise to many opinions given the context of the transformation of the city's traffic plan, a hot topic at the time of discussion. Pedestrian movements in external spaces and the adaptation of public transport are the subject of findings and also of proposals, while the disadvantages reported for housing give rise to few recommendations and prospects for adjustments. Much of the urban environments seems to be out of reach for older persons' capacities to influence adaptations yet desired.

Citizenship and social participation are variously regarded according to discussion groups and within the steering committee; a variety of definitions for the word 'citizenship' may be found. Thus, defined in a restricted fashion as participation in public consultations, the citizenship is heard mainly through elections. Nevertheless, a part of the persons consulted supports a less restrictive conception of social participation which extends it to forms of socialization and includes participation to activities organized by the city. The difficulties expressed during the focus groups about finding its place within the existing associations in the city—when retirement occurs or during life and health specific events—are given little credit, or are even rejected, in particular by the retirees' organizations that put forward the current range of activities and the diversity of volunteering. Possibilities for the senior population of contributing to the development of urban projects, deliberations, or even to decisions (in terms of planning, intermediate resources, etc.) are little considered.

The action plan will cover themes of the diagnosis following three categories: the environments, health/social services and social participation. The plan is based on collaboration with relevant professional services and puts forward current answers to coordinate, short-term improvements to be made as well as long-term projects. Proposed actions and those already committed are set forth in each theme; stakeholders and partners are identified as well as qualitative and quantitative indicators to be documented from now on or when things will be achieved. The most precisely defined areas are those for which the cross-cutting approach between municipal and public services has strengthened when joining the AFC (outdoor spaces and buildings; transportation and mobility; and communication and information).

The municipal election agenda will delay the process, and then will be halted for a time because the current municipal team will not be reelected. The duration of mandates for elected officials (six years) can thus have an impact on projects arising from consultations and encourage local decision-makers to prefer immediate actions. In this case, they are often preconceived notions from technical expertise at a distance from concerned populations. However, the new municipality will reactivate the AFC program at the end of 2014. A new elected official is "responsible for the aging of the population and seniors" and resumed the AFC components in proposing several cross-cutting actions on various topics including also an intergenerational housing project. The diagnosis will thus serve as a stepping stone for mediations previously initiated between services and associations and the use of topics for publicizing the new team's projects. Furthermore this new elected official has a strong visibility gained by the creation and extension of an association of seniors and families.

11.2.2.3 Case 3: Between Areas of Expertise and Activation of Seniors: Politics Fragmentation

The third urban situation contrasts with the previous two cases, as this latter city falls into the objectives of "Aging Well" but was not involved in the AFC approach. The presentation of this city serves as a tool to analyze differences between the WHO model and the French model. Located in the same department as the previous city and with a population twice as high, it is a university town which has several advanced training programs in the field of aging (medicine, law, management, sociology, psychology). Grey literature and several studies carried out by researchers and students include the territory and several dimensions related to the urban environment (housing, accommodation of disabilities, neighborhoods, home and institution support services, health, social participation, technological experiments, etc.). In addition, a number of data are produced at the city's request and the metropolitan area, either by their research departments—an observatory of population aging has been recently included in these departments—, or by commissioning a study from expert consultants. The city also has a number of professional networks and many associations involved in recreational activities, well-being and volunteering with retired people, people who are ill and/or isolated, etc. However, the fragmentation of competences in old age policies contributes to the blurring of priority objectives and methods for consultation in a field including multiple actors. The diversity of expertise is not enough to transform the diversity of specific and sectoral actions in long-standing efficient coordination.

Dispersion of Assignment of Competences

At first sight, the monitoring of old age policies does not appear within the assignment of an elected official in particular. Several mandates are affected by either action related to retirees: a deputy mayor is responsible for "social action," another elected official for "seniors" and a third one for the "third age club" (*Office des retraités*), while the issue of the "aging well" is under the responsibility of the first deputy mayor. Another deputy works on health issues, while disability and accessibility fall under yet another deputy mayor. The dispersion of elected officials' mandates frequently covers the one of municipal services, this proliferation of political actors in contiguous areas does not contribute to an increased collaboration between these partners. However, the old age policy must also confront with local policies (urban planning, equipment, etc.) that are often better rooted in the territory. In this context, the expertise for demographic, sociological, and psychological analyses encountered difficulty in asserting priorities regarding planning, housing, urban mobility. And this, especially since they are not supported by a cross-cutting approach on consultation, extended to different sectors and services and open to residents and citizens. Public disseminations of study results certainly form an arena for discussion, but they do not provide the continuity of the forum and do not generate consequent actions for this city which has experienced, also, periods of successful partnership collaborations in gerontology.

Despite the actions and allocated supports, the monitoring and priorities of the local policy of aging suffer from a blurring of their operationality. Thus, the calls for projects (alternative housing, multimedia, etc.) do not seem to correspond to seniors nor to professionals of the field of aging (Pennec 2012). This form of "governance by instruments" (Lascoumes and Le Gallès 2004) can only work if it is supported by a large number of stakeholders and after exchanging views, arguing, and compromising on the projects. For this city, while the tools (services, studies) are numerous in the field of aging as well as mandates of elected officials, the objectives and the means remain weakly developed for concerted actions.

Seniors' activities: between free choice and volunteers' mobilization

As in the other two cities studied, the local association of retirees, initiated from requests of retirees, developed a diversity of recreational and volunteering activities (reading assistance, crafts and repairs workshops, psychological support, etc.). There are collaborations with the social services (CLIC-CCSA) and with a few stakeholders in social and cultural sectors which enable the continuation of original projects and the inclusion in national actions (Monalisa). These proposals of activities between "recreational retirement" and "retirement solidarity" (Guillemard 2002) may be put in opposition when associations must make choices without priority guidelines first being set with local policies which alternate patronage and liberalism.

However, here, as in many cities, the mobilization of seniors is expected to maintain various forms of volunteering and it seems to be included in social actions toward seniors and isolated individuals. The city highlights and supports the action of groups who provide assistance, health, and socialization. This volunteerism for others seems to make seniors feel useful (*utilité sociale*), carried out for others, but also for their own "aging well," as one of the recommendations of prevention standards that are intended to them. However, these commitments into the city are more often seen as pure occupational activities than citizenship practices.

This third case study shows a context with diversified resources: the importance of existing studies and the diversity of their topics, a great number of services and various retiree associations and activities. Nevertheless, on this geographical area, the priority guidelines and the modalities for consultation and implementation do not benefit from good visibility. This is due in large part to the dispersion of assignments and overlaps of achievements from stakeholders involved in multiple partnerships with distinct timescales.

11.3 Unifying and Giving Priorities to Segmented Aging Policies

If the cities which are endorsing the AFC encounter difficulties to mobilize partners, develop and maintain consultations when achieving action plans, provisions of the WHO protocol nevertheless open the way to engaging collaborations and practical commitments. Beyond the three situations described, recurring elements will strengthen the synergy sought or hit the convergences of actions despite the number of available partners.

The "living together" cause and the values of social participation, active aging or of "aging well" are picked up by all the actors from social and health services up to various senior activities promoters, without forgetting the intensive advertising of insurances. But the dissemination of the rhetoric does not mean good news for cross-cutting actions nor for their adaptation to the diversity of populations, territories, or of inequalities of resources from individual and collective standpoints. Achievements are proving to be more difficult, especially when policies are built around the targeting of specific populations or when there is no clear leadership at the political level or the city's services level. Two points deserve to be discussed: the partitioning of fields affected by aging and formats assigned to "active aging" and citizen commitments.

11.3.1 Reaching a Cross-Sectoral Approach Instead of Working in Silos

The implementation of the AFC and the "Aging Well" approaches provides a magnifying glass of proximities and distances between sectors whose actions influence the life conditions of seniors. Within local policies, there is a hierarchy between decision-makers and areas of action, partly built by national guidelines. Several segmentations exist regarding health, housing or environment.

11.3.1.1 Domination of the Health and Social Sector and Medicalization of Aging

In France, healthcare actions are mainly under the responsibility of the government and are implemented by Health Regional Agencies and the departments. The successive policies, focused on the "dependent older adults" (distinction must be made from "persons with disabilities") and on pathologies (Cancer Plan, Alzheimer's Plan, etc.) have multiplied programs whose use is much more complex for users and adjustments for professionals are much difficult. Studying the implementation of the PAA, Réjean Hebert (2011), professor of geriatric medicine in Quebec, highlights it under a suggestive title "Misery and Greatness of the PAA in France." The action of municipalities can be confined to the incentive and the support of coordination and networking among services from health and social and from care sectors. According to the allocation of competences in municipalities, these sectors are collaborating to varying degrees; they consult each other in a more sustained manner when their responsibility is brought by a single elected official (health, disability, seniors). Divisions between areas such as disability and aging, or between medical and social sectors, were called into question and many projects aimed at their integration; it nevertheless remains to this day in an highly partitioned organization. In addition, many categorizations and representations of people (be treated, be cared or cured, be helped, be assisted, beneficiaries, residents, etc.) were conceived by health and social services professionals, without going through the effective participation of persons concerned beyond institutionalized settings. It is therefore these professionals who are first taking action, and then are solicited, to define and represent the "needs of seniors", leading to a certain medicalization of aging, with geriatric dominance. These deficit models of aging, in conflict with those of "active aging",are uneasy to be reversed.

11.3.1.2 Aging Almost Nonexistent in Adapting Urban Environments

Medicalization and social/health or dependency/disability distinctions are not the only sources of fragmentation of actions toward old age, but this segmentation is also noticeable in the transformation of urban environments and in the "city policy." In France, aging policies have pursued the objectives of aging in place for individuals by the allocation of hours for services (housework or home care) and of technical equipment. By contrast, the home has not been subject of a comprehensive policy including the adaptation of housing and urban environments that facilitate mobilities and sociabilities. Dwellings were adapted on an ad hoc basis, on a case-by-case basis regarding disabilities, based on means testing, and for certain housing projects by some social housing landlords. The adaptation of built and environmental spaces, which might facilitate a quality living environment accessible to all, was not carried out systematically. The necessary infrastructure rehabilitation (housing, transportation, buildings, and public spaces) has yet been the subject of research and many reports (Fondation de France 2007; Boulmier 2012). While it is already covered by the current legislation, the obligation to adapt and plan public spaces has regularly experienced derogations in their application. It must be stressed that these obligations are more often formulated in terms of disability compensation that in terms of matching needs and wishes of various audiences including the senior population. The local accessibility commissions include the representation of users, particularly "persons with disabilities" but do not take into account the "dependent older persons." Similarly, the "city policy" is mainly understood as addressing the "sensitive neighborhoods" and within the latter it is more specifically focused on groups such as young people, immigrants, travelers, etc. The local programing regarding urban planning, housing, mobility, and governance tools of elected officials, is struggling to integrate seniors who remain caught in the medical and social bind. However, the voice of the social and health sector is too often confused with "the voice of older people", and this sector does not collaborate much with other actors. Within municipalities, we found out that mandates of elected officials of these areas are often dispersed, that their powers are working remotely and that cross-cutting policies are quite rare. "On this point, there is still room for metropolitan areas to provide an alternative to retirement home and to aging in place service (confinement) and expand the range of housing and accommodations for everyone" (Fondation de France 2007, p. 15). For Catherine Collombet "aging represents not *a* challenge but a series of challenges of variable configuration, which public authorities must provide variety of answers" but she believes that France lags behind other countries from the fact that "public policies are a patchwork without clear strategies {… and} public services are spread over" (Collombet 2010, p. 154 and p. 168). However, given the extent of the rehabilitation of built areas, public authorities, communities, and organizations responsible for housing are now paying more attention to the various actions to be undertaken to define a new offer between the medicalization of accommodations and individual housing.

11.3.1.3 Low Level of Development of "Intermediate" Resources and Combined Services

Fact remains that, in our country, the antinomy constructed between aging in place services and seniors homes has not led to devise new intermediate forms of resources or to tackle the rehabilitation of built areas and environments. If new housing experiences are diversified, they are few in numbers and the sharp decrease of social residences (from CCSA—Public Social Welfare Centers and from associations) benefited residences from commercial private sector (Volant 2014). The weak support to the adaptation of individual dwellings has increased the needs of adapted housing and has widened social differences between individuals in terms of housing accessibility and environments. Also note that issues related to housing and health are updated under new features: in the name of social welfare, free choice, and even the principle of autonomy, responsibilities of individuals and moral and legal family obligations, are consistently repeted. In case of emergency and/or of limited resources, these elements reinforce the call for support to the family as reported in practices of (re)cohabitation between generations (Pennec 2015). Taking together the politics of care and of housing is becoming a major issue of old age policies.

On this issue, practices of municipalities range between intervention, primarily in social housing, and liberalism through the private market. Some municipalities have taken account of the tenants' and owners' requests to transform individual and collective spaces to intervene on environments (roads in particular). The adaptation of social housing is also scheduled in the long run with housing landlords and sometimes also negotiated with residents and relatives. However, local communities struggle to define a concerted policy with retirees who are predominantly owners (3/4) of their dwellings and who live in a detached house for 2/3 of households (Berger et al. 2008). For a large part of older owners, the "housing expenses" (repairs, taxes, electricity, etc.) can exceed their level of resources and require a dissaving. "New forms of impoverishment emerge among the population of small property-owner {…} in particular women whose retirement pensions remain notoriously lower to those of men, including for younger generations" (Ogg and Renaut 2010, p. 2). The number of poor owners–residents such as the one of poor tenants stands around 1 million for each group according to the National Housing Agency (Agence Nationale de l'Habitat 2012). Considering these data, the achievements highlighted such as "intergenerational" residences are of little concern to people. Alternative projects (grouped dwellings, cooperative housing, etc.) are not the subject of significant public support, and they can only be achieved from property of individuals. If some municipalities invite "project leaders" to submit alternative proposals, the process of calling for projects prevents from taking into consideration most vulnerable and most disadvantaged people. This type of local action may be close to liberalism as much as to the citizen participation put forward, when ways of achieving things are not accessible to the entire population. This type of liberalism is dominant in several cities if one considers the progress of private residences of the business sector.

Yet it is possible to do otherwise, on the condition of achieving intersectoral collaborative approaches. If we address these issues by leaving behind "the housing perspective" to consider alternatives in the organization of care, a few experiences deserve attention. They are emerging in the field of disability and in deficiencies due to aging. They consist in pooling the collective property of a territory, the private property of individuals and their rights to care and service benefits (Pennec 2015). Let us examine the two following cases. In Germany, care centers for people suffering from cognitive impairments are put in place in cooperative form between neighbors, parents and professionals. This organization at the neighborhood scale, coupled with care and community life, gives rise to innovation and establish strong links, not considered theoretically, as noted by Leenhart (2011) in several urban centers. These practices extend previous initiatives aimed at developing collective gatherings for all people by the extended access to places and social time (*temps sociaux*) in the presence of volunteers and professionals, such as the multigenerational centers studied by Erler (2009). Such cross-cutting actions may also be organized at home as shown by Rosenfelder (2014) in Strasbourg. From the adaptation of the private home of a resident, mediators are negotiating a joint project of care and sociabilities between the person and carers, parents and professionals. These mediations enable this individual home to become subsequently a place shared with other ill people, for care and sociabilities on a daily basis. This formula has spread to allow transition stays and rehabilitation within other private homes and to increase the pooling of more extensive services and more open on neighborhood structures.

11.3.1.4 Limited Cooperation Between Associations of Retired Persons

The last point on cross-cutting challenges concerns the emphasis put on actions and giving voice to seniors. Local programs for seniors address various issues such as presenting studies and expert reports, discussions with retirees about future projects. These community-initiated meetings can take place on an ad hoc basis or be the subject of regular consultations in the presence of professionals from various sectors. Finally, if these consultations sometimes rely on contributions from groups of retirees and/or neighborhoods, they are sometimes also collected by communities through Internet distant consultations. During consultations and public presentations of "Aging Well" and AFC projects, the forms of presentations may vary and may also show the segmentation of groups of stakeholders. Seniors' representatives refer to the mandate's area, the labor law for retirees, and users' rights related to services used. If the "right to citizenship/right to the city" and the "right to age" have been the subject of public meetings, sometimes on the government's initiative, they are less often put forward by associations. Yet the theme of citizenship now makes a notable comeback thanks to the public recommendation of a "civic/citizenship service" focused on seniors' volunteering. The plurality of logics advocated by seniors' representatives decreases and sometimes prevent the pooling of their proposals towards institutions and communities. For the latter, the purpose is to build a balance between representatives established through organizations such as users' committees (particularly in health sector) and the many retirees' organizations, while seeking to hear less structured speech from "the voice of ordinary people (lay perspective)." As stated by Dominique Argoud about seniors' representatives within recognized authorities, the exercise is perilous because:

> "The voice of ordinary people has a real power only if it transforms itself into an expert advice, thus avoiding being dismissed by other protagonists. But in doing so, it loses in part its subversive force by following the agenda of public authorities" (Argoud 2012, p. 68).

On the public stage, defections among seniors are emerging. This is the case, when the associations feel that they "do not measure up" with other retirees' organizations perceived as leaders or supported by the municipality, given the number of members or their seniority. It is the same when initiatives of local communities seem to be of less concern because of the irregular frequency of meetings or the distance from places to live and local concerns from residents. Finally, some retirees' associations keep their distance from others that they identify to social, and/or aging fields, such as the University of the Third Age (U3A) which, in France, rarely deal with aging and gerontology issues (Chamahian 2009; Pennec 2013). The situation of these U3As can be understood as a process of distinction with other retirees' organizations and proposals toward them, and it emphasizes the separate operation of these organizations. In comparison, the Spanish Community Centers, studied in Zaragoza by Monique Membrado, show extended collaborations between education, self-care, and sociabilities sectors as well as social work (Rouyer et al. 2008). The case of educational institutions is an example of working "in silo." Although facing a situation of strong demographic decline of their younger workforce, the reception of other categories of people remains limited and not very much innovative (Pennec 2013; Ogg and Phillipson 2013). In other countries such as Germany, Quebec, Brazil, etc., we observe less discriminating participations and more innovative educational approaches. Proposing the concept of "participatory university," Andre Rubião shows how some Brazilian projects are integrating social diversity. He stated that:

> The Center of citizenship (*le Pôle de citoyenneté*) {...} of the Federal University of Minas Gerais has been developing projects in the poor communities and the slums of Belo Horizonte for 10 years. The program is funded by the CNPq (National Council for Scientific and Technological Development) [and] the methodology [is] action-research (Rubião 2008, p. 119).

The Universities of Sao Paulo and Florianópolis welcome retirees and seniors. Courses are organized in collaboration with various people whose ages, social groups, and previous educational experiences are very diverse (Le Borgne-Uguen and Pennec 2014). Characteristics of the University of the Third Age (U3A) are analyzed by Formosa (2010) who questioned the pedagogical models by referring to feminist analysis, to critical gerontology and by considering social relationships (Formosa 2005). The issue of recognizing experiential knowledge of retirees is not about their activation instrumentalized through projects built for them but without them (the "aging well" plan: aging well, feeding well, enjoying well, dying well, etc.), but it concerns the increase of their empowerment on key elements for their own life and society. In several countries, programs of *Lifelong Learning in Later Life* have been developed. These experiences include some similarities with a few French initiatives but they struggle to survive because universities are not very involved; actually, U3A associations usually operate independently. The WHO programs such as "*Long Learning Cities*" and "Age-friendly Cities" give rise to innovative actions in a more confident manner in those countries whose policies are little partitioned and in those that attribute funding to cross-sectoral measures.

11.3.2 The Duty of "Successful Aging": Individual Responsibility and Seniors' Social Responsibility

The AFC approach also aims at strengthening the role of seniors by a co-construction of policies. However, in France, the normative frameworks of the "Aging well" rather advocate the individualization of prevention and the role of social usefulness by volunteering more than the participation of seniors in developing public actions.

11.3.2.1 Forms of "Active Aging" Serving the "Aging Well"

The concept of "active aging," developed by the WHO during the 1990s was initially centered on the "benefits of physical activity on health," presented under the designation "healthy aging" (WHO 1993). The orientation will be subsequently criticized by its individualized character which suggests that it belongs to everyone to ensure its own active aging. For Walker (2009), the mobilization of governments is required and he provides an understanding of active aging beyond the employability. As for the WHO, the concept will be extended to the optimization of health, participation, and security of citizens. In the scope of an inclusive society, "It is therefore in social terms, rather than individual, that AFC-Quebec considers participation, by enhancing, among other things, intergenerational bonds and, to a larger extent, the social interaction" (Paris et al. 2011, p. 31).

In France, "active aging" is often viewed as "aging well" and it has been briefly discussed, except by researchers who analyze it as a "polymorphic frame of reference" (Moulaert and Viriot-Durandal 2012; Guillemard 2013). Despite the high number of seniors who are underemployed, the retention of employees is excluded from the scope of questioning of old age policy-makers even though the blurring of time and transition toward retirement would allow to consider the continuities and blends with activities beyond employment. This occultation also prevails in cities where retirees have initiated production workshops with unemployed young people (equipment repair), interventions in schools (through the association *Lire et Faire lire* dedicated to support literacy initiatives) or in health services (hospitals, home-based), each time in coproduction with agencies and professionals.

Finally, action plans developed from diagnosis do not say much about collective expectations based on social distinctions, gender, relational capital, etc. The differential aging is little discussed, thus reinforcing dominant representations, aging stereotypes between two poles: active seniors and dependent seniors. In a simplified way, the commonly shared standard focuses on self-activation and seniors' activation, the younger ones but also the most advantaged (resources, health, social capital) and assistance, and/or solicitude, is directed to the dependent elderly.

11.3.2.2 From the Individual Prevention… to the Duty of Social Utility

The targeting of the senior population by a large number of stakeholders is the subject of more and more solicitations, essentially toward individuals. In addition to recreation, traveling, residences and services, private insurance companies focused on prevention (of dependency, health accidents) and more recently on job loss. Public policies also take part in recommendations dissemination in order to prevent individual aging for the adaptation of the living environment and the preservation of autonomy. The increased commercialization of these services and their potential self-financing draw up a new social configuration: aid and assistance toward the most insecure and frail; allocation of personal income and property assets for one's health and that of one's family members (Pennec 2005).

Moreover, concerning the availability of time from "inactive" older people, the maintenance of their social utility is desired, in formats sometimes codified by professional sectors. Complementing of paid work, volunteering may be subject to a certain professionalization through training and integration in teams, such as palliative care. These volunteer activities are put forward, more often when they are carried out through associations or when they are exerced through engagement (being president or secretary of an association is better seen than only being a member). Informal services within kinship and in the neighborhood or those networks of friendship are not very visible and therefore not enough recognized. Close relationship commitments, for which women's contribution is instrumental, are downplayed when it comes to developing local actions in terms of collective services focusing on mobility, sociabilities, and other support. By contrast, volunteer work from retirees is regularly requested to intervene with older, younger, and poor people. This situation has led several researchers to question the paradoxical nature of the duty of social utility (Gucher 2001) addressed to "inactive"retirees, whose a number of them has previously been submitted to the duty of an early exit from the workforce.

11.3.2.3 Citizenship Confined to Volunteering?

The AFC and "aging well" programs (in its "Living Together" dimension) put forward the desire to develop the participation of older individuals (seniors and older adults) to the definition of urban and territorial projects, and this, in an intersectoral manner that goes against the work in "silos." The exercise of citizenship is also mentioned but remains little defined and will not be the subject of specific action proposals. For all that, volunteers are rarely seen as experts, despite their quantitative and qualitative importance. The extent of their scope and the great number of associations contribute to mitigate or ignore their competencies. However, national efforts are made to federate these activities and to have them taken into account in public deliberations.

The building of a new figure of the "senior citizen" thus seems to be spreading: the "senior participants," actors of an individualized prevention, promise of *successful aging* and volunteer producers in the area of health and social services. "The duty to participate" addressed in "the city policies" to people of working-class neighborhoods (Carrel 2013) comes for seniors in a double or even triple viewpoint: self-care/self-maintenance, care/maintenance of ones' relatives and work for others in the city. Several dimensions are then to be considered, in particular the constraints of this normativity regarding the wide variety of retirement lifestyles, life course, and health trajectories.

11.4 Conclusion

The study of local policies has helped to locate different modalities according to which the French frame of references "Aging Well-Aging Together" and the one from the WHO's "Age-friendly Cities" have been disseminated and applied in urban environments. The objectives targeted by these two programs converge in part. This explains the overlapping of references made to the terminology of the two programs. For urban areas with the AFC label, the achievement of approaches varies when implementing the diagnosis and partnerships as well as in the options selected by action plans. In case 1, the approach produces effects of strengthening of actions fostering on aging integration. Several elements are then often put together: the authority of an elected official whose driving force helps promoting new modes of action, the valuation of "municipal social policy" and a long-term consultation with all involved stakeholders. Actions integrating aging are most often embedded with those concerning disability and sometimes more globally social cohesion. These cities have not always proceeded with the diagnosis step, preferring the results of previous partnerships carried out with service practitioners and their users.

In other urban contexts, incentives to initiate new synergies have been identified. After obtaining the "Aging Well" label, the AFC protocol has been engaged in the case 2 through the diagnosis in order to redefine local policies of aging and to ensure giving new impulse to social actions. At this stage, the process does not always make it possible to bring together the various stakeholders, the significant influence of health and social services model acting as an originally ignored but fully effective disincentive. The partitioning between professional sectors and other local policies (housing, transportation, disability, etc.) must then be pursued by long-term work in order to develop concerted action plans.

In case 3, as other cities have obtained the "Aging Well" label, the AFC protocol is not adopted. The case study does not deepen the compared analysis, between the AFC and non-AFC cities, nevertheless a few distinctive characteristics can be raised. In this city, the assignment of competences of several elected officials are involved and affect the actions in the field of aging, the global steering loses visibility as well as the orientation of the old age policy. The public presentation of the diagnosis phase by experts provides a temporary arena without building a forum for debates and development of projects in the framework of an extended cross-sectoral approach. Maintaining the division between "active" aging, the one regarding seniors, and aging targeted by assistance, maintains a long-standing perception of action, as well as overlaps and partitioning between services and actions.

The AFC's local appropriation contributes to develop a cross-cutting issue likely to establish a collective dynamic between the actors of the old age policies. This dynamic allows to offset or to stand by, areas of actions which "weight heavily" (equipment, urban planning, finance, etc.). The impact of a leadership (elected officials and administrative manager), on local priorities, is also a determinant factor when it comes to the level and duration of collective commitments. The progress of these programs and their achievements illustrate how the objectives can become obstacles, how they can progress, but ultimately introduce consultations and regular readjustments of coordinated actions. In addition, the cities involved in AFCs have strengthened their visibility at the national level, or even international, which induces a greater emphasis on aging policies within local policies. One of the advantages of the approach lies in the extension and the combination of themes by inserting them more closely in all local policies, those environments such as those of social participation.

But a series of challenges remain among which we distinguish between the difficulty of taking into account the diversity of seniors in the approaches studied on the one hand and on the other hand, the actual extent of seniors' power in these processes and beyond.

First: the diversity of seniors. If it is theoretically sought in the AFC approach, in practice this is often some types of seniors whose opinions are taken into account:

rather the younger old people, individuals in good health and with competences (particularly useful for volunteerism) in opposition to populations more fragile and more isolated. The AFC approach does not seem to question these partitions of the aging population.

Second: the power of seniors. The opinion of seniors has certainly been taken into account in some approaches that we observed. Nevertheless, we do not see any mention of these considerations in a more structural logic in the long-term. Moreover, it must be remembered that this “grey power at grassroots level” (Argoud 2012) exists in certain forms such as taking into account retirees’ organizations in a logic of clientelism or as taking into account users in an institution required by law or even through the very few Panels of the Wise put in place in France unlike other countries. However, beyond the “aging questions” research must also take into account social innovations like the “democracy conferences” (*assises de la démocratie*) and the “territories conferences” (*conferences des territoires*) (city 1) or ICT supporting social link (city 3) (Pennec 2012).

In conclusion, the observations made regarding French policies of aging show their focus on the issues of retirement, health, social actions, and recreation... to the exclusion of topics such as the participation in public life, culture, or even the “*Long-Life Learning.*” However, such prospects could better meet the two challenges to date, whether it is the issues of seniors’ diversity or those of the power of being old.

References

Agence Nationale de l’Habitat. (2012). *Prévenir et guérir les difficultés des copropriétés. Une priorité des politiques de l’habitat.* Paris.

Argoud, D. (2012). Le pouvoir gris à l’épreuve du territoire. *Gérontologie et société, 143*, 63–76.

Association des Maires de France. (2009). *Bien Vieillir Vivre ensemble.* Appel d’offre. Paris, 30 juin. Available: http://www.amf.asso.fr/document/fichier.asp?FTP=AMF_20090813_note_LABEL_BIEN_VIEILLIR.pdf&ID_DOC=9640&DOT_N_ID=7. Accessed December 8, 2014.

Berger, M., Rouge, L., Thomann, S., & Thouzellier, C. (2008). *Vieillir en pavillon: mobilités et immobilités des personnes âgées dans l’espace périurbain.* Université Toulouse-le-Mirail LISST-CIEU, Université Paris 1 LADYSS, Université Aix-Marseille 3 CIRTA. Rapport de recherche pour le Plan Urbanisme Construction Architecture.

Boulmier, M. (2012). *Bien vieillir à domicile: enjeux d’habitat, enjeux de territoires.* Paris: La Documentation française.

Carrel, M. (2013). *Faire participer les habitants? Citoyenneté et pouvoir d’agir dans les quartiers populaires.* Lyon: ENS Éditions, coll. «Gouvernement en question(s)».

Chamahian, A. (2009). Vieillissement actif et enjeux de la formation dans le temps de retraite. *Lien social et politiques, 62*, 59–69.

Collombet, C. (2010). «Habitat et urbanisme face au vieillissement». In V. Gimbert & C. Godot (Eds.), *Vivre ensemble plus longtemps.* Paris: Centre d’Analyse Stratégique. *Rapports & documents* (Vol. *28*, pp. 153–187). Paris: La documentation française.

Douguet, F., & Pennec, S. (2002). *Solitude et isolement chez les personnes âgées de 75 ans et plus.* Brest: Sufcep-Université Bretagne Occidentale. Coordination gérontologique Quimper. 115 p.

Erler, G. (2009). Des temps et des lieux pour la rencontre entre les générations: les centres multigénérationnels en Allemagne. *Informations sociales, 153*, 38–44.

Fondation de France. (2007). *Séniors, quelle intégration dans les documents de planification et d'urbanisme?* Paris: Fondation de France. Available: http://www.fnau.org/file/news/SENIORSquelleintegration.pdf. Accessed December 8, 2014.

Formosa, M. (2005). Femininism and critical educational gerontology. *Ageing International, 30* (4), 396–411.

Formosa, M. (2010). Life long Learning in later Life: The Universities of the Third Age. *Learning Institut Review, 5*, 1–12.

Garon, S., Beaulieu, M., Veil, A., Paris, M., & Bigonnesse, C. (2012). L'expérience québécoise du programme «Villes Amies des Aînés» de l'OMS: l'implantation dans 7 projets-pilotes. In J.-P. Viriot-Durandal, C. Pihet, & P.-M. Chapon (Eds.), *Les défis territoriaux face au vieillissement* (pp. 69–88). Paris: La documentation française.

Garon, S., & Veil, A. (2011). Les villes amies des aînées au Québec. Un mouvement de changement à large échelle en faveur des aînés. *Vie et Vieillissement, 9*(1), 6–12.

Gucher (dir.)., C. (2001). *Retraite et citoyenneté.Actualité d'une question paradoxale.* Grenoble: Presses Universitaires.

Guillemard, A.-M. (2002). De la retraite 'mort sociale' à la retraite solidaire. *Gérontologie et société, 102*, 53–66.

Guillemard, A.-M. (2013). Le vieillissement actif: enjeux, obstacles, limites. *Retraite et société, 65*, 17–38.

Hébert, R. (2011). Grandeurs et misères de l'allocation personnalisée d'autonomie en France. *Gérontologie et société, 138*, 67–82.

Lascoumes, P., & Le Gallès, P. (2004). *Gouverner par les instruments*. Paris: Presses de sciences Po.

Le Borgne-Uguen, F., & Pennec, S. (dir.; avec la collab. de Chauvin, K.). (2012). *Quimper: Ville Amie des Aînés-OMS. Diagnostic réalisé auprès des habitants et des professionnels.* Brest: ARS (EA 1749)—Université Bretagne Occidentale. 132 p.

Le Borgne-Uguen, F., & Pennec, S. (2014). *Carnets de terrain et d'entretiens auprès des universités: USP Sao Paulo et Florianópolis,* novembre.

Leenhart, H. (2011). De nouvelles formes d'organisation en Allemagne pour pouvoir vivre dans son quartier, même en cas de démence. *Gérontologie et société, 136*, 205–219.

Ministère des Affaires sociales, de la santé et des Droits des femmes. (2007). *Bien Vieillir—Vivre Ensemble* 2007–2009. Available: http://www.social-sante.gouv.fr/espaces,770/personnes-agees-autonomie,776/dossiers,758/plan-national-bien-vieillir,763/. Accessed December 8, 2014.

Moulaert, T., & Viriot-Durandal, J.-P. (2012). Le vieillissement actif sur la scène internationale: perspectives méthodologiques pour l'étude d'un référentiel polymorphe. *Les politiques sociales, 1–2*, 10–21.

Ogg, J., & Phillipson, C. (2013). L'offre de formation pour les 50 ans et plus dans le contexte britannique: un enjeu pour les universités dans l'avenir. *Gérontologie et société, 147*, 177–188.

Ogg, J., & Renaut, S. (2010). Vieillir chez soi: quels enjeux pour l'avenir?. *Cadr@ge, 11*. Juin. Paris: CNAV. 8 p.

Paris, M., Garon, S., Beaulieu, M., Veil, A., & Bigonnesse, A. (2011). Ville amie des aînés au Québec: l'apport de "vieillir en restant actif". *Vie et vieillissement, 9*(1), 28–32.

Pennec, S. (2005). Le travail filial et la préservation des biens de famille. *Enfances, Familles, Générations, 2*. Available: http://www.erudit.org/revue/efg/2005/v/n2/010915ar.htm. Accessed December 8, 2014.

Pennec, S. (2012). Vieillesse et usages des TIC. Analyse des réalisations et des logiques d'acteurs à l'échelle d'un territoire urbain. In J.-P. Viriot-Durandal, C. Pihet & P.-M. Chapon (Eds.), *Les défis territoriaux face au vieillissement* (pp. 137–156). Paris: La documentation française.

Pennec, S. (2013). Une université aux prises avec les demandes des retraités: entre éducation permanente, diffusion culturelle, prévention. *Gérontologie et Société, 147*, 147–161.

Pennec, S. (2015). La propriété questionnée par le vieillissement. In Denèfle, S. (dir.) *Pour penser des alternatives à la propriété.* Tours: Presses universitaires François Rabelais.

Petitot, C., Beard, J., Kalache, A., Plouffe, L., Cox, J., Powell, S., & Tahra, A.-M. (2010). Vers des environnements-amis des aînés. *Gérontologie et société, 132*, 229–242.

Rosenfelder, C. (2014). Résister par l'habitat. Le cas de l'association Agir Multi-générationnel (A2MG). *Émulations, 13*, 113–121.

Rouyer, A. (coord.) Clément, S., Membrado, M., Salord, T. (LISST), Mantovani, J., & Souchois, C. (INSERM-ORSMIP). (2008). *Droit de cité des personnes âgées et aménagement de l'espace public en ville. Représentations, modalités d'action et expériences urbaines en France, Allemagne et Espagne.* Paris: Ministère du logement et de la ville, Plan Urbanisme Construction Architecture.

Rubião, A. (2008). Une "université participative"? Réflexion à partir du programme "Pôles de citoyenneté". *Mouvements, 55*, 112–121.

Volant, S. (2014). L'offre en établissements d'hébergement pour personnes âgées en 2011. *Études et résultats, 877*. Paris: DREES.

Walker, A. (2009). Commentary: The Emergence and Application of Active Aging in Europe. *Journal of Aging and Social Policy, 21*(1), 75–93.

World Health Organization. (1993). *Health for all*. Copenhagen: WHO.

World Health Organization. (2002). *Active ageing, a policy framework*. Geneva: WHO.

World Health Organization. (2007). *Global age-friendly cities: A guide*. Geneva: WHO.

Chapter 12
A Belgian Case Study: Lack of Age-Friendly Cities and Communities Knowledge and Social Participation Practices in Wallonia

Thibauld Moulaert and Geneviève Houioux

12.1 Introduction

In February 2012, the Social Action, Equal Opportunities and Health Minister of the Walloon Region launched a project for "age-friendly cities, municipalities and region" towards municipalities, Public Social Welfare Centers (*Centres publics d'action sociale,* CPAS[1]) and associations. The call builds on the framework of the World Health Organization (WHO) Age-friendly Cities and Communities (AFCC) model initiated in 2007 (WHO 2007) according to active aging (WHO 2002).

[1]Public social welfare center (CPAS in French, OCMW in Flemish) is the provider of social services entitled to everyone in Belgium. Situated at the level of municipality, it provides social services so that everyone can live with dignity. It investigates what help is most appropriate given the personal or family circumstances and offers the appropriate means to meet people's needs.

T. Moulaert (✉)
REIACTIS, Dijon, France
e-mail: tmoulaert@gmail.com

T. Moulaert
School of Public Health, Université Catholique de Louvain, Louvain-la-Neuve, Belgium

T. Moulaert
Human and Social Sciences Institute, Université de Liège, Liège, Belgium

T. Moulaert
School of Social Work, University of Sherbrooke, Sherbrooke, QC, Canada

G. Houioux
Health Observatory of Hainaut (Observatoire de la Santé du Hainaut), Mons, Belgium
e-mail: genevieve.houioux@hainaut.be

G. Houioux
School of Public Health, Université Libre de Bruxelles, Brussels, Belgium

© Springer International Publishing Switzerland 2016

T. Moulaert and S. Garon (eds.), *Age-Friendly Cities and Communities in International Comparison*, International Perspectives on Aging 14,
DOI 10.1007/978-3-319-24031-2_12

The aim of the chapter is to show how this AFCC framework can serve to local actors to pursue their own objectives, even if they have a little or poor knowledge of what AFCC might produce. Through this objective, we want to answer these research questions: next to “active ageing in employment” (Moulaert 2012a), what kind of public policies can be produced by “active ageing”? How international referential for public policies such as “active ageing,” produced at international level like WHO or European one (Moulaert and Biggs 2013), is used by local stakeholders? Despite the finding and from a qualitative survey of 12 of the 60 cities that received funding from the Minister, this text shows that it remains possible to see signs of ‘social participation’ among seniors.[2] It reflects a certain Walloon attention to the ‘seniors’ social participation,’ to borrow one of the three pillars of the WHO’s definition of “active ageing”: “Active ageing is the process of optimizing opportunities for health, participation and security in order to enhance quality of life as people age” (WHO 2002, p. 12).

This attention is contextualized according to Walloon realities of the “*parole des vieux*” (“older people’s voice”) (Argoud and Puijalon 1999). This understanding arises with the existence of ‘municipality advisory councils of seniors’ (CCCA in French) or may appear as a practical case for some developments of any ‘Social Cohesion Plans’ (*Plans de Cohésion Sociale* or PCS[3]).

12.2 The Development of AFC[4] in Wallonia

For quite some time, the Walloon government has been interested in the effects of an aging population, as evidenced by its public policies and research mandates. While it is mainly directed toward demography or economy, the questions of ‘social

---

[2]For a literature review on social participation, see the work of Emilie Raymond and her colleagues; for example, Raymond et al. (2008).

[3]Since 2009, PCSs are plans cofunded by the Walloon Region (up to 75 %) and by the municipalities (at least 25 %) in order to improve ‘social cohesion,’ as defined and promoted by the Council of Europe since 2000. Supporting local partnership according four different topics (socioprofessional integration; access to a decent housing; access to health; (re)creation of social intergenerational and intercultural links), these actions aim to meet the respect of six rights (under regional competence): the right to a dignified salary; the right to health protection and social/medical care; the right to a decent housing and healthy environment; the right to work; the right to training; and the right to cultural and social growth. Source: Walloon Public Services. (2014). *Portail de la cohésion sociale*. (Online). Available from http://cohesionsociale.wallonie.be. Accessed July 20, 2014.

[4]As researchers, we refer to AFCC; Walloon Region government and local players (at least in 2012–2013) refer to AFC.

participation' and the recognition of seniors as full citizens are more recent. It is within this framework that the appeal of the Minister of Social Action, Equal Opportunities and Health was launched in February 2012. However, before the publication of this call, a tool already existed in the Walloon Region in favor of the social participation of seniors.

12.2.1 The Social Inclusion of Seniors Through the Creation of 'Municipality Advisory Councils of Seniors' (CCCA)

In 2006, under the Contract for the Future for Walloon/nes from 2005 and the Strategic Plan for the 'social inclusion' of seniors, the Minister of Home Affairs and Public Services invited the cities to implement 'municipality advisory councils of seniors' (CCCA). The Minister sought to structure a process that has existed for 20 years throughout Europe (Walker and Naegele 1999) and in Wallonia too, which concerns not only seniors, but all citizens (i.e., youth/children's or disabled people's councils), namely a growing interest for local and participatory democracy (Bacqué et al. 2005; Blondiaux and Fourniaux 2011).

If such councils were put in place during the 1990s, this legislation was intended to regulate processes that were specifically addressed to seniors (CCCA). It specifies objectives, mandate and responsibilities, without overlooking organizational arrangements, such as members age limit (i.e., a minimum age of 60 in 2006, 55 in 2012) or the possibility to use a person-resource, an expert in pension or in health promotion for example. According to Casman and Chaoui-Mezabi (2011, p. 9),

> the 2006 policy has reduced structural blur that characterized the CCCA and was joined by the Declaration of regional Walloon policy (2009-2014), which gives as priority inter alia to ensure a real place to elders through activation of their citizenship. Thus, it advocates the need to assess '*arrangements in place, define tags guaranteeing the proper functioning of the municipality advisory councils of seniors and their participation in the various municipal or para-municipal structures*'.[5]

Despite the 2006 policy, Casman and Chaoui-Mezabi (2011) observe a strong heterogeneity of practices of the CCCA from one municipality to another. Their study—the only one to date in the Walloon Region on the CCCA[6]—points to a

[5]All translations from French-speaking sources are from the authors.

[6]Their study was financed by the "Coordination des Associations Seniors" (Coordination of Seniors Associations), which is the meeting of all most representative associations of seniors in Wallonia. This association also supports a similar study on the Region of Brussels (Carbonnelle 2013).

series of dysfunctions in the CCCA's practices. They also made a series of proposals to amend their functioning. The minister in charge of local governance updated the 2006 policy in 2012 (improving the governance of the Council by clarifying the status of presidency; reducing the access age from 60 to 55, etc.) with the support of the Seniors Association's Coordination (CAS), sponsor of the study. Consequently, we see not only the Walloon government's interest in defining the place of seniors in local governance, but also an awareness of the forces and limits of such councils—political manipulation is a potential example. It is within this perspective that a specific call for projects appeared in early 2012.

12.2.2 The "Age-Friendly Cities, Municipalities and Regions" Governmental Call and the WHO Framework

The context of the regional Minister's call for projects was the "European Year of Active Ageing and Solidarity between Generations" in 2012. The AFC-Walloon program intended to register this European context of "active ageing." In reality, it is more precisely the framework of the WHO to which the call for proposals referred when its mission was to implement the first axis of the "Walloon Government's Orientation Note" from April 29, 2010. Indeed, the announcement of the call explicitly referred to the WHO vocabulary when it stated that

> Wallonia aims to support local authorities wishing to implement or to improve a policy encouraging active ageing by optimizing health, participation and security of older citizens, to improve their quality of life but also recognize seniors as a resource essential for their family, their community and the economy (Minister of Social Action, Equal Opportunities and Health, 2012, Call for projects, "Age-friendly Cities, Municipalities and Regions," February 3, p. 1).

The "objectives and methodology" refer to a methodological note annexed to the call. It mentions the WHO Guide (2007) as reference and also takes clear inspiration from the four elements of the approach of the 2010 WHO "Global Network of Age-friendly Cities and Communities" (GNAFCC). However, this comes with a serious limit, which is the timing of the actions supported by the call. While the WHO program presents a 5-year cycle, including 1 or 2 year(s) for consultations, the Walloon process can only take place in 1 year.

In the Walloon Region, the WHO's AFCC model had not found practical experience or research translation *until that time*. It sought to determine whether they proved to be crucial. Further developed in the results, needs and resources assessments were conducted in a relatively disparate manner from one municipality to another, due to a lack of time, experience and interest. Nevertheless, the invitation to mobilize the CCCA to do it was logically quite well followed, particularly in the 12 studied cases.

While the timing of the call might have been a strategic element for the minister's own campaign for further elections in 2014, at the municipality level, the timing for projects themselves (i.e., the submission of the projects was extremely short and included only one year funding, along with local elections during the period) was less optimal. Once the elections finished (i.e., after October 2012), the funding gave way to the significant impulse to register the project in 6-year terms. Conversely, some examples suggest that this agenda was indeed not optimal for all cities, particularly when the majority changed; because the CCCA has connections to the political local majority, this may raise some contradictions. It points out the *problem* between a project (potentially linked with research) agenda and political agenda (for similar discussion about the 'sustainability' dimension of AFCC, see the Australian chapter by Brasher and Winterton or the New York City chapter by Goldman et al.).

12.3 "For and by Seniors:" A Qualitative Study of 12 Walloon Municipalities

While we previously work on "active ageing in employment" (Moulaert 2012a) and on the production of "active ageing" at international level (Moulaert and Biggs 2013), this chapter aims to answer these questions: outside the main focus on 'work and employment,' what kind of public policies can be produced by "active ageing"? How international referential for public policies such as "active ageing," produced at international level like WHO or Europe (Moulaert and Biggs 2013) is used by local stakeholders?

Due to the quasi-absence of Belgian French-speaking studies on social participation and through the opportunity[7] to observe the early development of the AFCC approach in Wallonia, a qualitative exploratory study design was selected to meet the research goals. Diversity of cases was obtained via three criteria: size of municipalities, urban/rural character, and type of project (project submitted by the municipality, the social center (CPAS), the CCCA, or a mix of them). The cases were selected in each of the five Walloon provinces. Where possible, the projects that were evaluated positively or very positively by the selection committee (60 projects out of 172) were preferred. The selection was discussed and validated with the Cabinet of the Minister of Health and with the Ministry of Health (Table 12.1).

Within the 12 selected cities, three types of actors were chosen: elected politicians (the mayor, the deputy mayor in charge of aging or seniors/elderly, or the president or secretary of the CPAS), administrative staff from the CPAS or from the municipality (some being also members or managers/coordinators of a local social cohesion plan) and seniors (usually members or the president of the local CCCA). These types of actors have been identified as they generally appear in AFCC

[7]The first author of this chapter has been designated as expert to participate in the evaluation jury of the projects funded by the call of the minister.

Table 12.1 Selection of cities

Province	City	Degree of urbanization[a]	Main promoter of the project	Funding allowed, in €
Brabant Wallon	City A	D3	Municipality (Mun)	32,250
	City B	B2	Mun and CPAS	53,800
Hainaut[b]	City C	D3	Mun	40,000
	City D	B2	Mun	71,000
	City E	B3	CPAS	43,000
	City F	A	CPAS	61,500
Liège	City G	C3	CCCA and Mun	7750
	City H	C2	Mun	59,050
Luxembourg	City I	D3	CCCA	26,000
	City J	D1	Mun	50,450
Namur	City K	D2	Mun and CCCA	68,000
	City L	A	CPAS	80,000

[a]*Source* Federal Public Services Economy, Administrative Geography, from A for the biggest cities to D3 for the smallest villages
[b]The number of cities studied in Hainaut is due to the fact that we have benefitted from the voluntary collaboration of Geneviève Houioux from the Health Observatory of Hainaut

programs such as in the Quebec case (Garon et al. 2014). 47 individuals were interviewed in time spans that ranged from 20 to 120 min. Analysis of the interviews was conducted following a thematic analysis that referred to the AFCC and participatory democracy; selection of pertinent material was operated through systematic comparison of identified parts of interviews by both authors (Table 12.2).

The interviews were conducted individually or collectively at the discretion of the interviewees; the fact that they were conducted in groups brought up relevant information, and means, for example, that collaboration and trust were already instilled. On the other hand, a simultaneous conversation between an administrative civil servant and the newly elected president of the CCCA was—rather—an attempt to construct common relationships.

12.4 Results

Empirical observations relate to three elements: first, the profile of the respondents, second, the needs assessment of seniors and resources at their disposal; and, third, the central role of the administrative civil servants in the senior's social participation. These will help to answer the research questions, i.e., to understand how local stakeholders, even with little knowledge of the origins of 'active ageing' and of AFCC, re-used the AFCC framework to meet their own understanding of senior's social participation.

Table 12.2 Distribution of actors by cities

	Elected politician	Senior	Administrative staff of the city	Administrative staff of CPAS	Total
City A	1	1	1		3
City B	1	1	2*	2*	6
City C	1	1	2		4
City D	1	3*	1		5
City E	1	1		2	4
City F	1			2*	3
City G	1	1	1		4
City H	1	1	1		3
City I	1	1	2		4
City J	1	1	3*		5
City K	1	1*	1*		4
		1			
City L	1	1		1	3
Total	**12**	**14**	**14**	7	**47**

*Asterisks represent interviews that were conducted collectively; for example, in City B we met with the administrative civil servants of the municipality and the CPAS at the same time

12.4.1 Profiles of Respondents

Generally speaking, the respondents profile is quite different among the three categories (politicians, civil servants and seniors). It suggests that, behind the relationship between types of actors, there might be gender or age relations (or, more accurately, positioning in the life course) that come into play and can lead to misunderstandings.

12.4.1.1 The Elected Politicians

While there are the same number of men and female, their age ranged from 39 to 71 years. In some cases, there may be, therefore, identification with the target public, the seniors. The diversity is great, in terms of education, professional career, and current position (some are retired; others work part time; the rarest segment is still fully active).

We have identified a continuum of profiles of elected officials between two poles: on one hand, the initiator and, on the other hand, the consumer. In the call for projects, the first might see an opportunity to "manage" aging at the local level and develop new directions. On the other hand, he/she might see the call as an instrument at the service of his/her party against other party or political opponents. Such a continuum is also geared with the skills and personal capacities, relations, legitimation (is he/she re-elected?) of politicians. According to their position,they should also consider if such project will strengthen their position at local level, with the argument that they have been able to collect additional funding for municipal action. While funding has been central concern, an elected representative of City D noted the harshness of the negotiations of the municipality's budget, referring to the image of "*the balance of pharmacist.*" Being able to see an opportunity in AFCC might influence the sense of the balance.

12.4.1.2 The Administrative Civil Servants

While gender is not discriminant for the elected politicians in the sample, it is the opposite for administrative civil servants. The 21 interviewees only included one man.

The ages of the participants ranged from 24 to 55 years. Compared to the elected politicians, here there is not only younger people who were sometimes hired or promoted by politicians for whom they indirectly work, but there also is a more condensed range of age. When they are connected with the seniors of the city, their age may have an influence or, more accurately, their position in the life trajectory: younger workers sometime consider themselves to be facing "their" own seniors as an equivalent of "their" own family members.

Some administrative staffs who have either experience in local administration or prior experience can take a central position in the projects. They became operational link between the different actors with a key coordination role. In City J, an administrative staff manager defined her position during the first minutes of the interview ("*I am the liaison between the different teams, such as the PCS and the city council*"); as the interview continued, the centrality of her action for the maintenance of the project, throughout a change of elected politicians, became apparent.

As for the elected politicians, one may observe a continuum of profiles between two poles among administrative staff members: on one side, the network and initiator/facilitator agent and, on the other side, the executing agent. According to his (or, in the view of the sample, "her") personal and professional trajectory, the first type will take an active part in the project, while the second will be seen more like a follower. It appears that the first ones were apparently making more sense of the project with their own past and future. While the second pole might only consider the AFCC funding as a form of financial support ("*it is cash money*"), the first pole will indeed not deny the fact as a motivation, but it will be articulated with other preferences for action. In the end, the persons in the first role might consider themselves as the person who initiated the project or who "holds" it while, on the opposite, they will present themselves as a subcontractor of a superior (the elected politician or a member of the administration).

12.4.1.3 The Representatives of Seniors

As mentioned earlier in the chapter, access to CCCA has been reduced from 60 years in 2006 to 55 years in 2012. A similar trend is found toward more preventive actions in the field of employment policies where the age for some political measures has been lowered to 45 years (Moulaert 2012a). In the sample, most of the respondents were between 60 and 70 years old. Beyond this majority, 4 people between 75 to 85 years of age were profiled as *hyperactive* people who were ready to run out to the task, due to lack of future resources to replace them. While they may illustrate the "active" side of retirement, up to the limit and paroxysm (Katz 2000), they are often assimilated by administrative staff members as the "*active senior citizens*" for whom a city organizes "*full activities, holiday trips, dancing balls*" in contrast with the "*people who don't leave home, people who should not be disturbed within their daily routines.*" Certainly, the issue of isolation is recursive through the investigation and would rather affect the oldest old. But, in this case, how does one interpret the presence of people over the age of 75 in these activities? Are they exceptions? Or are they the first witnesses of deeper changes? While the danger of focusing only on "young seniors" exists within the "active ageing" concept (Boudiny 2013, Moulaert and Biggs 2013), it also makes empirical sense. The danger is that these groups of "active seniors" who are either young or old would create new intergenerational exclusions in which the generation of the younger literally chases that of their elders. In one of the cities that were observed, this phenomenon even happened as a political choice won by a different majority during the change of the forming of the CCCA. On the contrary, some CCCAs are aware of these potential tensions and build projects to reinforce solidarity between different seniors generations.

In the end, one can also draw a continuum around two poles of seniors: the involved/actor seniors versus the participant/consumer seniors. This can be illustrated by the role played and assumed by the CCCA, such a role being summarized between being the voices of seniors and being an activities provider like 'third-age clubs.' While often blurred, these two different sites of representations can be articulated. While the CCCA is officially a space for reflection, combined with actions, seniors associations that are embedded in their neighborhoods and take an active role in CCCA also represent places of activities. Even if they organize apparently very 'passive' activities like playing cards,[8] they might yet be sensors to feed the CCCA. In City I, local associations gathered in a "coordination of associations" and were simultaneously associated with a CCCA.

> The CCCA is more a reflection board issuing advice while the coordination aims at implementing. It is a bit like the parliament and the government, even the small hands of the government. For example, the principle of an activity is brought by the CCCA, but once we decide who goes to the bar, who does what, it goes to the coordination (Senior representative, City I).

---

[8]For a critical perspective supporting the importance of seeing active ageing in everyday life, see the work of Aske J. Lassen through the example of seniors playing the billiard (Lassen 2014).

While an individual who characterizes the first type of senior runs the risk of faltering if he/she does not feel heard by the city (seniors from City B explicitly mentioned it and it has been heard elsewhere, too), the latter runs the risk of failing to bring reflection to their local action. The first will not hesitate to be critical of the political actor, usually in a constructive way; the latter will have a greater tendency to be satisfied with what is put in place for them. The first will take an active role in the initiative of the project (as in City G), while the latter limit themselves to validating the program (some seniors from City H).

Eventually, the diversity of trajectory reveals a variety of knowledge and consideration about age and aging. Nevertheless, an important finding of the study is the poor knowledge of the AFCC perspective. If the Minister of Health's call adequately mentions the eight WHO's dimensions (2007) and emphasizes the role of the CCCA in the realization of the consultation of the needs and resources for seniors, only 3 people out of the 47 interviewed mentioned these areas and explicitly knew that there was some link to the WHO, many of them considering such framework made for "big cities."

12.4.2 The Needs and Resources Assessment as a Critical Element

While central since the original framework from WHO (2007) and systematically mentioned in international local practices as explored in the various chapters of this book, the "needs and resources assessment" in the early stages is undoubtedly one of the most critical points of the AFC-Wallonia. These assessments, which are quite disparate from one city to another, reveal the current lack of a standardized method. It clearly shows that such practice of assessment is still far from a series of actors identified in previous section.

12.4.2.1 Timing

The liminal question of the duration of the needs/resources assessments is central. Researchers in Quebec (Garon and Veil 2011) drew attention to this point: the start-up of an AFCC approach requires time. The participatory diagnosis performed in some case studies in Wallonia (Houioux et al. 2014) or other experiences of *peer research* as developed by the Vrij Universiteit van Brussel (Buffel et al. 2014) require at least a year. In practice, it might differ from this.

> It was although good to say that we started from an enquiry to answer to the call. {…} But be careful, here [the politicians] are not usually in favor of investigations. Do not create needs to which we cannot respond (Administrative staff PCS, City J).

12.4.2.2 Methods

Second, "needs and resources assessment" implies a systematic method. Admittedly, several PCS are familiar and have already conducted surveys similar to a "needs assessment." An administrative staff member said that she initiated a telephone survey. Elsewhere, one asked the population service to collect data's, another one spoke of a "satisfaction survey." In most of these cases, questionnaires were administered by mail or phone to all seniors that were 60 or 65-year old (or older) with no standard limit. However, no standardized method has been identified. One may question the value of these 'surveys' that are based on four or five questions, then 'analyzed' or 'updated in a graph' by individuals with poor or little statistical/research knowledge who evaluate the 'satisfaction' of seniors who are consequently perceived as 'users' or even 'consumers.'

Beyond the problematic relationship to 'investigations,' the access to the knowledge of social reality merits a special attention. Never in the 47 discussions has the possibility of collaboration with researchers been reported. Here the purpose is not to criticize local actors but, rather, to investigate the reasons that prevent them from having a research reflex *or* a rigorous methodological approach reflex to meet scientific standards. We can certainly discuss 'lack of time,' field's constraints or uncertain financial resources. Nevertheless, as researchers, one should wonder why such relations between the researchers and the local stakeholders appear to be underdeveloped and explore avenues that would benefit to both research and local actors.

12.4.2.3 Knowledge from the Field

The question of the existence/absence of a needs and resources assessments (when they were not purely and simply confused) has very frequently been resolved through the "knowledge from the field" or through the "personal experience."

> As [an] actor in the field, you already have a good broad knowledge. {…} It is also based on findings from practice. By working on the ground, I saw more and more old people with different requirements. And [I observed] that a lot of people want to come closer downtown here, where you have car parks, apartments, a shopping center, a hospital, banks…we also have such observations from home carers (Administrative staff of CPAS, City B).
>
> What has always interested me is the 3rd and 4th age. I did all of my [political] campaign on it. And according to my life experience [as a 60-year-old man] and professional [a GP]… due to my family and my wife's family, we are quite known in the region… (Elected politician, City K).

In a similar way, the seniors' opinion may be requested via a lighter consultation of CCCA. Here, the recurrent criticism is that its members "*only represent themselves.*" However, rigorous methods supporting a standardized perspective could be mobilized by the actors, as has been demonstrated by international research (Buffel et al. 2014; Garon et al. 2014) and as it will be explored in conclusion.

12.4.3 The 'Senior's Social Participation' in Practice and the Centrality of Administrative Staff as 'Intermediaries of Active Ageing'

If the senior's social participation is central in the AFCC approach, it is evident that what happened up to now in Walloon Region is less likely to be a standardized consultation process, but more an institutionalized form of recognition of "older people's voice" (Argoud and Puijalon 1999) through the CCCA. Beyond the recognition of CCCA by the local authorities,[9] what is observed here is the key role played by administrative staff in organizing such CCCA. These members of the administration, when involved in the process, illustrate the hypothesis of "active ageing intermediaries" (Moulaert 2012b), that is, actors who are concretely operationalizing a theoretical model of "active ageing." For example, the already quoted Administrative manager of City J illustrates such a role: "*I am the liaison between the different teams, such as the PCS and the communal college.*" As the interview continues, her key role in maintaining coherence to the project while new politicians have been elected was highlighted. This role played by administrative staff members is not that far from the role played by municipal actors from the steering committee of the AFCC Quebec experiences (Garon et al. 2014).

The role of 'intermediary of active ageing' is particularly significant in the presence of involved/actor seniors. They will be mobilized at different stages during the project (at the writing of the project; during the setting of benches; when the municipality decides to edit a "senior guide," etc.). When playing a role of intermediation, the agents as facilitators carefully link all of the players with the seniors on one side and politicians on the other. Drawing on their professional or personal experience, they seek to operationalize the public action desired by the first and decided by the latter. This situation emerges particularly well during the installation of benches at the request of older people.

In the different situations encountered, the choice of benches is an exemplary case. At the start of the call, when the demand for benches is to be found in the draft proposition, it is administrative staff members who identify the different types of benches. Later in the project, the negotiation of location of benches is done in collaboration with seniors, based on their lived experience of territorial space and neighborhood. However, tensions may arise. In a rural municipality, when it comes to the final choice, the head of the PCS agrees to question, and finally follow, the opinion of seniors.

> Reflection is done with the seniors, but the bulk of the work is for the administration. For example, one should always accompany the benches. For the benches, one should start with a state of the art [of available types of benches]. We must, therefore, accompany seniors. {…}We bring the material [and] prepare it, but the reflexion is for them. And the decision,

[9]Because the recognition of CCCA is optional at the municipality level, only half of Walloon municipalities have such councils (Casman and Chaoui-Mezabi 2011).

it is them who take it. The driver is still us. It is not them who will set a date for a meeting, etc. (Administrative staff PCS, City I).

For the benches, based on the budget, I took the technical details. And I suggested it to them. But it is they who decide what we do, where we do. In short, we keep control because there are administrative rules and because you need the collaboration of the Department of Work, etc. We manage it, but we give them the timing. I integrate them into the majority of the project so that they understand why we may sometimes accelerate [and] why we must sometimes wait. {…} For the choice of the benches, I personally wanted to put [in] a 1-person bench; this is what I identified. But in discussing with the seniors working group, they preferred a friendlier, rounded bench. It was a matter of [being] more affordable, but [it] can sit more people. And really, I had not seen it, that was the price and the usability for several people. Then, for elsewhere, [they wanted] classic benches; they asked that there be a seat back with a comfortable and strong wood seat. While on the central place, [since] it comes with an urban revitalization project, they agreed on a metal bench. It was really their choice, their application. Because I would never have thought about it. And so, in the tracking they do, they know what furniture will be at each location (Administrative manager PCS, City I).

More than an illustration of the *bottom-up* argument, these quotes show how the administrative manager builds a social participation approach by combining a bottom-up perspective with a top-down approach. Such a dynamic process also suggests, as illustrated in the second quote, that administrative staff in charge of the project might also transform and adapt their own vision in connecting a vision (given here by the call of the Minister of Health in direction of AFC) and everyday life practice. In doing so, they confirm the need not only for more empirical research of "active ageing" (Lassen 2014) but also for connecting such perspectives with theoretical discussions of the notion (for further proposition in that direction, see the chapter of Moulaert, Boudiny and Paris in this book).

12.5 Conclusion: Toward a New Age-Friendly Cities and Communities Agenda in Wallonia

This chapter aims to better understand how the AFCC framework has served to local actors in Walloon Region to pursue their own objectives. The contribution shows that field workers and politicians have little or poor knowledge of what AFCC might produce and demand, the relative practice of 'needs and resources assessment' being a strong example of actual challenges. While this chapter has been based on a qualitative study of 12 cities that received a subsidy in 2012, further development in Wallonia might need a more structured framework. This is summarized in the following question. Under what conditions can we consider *today and tomorrow* a realistic approach for AFCC in Wallonia?

Two basic conditions emerge from the literature and international practices: the first concerns the place of research, while the second pertains to the inclusion of existing approaches in a situated context.

The first condition calls for the presence of a research component within the program; absent in Wallonia in 2012, it is a key element of many AFCC experiences

—either in Belgium (the city of Brussels entered the AFCC in 2009, with the support and research conducted by the team of Professor Verté of the VUB) or abroad, as illustrated in this book. Nevertheless, the presence of researchers is not a sufficient condition. Elsewhere (Moulaert and Garon 2015), we further defend the hypothesis that such researchers should not only be able to position themselves on academic issues, but they should also articulate knowledge *and* action (rather than valuing the first at the expense of the second).

This research might also focus on communication issues which, so far, are an undeveloped theme in the field of the AFCC. Indeed, communication is one of the eight themes explored by the WHO. Indeed, it has been put forward as a concrete local tool in the fight against the stereotypes associated with age, particularly in Quebec (Garon et al. 2014). But one should better discuss the idea that one of the strengths of the "age-friendly" approach is its ability to "brand" local policies. When communication becomes a tool of legitimation, a more detailed study deserves to be developed from the lessons of the sociology of innovation on the importance of the spokesman (Akrich et al. 1988) or from a "*sociologie du packaging*" (Cochoy 2002). Applications of the diffusion of innovation in social work theory (Dearing 2009) that stress the importance of the visibility of initiatives, at times with different audiences, might also generate great support.

The second condition recalls the need to (re)start from existing situations rather than wanting to implement a program *ex nihilo*. Clearly, it is not the idea of taking a model (such as the WHO model) and applying it without reflection in a local territory such as Wallonia. To monitor such a transfer from international space of production of knowledge such as WHO or European level (Moulaert and Biggs 2013), social sciences research might be appropriate.

In Wallonia, the existence of the PCS (for composing the territorial diagnosis) of the CCCA (to ensure the legitimacy of the seniors' voice) and the potential role of administrative staff members as "intermediaries," could be mobilized. Coming to an end, we suggest making one step further by bridging such existing practices—with their potentials and limits presented in the chapter. We present "participatory diagnosis" (Houioux et al. 2014) as a potential inspiration to meet the challenges of AFCC in Wallonia.

In response to the call of several CCCAs from the province of Hainaut, South West of Belgium, through municipal civil servants or elected officials who wish to understand the needs of seniors in their municipalities, the Health Observatory of Hainaut (*Observatoire de la Santé du Hainaut, OSH*) developed a participatory approach of collecting data (Houioux et al. 2014). This position considers older people as actors of their health, well-being and quality of life and finds its origin beyond AFCC, that is within Ottawa Charter (WHO 1986) for health promotion and its values of community participation.

The first steps of "participatory diagnosis" were in place 5 years ago. If these requests were initially made by quantitative surveys mailed to seniors in a city, then the opportunity to operationalize the principles of citizen participation for seniors quickly appeared. Indeed, senior citizens wished that their voices were present in the CCCA. They are able to offer time as well as their capacity to carry out this type of investigation. The procedure may vary from one city to another, but it is based on semi-directed interviews in accord with a grid of a dozen open questions that relate to the well-being and quality of life, the satisfaction of needs, the knowledge of services and suggestions. Content and procedures are adapted with the members of the CCCA to better match the realities of the seniors from the municipality. The formulation of hypotheses, the selection of the information to be collected, the drafting of the harvesting tool data, the recruitment of the sample, interviews with seniors at home, the entry of data, analysis, the drafting of a report and the development of recommendations to be transmitted to the local authorities are *all* led by members of the CCCA with the support of the OSH. The survey protocol is designed and accomplished from the beginning to the end by the seniors of the CCCA with the help of the Observatory. Only the sampling based on the structure of the population is the responsibility of the latter. The CCCA, which is committed to the approach, is "coached" for a 6-week period by a health promotion professional. The process extends over a year. At the same time, identical content surveys are conducted among professionals and the city's services for seniors.

"Participatory diagnosis" are not the ultimate answer to the various challenges faced by AFCC in Wallonia. Nevertheless, it represents an empirically based strategy to better taking into account the 'needs and resources' for seniors at a local level and a structured scientific method that has been missing until now in AFCC in Wallonia.

References

Akrich, M., Callon, M., & Latour, B. (1988). À quoi tient le succès des innovations? 2: Le choix des porte-parole. *Gérer et Comprendre, Annales Des Mines, 12*, 14–29.

Argoud, D., & Puijalon, B. (1999). *La parole des vieux. Enjeux, analyse, pratiques*. Paris: Dunod.

Bacqué, M.-H., Rey, H., & Sintomer, Y. (2005). La démocratie participative, un nouveau paradigme pour l'action publique ? In M.-H. Bacqué, H. Rey, & Y. Sintomer (Eds.), *Gestion de proximité et démocratie participative. Une perspective comparative* (pp. 9–46). Paris: La Découverte.

Blondiaux, L., & Fourniau, J.-M. (2011). Un bilan des recherches sur la participation du public en démocratie : beaucoup de bruit pour rien ? *Participation, 2011/1*(1), 8–35.

Boudiny, K. (2013). "Active ageing": From empty rhetoric to effective policy tool. *Ageing and Society, 33*(6), 1077–1098.

Buffel, T., McGarry, P., Phillipson, C., De Donder, L., Dury, S., De Witte, N., & Verté, D. (2014). Developing age-friendly cities: Case studies from brussels and manchester and implications for policy and practice. *Journal of Aging and Social Policy, 26*, 52–72.

Carbonnelle, S. (2013). *Enquête sur les Conseils Consultatifs Communaux des Aînés (CCCA) en Région de Bruxelles-Capitale réalisée par la Coordination des Associations de Seniors (CAS). Présentation des résultats et analyse*. Bruxelles: CDCS.

Casman, M.-T., & Chaoui-Mezabi, D. (2011). *Rapport d'étude des Conseils Consultatifs Communaux des Aînés (CCCA) en Région Wallonne*. Université de Liège.

Cochoy, F. (2002). *Sociologie du packaging*. Paris: Presses Universitaires de France.

Dearing, J. W. (2009). Applying Diffusion of Innovation Theory to Intervention Development. *Research on Social Work Practice, 19*(5), 503–518.

Garon, S., Paris, M., Beaulieu, M., Veil, A., & Laliberté, A. (2014). Collaborative partnership in age-friendly cities: Two case studies from Quebec, Canada. *Journal of Aging and Social Policy, 26*(1–2), 73–87.

Garon, S., & Veil, A. (2011). Les Villes amies des aînés au Québec : un mouvement de changement à large échelle en faveur des aînés. *Vie et Vieillissement, 9*(1), 6–12.

Houioux, G., Vanlierde, A., & Storme, S. (2014). La participation n'a pas d'âge. Accompagnement d'un diagnostic de la qualité de vie et de la santé des aînés par un Conseil consultatif communal des Aînés. *Éducation Santé, 299*, 7–11.

Katz, S. (2000). Busy bodies: Activity, aging, and the management of everyday life. *Journal of Aging Studies, 14*(2), 135–152.

Lassen, A. J. (2014). Keeping disease at arm's length—how older Danish people distance disease through active ageing. *Ageing & Society, FirstView*, 1–20. (Online). Available from http://doi.org/10.1017/S0144686X14000245. Accessed October 20, 2014.

Moulaert, T. (2012a). *Gouverner les fins de carrière à distance. Outplacement et vieillissement actif en emploi*. Bruxelles: Peter Lang.

Moulaert, T. (2012b). Pourquoi les francophones préfèrent-ils la sociologie du vieillissement à la gérontologie critique ? *Gérontologie et Société, 142*(3), 81–99.

Moulaert, T., & Biggs, S. (2013). International and European policy on work and retirement: Reinventing critical perspectives on active ageing and mature subjectivity. *Human Relations, 66*(1), 23–43.

Moulaert, T., & Garon, S. (2015). Researchers behind policy development: Comparing "Age-Friendly Cities" models in Quebec and Wallonia. *Journal of Social Work and Practice, 29*(1), 23–35.

Raymond, É., Gagné, D., Sévigny, A., & Tourigny, A. (2008). *La participation sociale des aînés dans une perspective de vieillissement en santé. Réflexion critique appuyée sur une analyse documentaire*. Québec: Direction de santé publique de l'Agence de la santé et des services sociaux de la Capitale-Nationale, Institut national de santé publique du Québec, Centre d'excellence sur le vieillissement de Québec et Institut sur le vieillissement et la participation sociale des aînés de l'Université Laval.

Walker, A., & Naegele, G. (1999). *The politics of old age in Europe*. Buckingham, England: Open University Press.

WHO. (1986). *Ottawa Charter for Health Promotion*. WHO.

WHO. (2002). *Active ageing. A policy framework World Health Organization to the Second United Nations World Assembly on Ageing*. Geneva: WHO.

WHO. (2007). *Global age-friendly cities: A guide*. Geneva: WHO.

Chapter 13
Whose Responsibility? Challenges to Creating an Age-Friendly Victoria in the Wider Australian Policy Context

Kathleen Brasher and Rachel Winterton

13.1 Introduction

Utilizing one Australian state (Victoria) as a case study, this chapter explores how the implementation of the World Health Organization (WHO) Global Age Friendly Cities and Communities (AFCC) model can be impacted negatively by a lack of vision for AFCC within government policy. Over recent years, the WHO Global Network of Age-Friendly Cities and Communities has been gaining traction across international contexts. The AFCC model engages cities to be more age-friendly through promoting the establishment of environments to enable active ageing by optimizing older adults' opportunities for health, participation, security and life-long learning (World Health Organization 2007). In Australia, the provision of age-friendly cities is a critical concern, given its high levels of urbanization and growing rates of population ageing (Davis and Bartlett 2008). However, while two Australian cities participated in the research guiding the development of the WHO AFCC Guide (World Health Organization 2007), the uptake of the AFCC model nationwide has been relatively stagnant (Ozanne et al. 2013). This is particularly the case in Victoria, despite the initial provision of $1.4 million in funding from state government to lead the implementation of AFCC within Victorian local government areas (Municipal Association of Victoria 2009).

This chapter argues that the limited uptake of AFCC in Victoria can be attributed to the inability of various levels of government to foster coherent policies

K. Brasher (✉)
This day & age, PO Box 4034, Box Hill South, VIC 3128, Australia
e-mail: kathleen@thisdayandage.com.au

R. Winterton
Research Fellow, John Richards Initiative, Australian Institute for Primary Care and Ageing, La Trobe University, PO Box 821, Wodonga, VIC 3689, Australia
e-mail: R.Winterton@latrobe.edu.au

© Springer International Publishing Switzerland 2016

T. Moulaert and S. Garon (eds.), *Age-Friendly Cities and Communities in International Comparison*, International Perspectives on Aging 14,
DOI 10.1007/978-3-319-24031-2_13

supporting the implementation of AFCC. A key element of the WHO age-friendly framework is its bottom-up, participatory approach, where older people's own experiences provide a basis for government policies (World Health Organization 2007). This is contrasted with a 'top-down' approach, where local authorities or governments take a significant role in determining guidelines for the age-friendliness of a community (Lui et al. 2009). However, many models indicate that the intersection of both top-down and bottom-up input is crucial in developing age-friendly settings (Lui et al. 2009; Menec et al. 2013). Thus, this chapter draws on Australian and Victorian policy, practice and academic literature to critically assess the factors that have inhibited the implementation of the AFCC model within Victoria, and to demonstrate how a lack of vision and coherence within top-down approaches can inhibit the development of 'bottom up', localized age-friendly initiatives.

13.2 Background

13.2.1 The Context

Victoria is Australia's most southern mainland state, with a population of approximately 5.3 million (Australian Bureau of Statistics 2013). While it is the smallest mainland state, it is also the second most populous (Australian Government 2013) and has the highest population density among the Australian states at 25 persons per km^2 (Australian Bureau of Statistics 2014a). However, it is also the most urbanized, with the capital city, Melbourne, home to more than 4 million people (Australian Government 2013). Thus, while Greater Melbourne comprised 76 % of Victoria's overall population in 2012 with a population density of approximately 430 people per km^2 (Australian Bureau of Statistics 2014a), this is mediated by the sparseness of Victoria's rural and regional locations. Some rural and remote regions in the state's west and north-west agricultural regions have population densities of less than 0.5 persons per km^2 (Australian Bureau of Statistics 2014b).

Within this demographic and geographic context, Victoria is experiencing population ageing consistent with both international and national trends. In 2011, 14.2 % of Victorians were aged 65 years and over, consistent with the proportion of Australians in this age cohort (14 %). However, the state's population is also ageing, with its percentage of people aged 65 years or above increasing from 13.6 % in 2006 (Australian Bureau of Statistics 2013). Again, there is diversity between Victoria's urban and rural regions in this regard. While 13 % of the Greater Melbourne population was aged over 65 years in 2011, this figure was higher outside the urban centre, with 17 % of the population in Victoria's rural and regional areas aged 65 years and above. In a number of Victorian rural communities, approximately one-third of the population were over 65 years of age (Australian Bureau of Statistics 2012).

Further, this population ageing is predicted to increase, with the aged 65 and over population in Victoria predicted to triple by 2051 as a result of the current size of the 45–65 cohort and increasing life expectancies. As a proportion of the general population, a recent state government report indicated that by 2051, the percentage of Victorians aged 65 and over will rise to 21 % (Victorian Government 2014). Thus, with older people increasing in number and comprising a greater percentage of the population, the need to cater more strategically for an ageing population is paramount. From the perspective of planning and service provision for ageing populations in Victoria, local government has a strong remit (Australian Local Government Association 2006). In respect to local governance, Victoria is segmented into 79 municipal districts often referred to as Local Government Areas (LGAs) or councils which are separate Organizations established under the Local Government Act 1989. Demographically and geographically they are varied, with populations ranging from 3200 to over 253,000 people, and land area fluctuating from 8 to 22,000 km^2 (Municipal Association of Victoria 2011).

13.2.2 The Initiative

The early impetus for AFCC in Victoria was provided in 2004 with the release of the *Positive Ageing: A Strategy for Current and Future Senior Victorians* plan by the Department for Victorian Communities, which was intended to contribute directly to the Victorian Government's wider social action policy (Office for Senior Victorians 2005). While this strategy was located within the positive ageing discourse, it also drew upon age-friendly rhetoric following visits to Melbourne by Dr. Alex Kalache, the then Director of Ageing and the Life Course Programme at the WHO. The key aim of the strategy was to maximize the quality of life and social recognition of older Victorians. To achieve this, the *Positive Ageing Strategy* was underpinned by five key principles: confidence in the rights of older people to be upheld; their autonomy accepted and dignity respected; certainty that older people are valued and listened to for their past, current and future contributions; opportunities for older people to fully participate in their communities; access to information, support, and services to maximize their independence and maintain health and wellbeing; and government services and communities which are responsive to the particular needs and interests of older people and which recognize the increasing diversity of our community. Over a 4-year period (2005–2009), $5.1 million was committed to this strategy to support positive ageing initiatives (Office for Senior Victorians 2005).

As part of this package, $1.4 million was allocated over 4 years to fund partnership projects between local government and the community to develop neighborhoods that were more age-friendly. This program, titled the *Positive Ageing in Local Communities* project, was a joint initiative between the Municipal Association of Victoria (MAV), the state's legislated peak body for local government, and the Victorian branch of the Council of the Ageing (COTA), the nation's peak advocacy organization for older people. The aims of this project were twofold: to build local government capacity in planning for population ageing, and to provide leadership in promoting age-friendly communities (Municipal Association of Victoria 2009). To do so, three categories of demonstration projects were funded: the development of positive ageing strategies, components of positive ageing strategies, and implementation of actions identified in existing positive ageing strategies (Project Partnerships and Community Planning and Projects 2008). Thirty-eight projects were funded, involving 42 of Victoria's 79 councils, with a Positive Ageing Advisor funded to work with councils to address positive ageing issues (Municipal Association of Victoria 2009). In the evaluation of the Positive Ageing in Local Communities project, results indicated that the project assisted in raising awareness of positive ageing within local government's planning agendas. While only 16.5 % of Victorian councils had an ageing strategy prior to the commencement of the project, this increased to 44.3 % by 2008. Only 7.6 % of councils were not in the process of developing a strategy for implementation all of which were smaller rural councils (Project Partnerships and Community Planning and Projects 2008).

In the second year of funding, COTA argued for Victoria's involvement in the WHO Global Age Friendly Cities initiative, facilitating meetings between Dr. Kalache and the MAV. Funding was then provided by the *Positive Ageing in Local Communities Project*, through collaboration between the MAV, COTA and the University of Melbourne for one urban municipality to participate in the international WHO AFCC project in 2007. Upon the launch of the WHO's *Global Age-Friendly Cities Guide* in late 2007, the Guide and accompanying resources were actively promoted to local governments to be incorporated into their positive ageing strategies. However, a further evaluation undertaken by the MAV exploring the use of the WHO AFCC guide and checklist within local government indicated that the take-up was not comprehensive. Of the 58 Victorian councils that responded to the survey, 30 % were not aware the resources existed, and one-quarter of those who were aware, did not used the guide or checklist in any form. For those who used the tools, they were primarily used as a reference document for positive ageing strategic plans (24 councils), to guide discussion with older persons' reference groups (15 councils), or to undertake age-friendly assessments from the perspective of councils. Only 12 councils used the tools to engage with older people in assessing the age-friendliness of communities, while only eight councils used the tools for all of these objectives (Municipal Association of Victoria 2009).

13.2.3 Research Approach

In light of the intermittent development of the age-friendly communities framework in Victoria (Ozanne et al. 2013), the subsequent sections highlight some key factors that have contributed to the lack of uptake of AFCC in the Victorian context.

To achieve this, we reviewed the Australian policy literature and academic material pertaining to health, ageing and/or age-friendliness. The review was undertaken in two stages: first, a manual search of federal, state and local government and policy websites, in conjunction with a broader web-based search was carried out. Second, a Google Scholar search to locate academic, peer-reviewed articles pertaining to the use of age-friendly in Australia was completed.

A content analysis of this literature was then undertaken to identify factors at the federal, state and local government level which had either been acknowledged as having a direct impact on the uptake of AFCC, or created policy contexts which were detrimental to the uptake and development of AFCC in Victoria. These findings were then juxtaposed against critical success factors identified in an analysis of the international AFCC literature: strong government leadership; a positive vision of ageing and of older people as a resource; consistency across levels of government; multi-sectorial partnerships; and government support for the active involvement of older people in process of implementation.

13.3 Factors Influencing Lack of AFCC Uptake in Victoria, Australia

International research has demonstrated that strong and effective leadership and vision across all levels of government (local/municipal, provincial/state and federal) are required to encourage, promote and facilitate AFCC to ensure consistency and coordination of effort, and to develop policies and services that promote active ageing (Spina and Menec 2013; Menec et al. 2013; Fitzgerald and Caro 2013). However, the adoption of AFCC in Victoria has been primarily impacted by a lack of strategic vision, leadership and direction across varied levels of government. This vision relates to both population ageing generally and the adoption of the AFCC framework more specifically.

Australia, like many geographically large countries, is a federation, with three levels of governance—federal parliament, state and territory parliaments, and local councils—allowing decisions to be made by the most appropriate level of government. Each level has its own responsibilities, although in some cases these responsibilities overlap. The Council of Australian Governments (COAG) is the peak intergovernmental forum in Australia to promote collaboration and agreement between levels of government (Council of Australian Governments 2012). However, despite two Australian cities participating in the initial research and development of the WHO AFCC guide (World Health Organization 2007), neither has single voice of leadership has been heard in public policy nor have those with power been invested in the WHO approach. This has resulted in a lack of multi-sectorial partnerships, both between and within different sectors of government, advocacy and community Organizations, academics and older people themselves. These issues are discussed in the subsequent section.

13.3.1 Lack of Vision

13.3.1.1 Lack of Federal Support for AFCC

A major issue inhibiting the implementation of AFCC in Victoria is the lack of vision, and subsequently direction, emerging from top-down governance structures. While government policies relating to ageing are critical for the successful implementation of AFCC (Garon et al. 2013), ageing policy in Australia has been developed from a top-down perspective, with policy agendas set by government and bureaucrats (Encel and Ozanne 2007). Significantly, in the Australian context, there is currently no consistent federal or state policy leadership relating to age-friendliness as exists in other countries (Ozanne et al. 2013). While population ageing was a prominent concern for the Coalition government that led Australia from 1996 to 2007 (Everingham et al. 2010), Australia has not had a strategic plan for ageing since the 2000 *National Strategy for an Ageing Australia*, which highlighted the need for "policies that support continued economic and social contributions by older people" (Department of Health 2000).

In subsequent years, no specific vision for older Australians emerged on the policy front. With the reign of the Rudd/Gillard Labor government from 2007 to 2013, wider population responses were instituted, such as the Social Inclusion Unit, which sought to ensure that all Australians must have the 'resources, opportunities and capability to learn by participating in education and training; work by participating in employment, in voluntary work, and in family and caring; engage by connecting with people and using their local community's resources; and have a voice so that they can influence decisions that affect them' (Social Inclusion Board 2008). While this unit was disbanded in 2013 by the incoming Coalition government, key criticisms relating to the Social Inclusion Unit's platform were its lack of attention to the specific issues faced by older people, its failure to identify older Australians as a priority group, and its lack of commitment to the age-friendly discourse (Lui et al. 2009, 2011). As the Australian Productivity Commission—the federal government's independent research and advisory body—recently noted, while the need for "timely action to address the consequences of demographic change" was observed in 2005, "the discussion of the possible opportunities and policy challenges presented by an ageing population seems to have waned" (Productivity Commission 2013, p. 3).

13.3.1.2 Lack of Direction from State and Local Government

Limited direction has also been an issue at the state and local government level. While the Australian Local Government Association (ALGA) provided some early direction to LGAs for age-friendly implementation through their 2006 report and checklist, *Age-Friendly Built Environments: Opportunities for Local Government*, this was restricted solely to built environment characteristics (Australian Local

Government Association 2006). This is problematic, as international studies have noted that strong municipal leadership is a critical success factor in successfully implementing AFCC (Garon et al. 2013; Lehning 2013; Buffel et al. 2013). This is also in contrast to other Australian states, such as South Australia, who have demonstrated strong policy support for active ageing. Their strategic plan, *South Australia's Communities for All: Our Age Friendly Future* was released in 2012, which provided guidelines and toolkits for state and local government to implement age-friendly communities. Significantly, it was based on the WHO AFCC checklist, with strategic amendments to reflect the South Australian context (Government of South Australia 2012). However, similar progress in Victoria has been much slower.

In early 2011, the incoming Victorian Coalition government established an inquiry into the opportunities for participation of Victorian seniors (Family and Community Development Committee 2012) which acknowledged the limitations associated with AFCC in Victoria, stating that:

> While progress has been made in many local areas, the Committee found that there is no ongoing, coordinated or consistent commitment at a state or local level to develop a cohesive, statewide age-friendly strategy. The Committee heard that despite early progress, there is a need to sustain momentum in Victoria. Local councils indicated they are seeking the leadership and support of a statewide age-friendly policy to sustain the effort (Family and Community Development Committee 2012, p. xxiv).

This statement referred primarily to the actions associated with the Victorian Labor Government's 2005 *Positive Ageing: A Strategy for Current and Future Senior Victorians* plan, which was informed by the WHO Active Ageing policy and invested $1.4 million into local government. As previously described, this strategy financed the *Positive Ageing in Local Communities* project, which provided funding for a Positive Ageing advisor based at the MAV, and for 31 demonstration projects in three categories to encourage local governments to address positive ageing issues. However, within these funded projects, there was no clear whole-of-government support or direction for AFCC. Significantly, while the 2009 evaluation of the *Positive Ageing in Local Communities* initiative had noted a greater awareness of 'positive ageing' at a strategic level, there was no evidence of leadership or involvement from the executive level at the MAV or the executive level within individual councils through this funding. The subsequent review of the uptake of the AFCC in local government expressed the value of the framework to staff in aged care or community services within council; however, the need for state and federal governments to provide a 'more integrated government strategy for supporting positive ageing and age-friendly communities across Australia' was noted (Municipal Association of Victoria 2009). This has similarly been observed in other Australian research with seniors and service providers, who advocated that the role of local government was significant in both creating an age-friendly framework and in facilitating actions at a community level (Everingham et al. 2010).

Within the *Positive Ageing in Local Communities* review, it was also noted that while local government was active in incorporating age-friendly planning principles throughout Victoria's municipalities, there had been little in the way of common criteria about what age friendly communities in Victoria should look like. Issues were also noted in relation to the capacity of rural and regional Victoria to implement age-friendly strategies and planning. In response to these issues, a key recommendation for the Victorian Government to liaise with federal government to promote the development of a national age-friendly strategy was made by the Family and Community Development Committee in its report. It also recommended that the Victorian Government seek to join the World Health Organization Global Network of Age-friendly Cities, and support local government authorities to achieve the same goal (Family and Community Development Committee 2012).

As a result of this lack of direction from federal and state government, and the subsequent impact on local government policies, responsibility for the age-friendly agenda in Victoria has largely been devolved to the Positive Ageing Policy Advisory Group through the MAV. Given the lack of policy support from governments in the form of funding and resources, this has primarily taken the form of capacity-building and lobbying (Ozanne et al. 2013).

13.3.1.3 The Impact of Government Change on AFCC and Active Ageing Discourses

The lack of leadership and direction for the ongoing implementation of AFCC has been significantly impacted by changes of government at both federal and state level during the past decade in Australia. These changes have resulted in competing discourses within policy, and a lack of policy support for AFCC implementation at all levels of government. Discourse analyses of Australian ageing policy over the past two decades indicate that governments at all levels have traditionally pursued policies of ageing that seek to address economic deficits associated with population ageing (Aberdeen and Bye 2013). However, the 2013 change from a Labor to Coalition federal government in particular has promoted a policy discourse around population ageing which is largely opposed to the AFCC approach. While the AFCC active ageing approach emphasizes the valuable contributions that older people can make to society and recognizes them as an asset (World Health Organization 2002), policy discourses promoted by the current federal Coalition government have largely promoted negative views of ageing, associating ageing with decrepitude, frailty and the cause of stress on national budgets. Population ageing is seen as a 'culprit', albeit 'a virtuous one', with the views of older people themselves viewed at best ambivalently (Productivity Commission 2013).

As Everingham et al. (2010) have noted, this shift in welfare policy toward promoting individual responsibility for independence in older age, coupled with growing marketization, has resulted in a strong emphasis on the financing of aged care services within ageing policy discussions in Australia. Thus, within the past 5 years, the Federal Government's primary focus for significant policy reform regarding population ageing has occurred in the aged care sector. In 2010, the Productivity Commission was asked to report on aged care, releasing its report,

Caring for Older Australians, in 2011. The *Living Longer Living Better* reforms, with significant change to the aged care sector over a 10-year period, were signed into legislation in June 2013 (Parliament of the Commonwealth of Australia 2013).

In March 2010, the federal Labor Government set up an Advisory Panel on the Economic Potential of Senior Australians, mandating the Panel to produce a comprehensive policy review on the opportunities and challenges associated with an ageing Australia. The Government's 2011 response to the Panel's third report, *Realizing the economic potential of senior Australians: turning grey into gold* (Advisory Panel on the Economic Potential of Senior Australians 2011) led to the establishment of an ongoing Advisory Panel on Positive Ageing to produce an Active Ageing Agenda, an overall strategy to maximize the potential of an ageing population. Moreover, the Government indicated its intention to invest in the WHO AFCC approach, recommending that the federal government:

> develops a national framework to recognise the life course approach to Active Ageing, based on World Health Organization (WHO) Active Ageing principles and including wellness, age-friendly environments, availability and accessibility of effective health care, and active participation in all aspects of community life (Government Response to the Final Report of the Advisory Panel on the Economic Potential of Senior Australians 2012, p. 1).

In doing so, the Panel proposed that the inter-government COAG forum develop an agreement to implement the WHO Active Ageing principles of health, participation and security. It was suggested that COAG's five strategic themes enabled collaborative work on positive ageing issues, with the states and territories possessing their own positive/active ageing strategies which were informed by the World Health Organization's principles or similar principles (Advisory Panel on the Economic Potential of Senior Australians 2011). However, in the first month of the new Coalition government's election in September 2013, it dismantled the Panel less than 8 months from the completion of its blueprint, shifting policy responsibility for ageing to the Department of Social Services. As the new Prime Minister, Tony Abbott, noted, 'Many of these non-statutory bodies have outlived their original purpose or are not focused on the government's policy priorities' (O'Keefe and Egan 2013).

In May 2014, the federal Coalition government released the National Commission of Audit Report and brought down its first budget, using both to signal a reframing of the public policy debate on ageing to one of a fiscal problem associated with service provision, with the solution to be found in economic policies based on longer workforce participation and a curtailing of the age pension. In the media release for the National Commission of Audit Report, the Commission chair, Tony Shepherd, set the tone for the government position on population ageing, declaring that:

> the Commission recognizes the unfairness of saddling today's children with our debts. With an ageing population there will be fewer people of working age to look after the retired. They should not inherit our debt as well as the burden of looking after us (National Commission of Audit 2013, Media Release).

Issues associated with an evolving policy environment have also been observed at the state government level (Ozanne et al. 2013), which have similarly restricted the take-up of AFCC at a local government level. While the state Labor government released its *Ageing in Victoria: A Plan for an Age-Friendly Society* in 2010 to address some of the issues identified within the *Positive Ageing in Local Communities* project (Department of Planning and Community Development 2010), the plan was abandoned shortly after its release due to a change in government by the present Coalition government.

13.3.1.4 Competing 'Age-Friendly' Discourses

The competing discourses present within the relevant policy and academic literature in relation to defining 'age-friendly' have also proved problematic in terms of providing a comprehensive vision of what AFCC should look like. Where Menec et al. (2011) have indicated that the term 'age-friendly' has been gaining worldwide momentum internationally within the academic and policy literature, its use and application in the Australian and Victorian context has been largely inconsistent. Within federal and state-level policy approaches to date, the AFCC paradigm has often been subsumed under alternate conceptual approaches to ageing. Terminology such as active ageing, positive ageing and healthy ageing have used interchangeably by various levels of government (Everingham et al. 2010), with the age-friendly concept recognized only in the context of policies focused on these concepts. A key example relates to the recent federal *Living Longer Living Better* reforms, which incorporates a positive ageing agenda that will create "policies to harness these demographic and societal shifts and aims to enable older Australians to live more active and financially secure lives in retirement" (Commonwealth of Australia 2012, p. 34). Active ageing is thus encompassed under the positive ageing agenda, with no mention made of AFCC (Commonwealth of Australia 2012).

Similar issues exist within Victorian state and local government policy. As Ozanne et al. (2013) have noted, rapid urban growth has resulted in increased emphasis on the creation of liveable, inclusive and sustainable communities, with only marginal recognition of what an age-friendly community might encompass. Within local governments, ageing strategies are badged as 'positive ageing' plans, as their development was financed through the 2005 state government funded *Positive Ageing in Local Communities* initiative. Further, given that the WHO AFCC guide was released during the latter stages of the project implementation, the AFC framework has been largely retrofitted within already prepared council positive ageing strategies rather than used in its entirety, with varying levels of commitment to the AFCC framework.

These competing discourses have also been an issue in relation to the academic literature, with little emphasis on the concept of age-friendly within the Australian ageing or community development literature. This is in stark contrast to countries such as Canada, the United States and the United Kingdom, who have a considerable empirical and conceptual literature exploring various issues and applications of the WHO AFCC framework (for examples, see Menec et al. 2013; Keating et al. 2013; Lehning 2013; Buffel et al. 2013). While international researchers have engaged closely with the active ageing concept and the AFCC framework, to date this has not occurred in Australia. Rather, concepts such as healthy ageing (e.g. Davis and Bartlett 2008), productive ageing (Davis et al. 2012) and ageing well (e.g. Warburton et al. 2008; Everingham et al. 2010) appear to be prevalent in academic work exploring the ability of communities to provide environments that enhance quality of life for older Australians. Further, in discussing 'age-friendly' within Australian ageing research, the trend has been to utilize the concept as a descriptor, rather than as a comprehensive framework in its entirety. Only a small number of Australian studies have explicitly utilized the AFCC framework or contextualized their results within the WHO AFCC discourse (for examples, see Everingham et al. 2009; Ozanne et al. 2013). As has been the case in policy, AFCC concepts have primarily been 'retrofitted' to findings, rather than research being designed explicitly with the AFCC framework in mind. While this by no means indicates that these findings are not applicable to or relevant to age-friendly approaches, questions must be raised as to the palatability of findings that are not directly situated within the AFCC discourse within policy circles.

13.3.2 Lack of Multi-sectorial Partnerships

13.3.2.1 Lack of Partnerships Across Government Departments

Given the limited strategic direction for AFCC at the federal, state and local level, the formation of multi-sectorial, whole-of-government partnership responses to guide its implementation have also been fragmented. The WHO, through the cycle of continual improvement, encourages communities to build partnerships with a broad range of organizations including nongovernmental organizations and academic institutions (World Health Organization 2014). While international research has similarly noted that partnerships across different sectors are fundamental to the successful implementation of age-friendly environments (Lui et al. 2009; Green 2013; Garon et al. 2013; Buffel et al. 2013), this process has been relatively disjointed in Victoria. Significantly, the 2000 *National Strategy for an Ageing Australia* has since been critiqued for its lack of attention to multi-sectorial planning and involvement by other spheres of government (Everingham et al. 2010).

The policy directions presented in the AFCC program are a major paradigm shift through the focus on place and community rather than the economic burden of ageing, and a whole-of-government response rather than from health and welfare services alone. However, for state and local government in Victoria, the development of policy, program and service responses has remained confined within ageing and aged care, or ageing and disability sections of governments. This was noted in the *Inquiry into Opportunities for the Participation of Senior Victorians*, with the report suggesting the imperative to shift from the current focus on health service delivery as the exclusive policy domain of government, to a whole-of-government strategy across the life course through "cross-departmental coordination within Victoria and partnership across levels of government" (Family and Community Development Committee 2012, p. 59). Within local government, there is no evidence of councils adopting a comprehensive, integrated approach to AFCC. The evaluation of the *Positive Ageing in Local Communities* project noted the limited engagement of staff from other departments within the project, with only three councils reporting the involvement of all sections of council. Mostly, strategic plans were, in effect, aged and disability service plans, not 'whole of council' plans. Importantly, local governments appear not to have sought to work in partnership with other Organizations in their communities, nor was there evidence of a regional or joined up response for those issues beyond the remit of a single local government (Project Partnerships and Community Planning and Projects 2008).

The key mechanism for ensuring the establishment of genuine partnerships is the creation and composition of a steering or advisory committee to achieve the vision of an age-friendly community, whether at national, state or local level. For example, the Canadian government, through its milestones approach, mandates the establishment of a steering or advisory committee that includes the active involvement of older people (Public Health Agency of Canada 2012). In Victoria, there was neither over-arching steering committee established for the *Positive Ageing* strategy, nor specifically for AFCC. Neither was there direction given to local governments on the process of implementation for the development of a positive ageing strategy, hence the involvement of older people varied significantly (Municipal Association of Victoria 2009).

13.3.2.2 Insufficient Involvement of Older Adults

The WHO Guidelines, supported by the Madrid International Plan of Action on Ageing, view the involvement of older residents of community as essential to the development of AFCC. This involvement moves beyond consultation to include 'participation in all established processes to set priorities, propose solutions and monitor implementation' (World Health Organization 2014). While all councils who took part in the evaluation of the *Positive Ageing Strategy* sought the views of older people, most did so only through surveys. Only three councils actively involved older people in preparing their positive ageing strategy by employing older residents as community researchers (Project Partnerships and Community Planning and Projects 2008). Further, within the 31 demonstration projects funded by the state government, less than two thirds (61.2 %) had a steering or reference group that included at least one older person. Thus, only 1 % of the 10,782 older people involved in the 31 demonstration projects were involved in project working groups or steering committees where they had direct influence on decision making. While council staff who responded to the evaluation reported that the involvement of older people was significant in "having a real influence on project decision making", there was no evidence of such change (Project Partnerships and

Community Planning and Projects 2008, p. 75). There were also no terms of reference or guidelines for the involvement of older people on these committees to assess how empowering such processes are for older people. Importantly, older people in some councils reported a less than positive response to council consultation processes in AFCC (Municipal Association of Victoria 2009), suggesting that without an active, legitimate involvement in the creation of AFCC, such approaches are seen as tokenistic.

13.3.2.3 Limited Engagement with the Academic Sector

From the perspective of partnerships, there has also been limited engagement of the academic sector in the Victorian context. Evaluations of international age-friendly initiatives have highlighted the importance of partnerships between university researchers and provincial governments to support municipal governments and communities in relation to assistance with surveys, conducting consultations and developing action plans. However, these partnerships have been contingent on federal research funding to support university involvement in AFCC initiatives (Menec et al. 2013; Buffel et al. 2013). One Victorian study exploring the development of age-friendly environments in the growth corridors of Melbourne has noted that while partnerships between universities and state health departments have been crucial in implementing the mandate, without an explicit age-friendly framework, competing discourses in relation to narrow service and planning focuses become problematic (Ozanne et al. 2013). Thus, questions must be raised as to why researchers have not become engaged with age-friendly research, or, to this end, critiqued the lack of commitment to the WHO framework in the wider Australian context.

13.4 Age Friendly Victoria Through the Bottom-Up Approach

Research has indicated that age-friendly initiatives are best implemented at the provincial/state level (Menec et al. 2013), yet as Ozanne et al. (2013) have observed, responding to population ageing from the perspective of AFCC has not yet sufficiently been incorporated into existing Victorian planning and urban development models. Without a coordinated, comprehensive approach to the implementation of AFCC by government, COTA Victoria has continued to develop its bottom-up approach at the state level.

In 2012, COTA Vic received $300,000 from the philanthropic Lord Mayors Charitable Foundation to establish *Age Friendly Victoria*—a program of education, training and resourcing of groups of older people in their own communities to work with the AFCC framework. COTA Vic has developed a peer-education model with its volunteers reaching out to groups of older people across the state to inform them about the AFCC model, and to teach them how to undertake a community audit and develop a plan of action. Through a partnership approach, COTA is assisting older people to develop local networks across academia, government, business and civil society to bring about positive change in their community. The objective of the project is to enable 20 local governments to be ready to be accredited by WHO as Age-Friendly Cities and Communities by mid-2014. This funding has also enabled COTA to respond to requests from individual local governments for advice on AFCC. A framework to guide the monitoring and evaluation of the COTA *Age Friendly Victoria* process was developed in late 2013, in consultation with academic and government partners, COTA and older people themselves.

However, as international age-friendly evaluations have noted, bottom-up approaches must be coordinated with support and collaborations from higher-level governance structures if AFCC initiatives are to be successful (Menec et al. 2013). Indeed, COTA Vic has no direct influence on local governments' willingness or ability to seek accreditation with the WHO.

13.5 Conclusion

Valuable lessons can be learned from the previous attempts at implementing AFCC in the Victorian context, as this chapter has demonstrated. The limited success in Victoria to date can be attributed to a lack of strategic direction from federal and state governments in the form of policies, funding, leadership, reporting criteria and collaborative partnerships. At a federal level, the transformation of Australia's initial involvement in the AFCC initiative into knowledge to inform the development of policies at government level stalled when successive federal governments failed to provide the leadership required to co-ordinate an AFCC program. While all levels of government have acknowledged the benefits of active ageing, this has not translated to a willingness to accept responsibility for implementation. Specifically, no direct policy commitment has been made to AFCC at a state or federal level, as a result of two primary factors: the increasing emphasis on the fiscal implications of population ageing and the provision of aged care services within policy; and the evolving policy environment. Where recommendations have been made to address the lack of attention to AFCC within commissioned enquiries or reviews, these have in most cases been abandoned with changes in government, where a new policy discourse has emerged. This lack of continuity has resulted in inconsistent and piecemeal use of 'age-friendly' terminology, and fragmentation of policies and programs that have not coalesced into a wider AFCC strategy within Victoria. While local government are attempting to implement AFCC within their municipalities with the assistance of support from COTA and MAV, the legitimacy and sustainability of their actions are threatened where they are not accountable to a wider network or lack the policy support to move forward.

Moreover, this poor leadership and strategic direction has resulted in limited collaboration between different sectors, where no governance or steering structure has been established to direct this process. In the Victorian context to date, this has resulted in limited and inconsistent engagement of older people in steering the AFCC process across different municipalities, difficulties in implementing a whole-of-government approach beyond the remit of government ageing and aged care departments, and in engaging academics. It can be suggested that the lack of engagement of Australian academics with the AFCC initiative to date reflects current government priorities toward population ageing, and therefore funding research. However, Australian age-friendly research is needed to better inform policy, as well as implementation strategies at the state and local level.

In conclusion, to ensure the continued sustainability and successful implementation of AFCC in Victoria, and in other Australian states, top-down direction is needed in the form of policy commitment and support of 'bottom up' initiatives, such as the COTA *Age Friendly Victoria* initiative. Some coordinated efforts nationwide are beginning to emerge to support these programs, including the first Australian Age Friendly Cities and Communities conference hosted by the Australian Capital Territory state government, and the subsequent formation of the Australian Network of Age-Friendly Cities and Communities (Curtis 2013). However, the sustainability of these initiatives and collaborations will be contingent on a paradigm shift within federal and state governments to reflect a vision where older people are recognized as an asset rather than a liability, and are recognized in policies other than those relating to aged care and disability planning. To achieve this, continued input and advocacy is required from local government, advocacy Organizations, academics and older people themselves.

References

Aberdeen, L., & Bye, L. A. (2013). Challenges for Australian sociology: Critical ageing research-ageing well? *Journal of Sociology, 49*, 3–21.

Australian Bureau of Statistics. (2012). *3235.0—Population by Age and Sex, Regions of Australia, 2011* [Online]. Available: http://www.abs.gov.au/ausstats/abs@.nsf/Products/3235.0~2011~Main+Features~Victoria. (Accessed July 20, 2014).

Australian Bureau of Statistics. (2013). *2011 Census QuickStats—Victoria* [Online]. Available: http://www.censusdata.abs.gov.au/census_services/getproduct/census/2011/quickstat/2?opendocument&navpos=95. (Accessed July 18, 2014).

Australian Bureau of Statistics. (2014a). *3218.0—Regional Population Growth, Australia, 2012–13* [Online]. Available: http://www.abs.gov.au/ausstats/abs@.nsf/Latestproducts/3218.0Main%20Features52012-13?opendocument&tabname=Summary&prodno=3218.0&issue=2012-13&num=&view=.

Australian Bureau of Statistics. (2014b). *West Wimmera (S) (LGA)* [Online]. Available: http://stat.abs.gov.au/itt/r.jsp?RegionSummary&region=26890&dataset=ABS_NRP9_LGA&geoconcept=REGION&datasetASGS=ABS_NRP9_ASGS&datasetLGA=ABS_NRP9_LGA&regionLGA=REGION&regionASGS=REGION. (Accessed July 16, 2014).

Australian Government. (2013). *Victoria—climate change impacts in Victoria* [Online]. Commonwealth of Australia, Department of the Environment Available: http://www.

climatechange.gov.au/climate-change/climate-science/climate-change-impacts/victoria. (Accessed July 16, 2014).

Australian Local Government Association. (2006). *Age-friendly built environments: opportunities for local government*. Deakin: Australian Local Government Association.

Buffel, T., McGarry, P., Phillipson, C., De Donder, L., Dury, S., De Witte, N., et al. (2013). Developing age-friendly cities: Case studies from brussels and manchester and implications for policy and practice. *Journal of Aging and Social Policy, 26*, 52–72.

Commonwealth of Australia. (2012a). *Government Response to the Final Report of the Advisory Panel on the Economic Potential of Senior Australians*. Canberra: Commonwealth of Australia.

Commonwealth of Australia. (2012). *Living Longer. Living Better.* Canberra: Australian Government, Department of Health and Ageing.

Council of Australian Governments. (2012). *About COAG* [Online]. Available: https://www.coag.gov.au/about_coag. (Accessed July 31, 2014).

Curtis, K. (2013). Recipes for successful age-friendly communities. Australian Ageing Agenda, 11 October.

Davis, S., & Bartlett, H. (2008). Healthy ageing in rural Australia: Issues and challenges. *Australasian Journal on Ageing, 27*, 56–60.

Davis, S., Crothers, N., Grant, J., Young, S., & Smith, K. (2012). Being involved in the country: Productive ageing in different types of rural communities. *Journal of Rural Studies, 28*, 338–346.

Department of Health. (2000). National Strategy for an Ageing Australia [Online]. Available: http://www.health.gov.au/internet/main/publishing.nsf/Content/health-budget2000-fact-acfact1.htm. (Accessed July 29, 2014).

Department of Planning and Community Development. (2010). *Ageing in Victoria: A plan for an age-friendly society 2010–2020*. Melbourne: State Government of Victoria.

Encel, S., & Ozanne, E. (2007). The politics of ageing. In A. Borowski, S. Encel, & E. Ozanne (Eds.), *Longevity and social change in Australia* (pp. 296–315). Sydney: University of New South Wales Press.

Everingham, J., Bartlett, H., Lui, C., Warburton, J., & Cuthill, M. (2010). Rhetoric to action: Local perceptions of policy goals and actions on population ageing. *Gerontological Social Work, 53* (8), 760–775.

Everingham, J.-A., Petriwskyj, A., Warburton, J., Cuthill, M., & Bartlett, H. (2009). Information Provision for an Age-Friendly Community. *Ageing International, 34*, 79–98.

Family and Community Development Committee. (2012). *Inquiry into opportunities for participation of Victorian seniors*. East Melbourne, VIC: Parliament House.

Fitzgerald, K. G., & Caro, F. G. (2013). An overview of age-friendly cities and communities around the world. *Journal of Aging and Social Policy, 26*, 1–18.

Garon, S., Paris, M., Beaulieu, M., Veil, A., & Laliberté, A. (2013). Collaborative partnership in age-friendly cities: Two case studies from Quebec, Canada. *Journal of Aging and Social Policy, 26*, 73–87.

Government of South Australia. (2012). *South Australia's Communities for all: Our age-friendly future—age-friendly neighborhoods: Guidelines and toolkit for local government*. Adelaide: Government of South Australia.

Green, G. (2013). Age-friendly cities of Europe. *Journal of Urban Health, 90*, 116–128.

Keating, N., Eales, J., & Phillips, J. (2013). Age-friendly rural communities: Conceptualizing "Best-Fit". *Canadian Journal on Aging, 32*(4), 319–332.

Lehning, A. J. (2013). Local and regional governments and age-friendly communities: A case study of the San Francisco Bay Area. *Journal of Aging and Social Policy, 26*, 102–116.

Lui, C., Everingham, J., Warburton, J., Cuthill, M., & Bartlett, H. (2009). What makes a community age friendly? A review of international literature. *Australasian Journal on Ageing, 28*, 116–121.

Lui, C., Warburton, J., Winterton, R., & Bartlett, H. (2011). Critical reflections on a social inclusion approach for an ageing Australia. *Australian Social Work, 64*, 266–282.

Menec, V., Means, R., Keating, N., Parkhurst, G., & Eales, J. (2011). Conceptualizing age-friendly communities. *Canadian Journal on Aging, 30*, 479–493.

Menec, V. H., Novek, S., Veselyuk, D., & Mcarthur, J. (2013). Lessons learned from a Canadian Province-wide age-friendly initiative: The age-friendly Manitoba Initiative. *Journal of Aging and Social Policy, 26*, 33–51.

Municipal Association of Victoria. (2009). *The World Health Organization Global Age-Friendly Cities Guide and Checklist: A review of their use by local government.* Melbourne: Municipal Association of Victoria.

Municipal Association of Victoria. (2011). *Jobs in Victorian local government.* Melbourne: State Government of Victoria.

National Commission of Audit. (2013). *National Commission of Audit Releases Review of the Activities of the Commonwealth Government* [Online]. Available: http://www.ncoa.gov.au/media-release.html. (Accessed July 28, 2014).

O'Keefe, D., & Egan, N. (2013). Advisory panel on positive ageing scrapped. *Australian Ageing Agenda,* 8 November.

Victorians, Office For Senior. (2005). *Positive ageing: A strategy for current and future senior Victorians*. Melbourne: Department for Victorian Communities, Victorian Government.

Ozanne, E., Biggs, S., & Kurowski, W. (2013). Competing frameworks in planning for the aged in the growth corridors of Melbourne. *Journal of Aging and Social Policy, 26*, 147–165.

Parliament of the Commonwealth of Australia. (2013). Aged Care (Living Longer Living Better) Bill 2013. In Representatives, H.O. (ed.). 76. Canberra: Parliament of Australia.

Commission, Productivity. (2013). *An ageing Australia: Preparing for the future*. Canberra: Australian Government.

Project Partnerships and Community Planning and Projects. (2008). *MAV COTA Positive Ageing in Local Communities. Project Evaluation Prepared for Municipal Association of Victoria and Council on the Ageing Victoria*. Victoria.

Public Health Agency of Canada. (2012). *Age-friendly communities* [Online]. Available: http://www.phac-aspc.gc.ca/seniors-aines/afc-caa-eng.php. (Accessed August 4, 2014).

Social Inclusion Board. (2008). *Social inclusion principles for Australia.* Canberra: Social Inclusion Board, Department of Prime Minister and Cabinet.

Spina, J., & Menec, V. H. (2013). What Community Characteristics Help or Hinder Rural Communities in Becoming Age-Friendly? Perspectives from a Canadian Prairie Province. *Journal of Applied Gerontology, September 9.* [Online]. Available: doi:10.1177/0733464813496164. (Accessed August 4, 2014).

Victorian Government. (2014). *Victoria in Future 2014: Population and household projections to 2051*. Melbourne: Department of Planning and Community Development.

Warburton, J., Everingham, J., Cuthill, M., & Bartlett, H. (2008). Achieving effective collaborations to help communities age well. *Australian Journal of Public Administration, 67*, 470–482.

World Health Organization. (2002). *Active ageing, a policy framework*. Geneva: WHO.

World Health Organization. (2007). *Global age-friendly cities: A guide*. Geneva: WHO.

World Health Organization. (2014). *Ageing and the lifecourse: Network cycle of continued improvement* [Online]. Available: http://www.who.int/ageing/age_friendly_cities_process/en/. (Accessed August 4, 2014).

Chapter 14
Toward an Age-Friendly European Union

An Interview with Anne-Sophie Parent and Julia Wadoux, AGE Platform Europe

Anne-Sophie Parent and Julia Wadoux

Thibauld Moulaert and Suzanne Garon: *Can you tell us what is AGE Platform Europe, its mission, vision and main work priorities?*

Anne-Sophie Parent/Julia Wadoux: AGE Platform Europe (AGE) is a European network composed of 150 organizations representing directly more than 40-million senior people in the European Union (EU). AGE aims to voice and promote the interests of the 190 million citizens aged 50+ living in the EU and to raise awareness of the issues that concern them most. AGE was established in 2001 to give a voice to older people in relevant EU policy debates through the active participation of their representative organizations in the platform's governing bodies and expert groups. Since its creation, AGE's work has been funded by grants from EU programs and membership fees paid by its member organizations.

AGE's vision is that of a society "for all ages" based on solidarity and cooperation between generations, a society where everyone is empowered to play an active part in society and enjoys equal rights and opportunities in all stages of life regardless of age, sex, race or ethnic origin, religion or belief, social or economic status, sexual orientation, physical or mental condition, or need for care.

AGE aims at fostering a more positive approach to demographic change and at promoting innovative and supportive solutions to respond to the needs of Europe's aging population. For that reason, AGE's work focuses on a wide range of policy areas that impact on older people: antidiscrimination, active aging, social protection, pension reform, social inclusion, health, research, accessibility of public transport and of the build environment, and new technologies in support to active and healthy aging.

A.-S. Parent (✉) · J. Wadoux
AGE Platform Europe, Rue Froissart 111, 1040 Brussels, Belgium
e-mail: annesophie.parent@age-platform.eu

J. Wadoux
e-mail: julia.wadoux@age-platform.eu

© Springer International Publishing Switzerland 2016
T. Moulaert and S. Garon (eds.), *Age-Friendly Cities and Communities in International Comparison*, International Perspectives on Aging 14,
DOI 10.1007/978-3-319-24031-2_14

TM/SG: *Why did AGE decide to promote Age-Friendly Environments (AFE) at EU level*?

ASP/JW: According to Eurostat, the statistical office of the EU, the number of people aged 60 and over living in the EU will increase by about two-million persons per annum in the coming decades, while the working age population will start to shrink as a result of lower fertility rates among post baby boom generations. By 2060, those aged 65 and over will become a much larger share (rising from 17 % in 2010 to 30 % of the population), and those aged 80 and over (rising from 5 to 12 %) will almost become as numerous as the young population in 2060 (European Commission 2012). This will result in an increasing number of very old and dependent persons and fewer young persons to care for them (relatives or healthcare professionals).

In line with the World Health Organization (WHO) holistic approach to active aging, AGE members believe that the best way to help Europe overcome this unprecedented demographic challenge is to empower older people to remain actively engaged in the labor market and in their communities for as long as possible. This requires action to combat age discrimination and to promote the concept of "AFE" in all areas of life: transport, the built environment, urban planning, health and long-term care services, mainstream goods and services, etc. Removing existing barriers that prevent older people from participating on an equal basis with other citizens will enable everyone to age in better health and to live independently for much longer. This will also help lower the pressure on public health and care budgets and will enable our societies to better cope with demographic aging in ways that are sustainable and fair to all generations.

TM/SG: *Is AGE cooperating with the WHO Global Network of Age-Friendly Cities and Communities (GNAFCC) based in Geneva WHO Headquarters and WHO Europe Healthy Cities network*?

ASP/JW: Yes, in fact our cooperation with WHO started in 2008 when AGE signed an agreement with the WHO in which we committed to promote the concept of AFE across the EU. In 2015, AGE has joined the WHO GNAFCC as institutional affiliate program. We also cooperate with WHO Europe because the Global Network of Age-friendly Cities and Communities is coordinated at WHO Headquarters and managed in close collaboration with all WHO Regional Offices. This enables us to build synergies also with WHO Europe Healthy Cities Network and their Strategy and Action Plan for Healthy Aging in Europe 2012–2020.

We share WHO views that the physical and social environments are key determinants of whether people can remain healthy, independent, and autonomous long into their old age. AGE members are convinced that promoting active and healthy aging through AFE is one of the most effective approaches for responding to demographic aging and should help older people live longer in better health and free of severe limitations in their daily activities that may otherwise result from their health problems or disabilities. In our view, the problem is not the long life expectancy that we have achieved in the last century thanks to progress made in the medical field and better nutrition and hygiene, but rather the long years many of us spend in poor health at the end of our lives. The so-called "healthy life expectancy" is unfortunately not improving in parallel with life expectancy. Economists' alarmist long-term projections on the impact of aging on public budgets are based on the assumption that this trend will continue in the coming decades and, with the expected sharp increase in the number of very old and dependent people, the demand on health and long-term care services will explode.

Because we are convinced that demographic aging can be turned into an opportunity if the challenges we are facing today are addressed properly, AGE decided to promote the WHO GNAFCC holistic approach and methodology to encourage public authorities and the private sector to adapt the way our society is organized to the needs of our aging population and support longer, healthier, and more autonomous lives. Our aim is to empower older people to keep control of their own life for much longer, improve their quality of life in old age, and release the growing pressure on traditional care and assistance. We know through the daily experience of our member organizations that, depending on the environments they live in, older persons can become dependent on others for their daily activities quite early in their life or, on the contrary, continue to live independently until very old age if their environment is supportive.

In our view, the promotion of AFE across the EU should also boost the economy through demand for innovative solutions which should result in job creation for both young and older workers and help to rise the effective retirement age, thus safeguarding our social protection models so dear to European citizens' heart.

TM/SG: *Can you tell us in more details how AGE managed to introduce the concept of AFE at EU level?*

ASP/JW: Following the signature of our agreement with the WHO GNAFCC, we started to promote WHO's AFEs' approach and methodology in all relevant EU policy debates and discussions on EU funding priorities. Soon, the concept gained visibility among EU leaders as well as local and regional authorities involved in EU projects or networks of regions devoted to aging issues.

AGE members are convinced that older people can and should be key drivers for change in this area. This is why they decided that AGE should use the momentum of the European Year 2012 on Active Aging and Solidarity between Generations (EY2012) to launch a large campaign entitled "Towards an age-friendly European Union by 2020." The EY2012 gave us the opportunity to build support for the concept of AFE among a large group of stakeholders who share our vision, and together we managed to import this approach in the relevant EU processes.

For example, we seized the opportunity of the launch of the European Innovation Partnership on Active and Healthy Aging (EIP AHA) by the European Commission end of 2011 (http://ec.europa.eu/health/aging/innovation/index_en.htm). The EIP AHA was the first of such partnership to be launched under the so-called Europe 2020[1] Innovation Union flagship initiative, the "EU strategy to create an innovation-friendly environment that makes it easier for great ideas to be turned into products and services that will bring our economy growth and jobs." EIPs are not funding programs but a new approach adopted by the European Commission to boost EU research and innovation.

As explained on the above-mentioned EIP AHA website, its objective is to increase by two years the average healthy lifespan of Europeans by 2020 and to help older people remain healthy and autonomous as they age through:

- enabling EU citizens to lead healthy, active, and independent lives while aging;
- improving the sustainability and efficiency of social and health care systems;
- boosting and improving the competitiveness of the markets for innovative products and services, responding to the aging challenge at both EU and global level, thus creating new opportunities for businesses.

The EIP AHA seeks also to develop more efficient health, long-term care, and social services for Europe's aging populations, and to support innovative coproduction models for active and healthy aging across the EU. This should support evidence-based and cost-efficient health systems that will help reduce the cost of population aging on public health and social budgets, a key objective of the EU Europe 2020 Strategy for smart, sustainable, and inclusive growth.

AGE Platform Europe was actively involved in the Steering group (SG) of the EIP AHA and succeeded—after long discussions—in getting the promotion of AFE in support to active and healthy aging adopted as one of the EIP AHA first key actions addressed by six Action Groups. As a result, the so-called D4 Action Group was established and is since devoted to AFE.[2]

TM/SG: *Through this D4 Action Group, what are you concretely trying to achieve with your EU campaign on AFE compared to initiatives already taken by local stakeholders*?

ASP/JW: Promoting innovative AFE may sound the obvious thing to do. Yet one can wonder why so many solutions supporting active and healthy aging, which have already been developed as pilots across Europe, remain isolated and do not get scaled up, not even in their own country. Their impact on addressing Europe's demographic challenge is thus limited because, except in a very few countries that have a national program on aging (e.g. Slovenia, France, Poland), the various governance levels that need to work together do not necessarily coordinate their action. So far public authorities, the private sector, the research community, and civil

[1]Europe 2020 is the European Union's policy framework for the period 2010–2020. For more information on Europe 2020. [Online]. Available: http://ec.europa.eu/europe2020/europe-2020-in-a-nutshell/index_en.htm. [Accessed December 23, 2014].

[2]EIP AHA Market Place Action Groups. [Online]. Available: https://webgate.ec.europa.eu/eipaha/actiongroup. [Accessed December 23, 2014].

society organizations have not been able to make the best of all the opportunities that exist in the framework of the EU policy processes and funding programs.

This is why within the EIP AHA D4 Action Group, in May 2012 AGE committed to set up and moderate an EU virtual network on AFE to connect stakeholders (public authorities, non-governmental organizations (NGOs), service providers, industry, and researchers) interested in the promotion of AFE at local, regional, national, and EU level. We committed also to collect good practices and to work toward the launch of a Covenant on Demographic Change to support the scaling up of innovative AFE across the EU.

In December 2012, the European Commission launched a call for proposals for the setting up of a thematic network on AFE and launch of a Covenant on Demographic Change under the ICT Policy Support Program (ICT PSP), offering us the possibility to apply for EU funding to implement our commitment with a larger group of partners. Our project proposal was accepted by the European Commission and the project called AFE-INNOVNET—for **A**ge-**F**riendly **E**nvironments **Innov**ation **Net**work—started in February 2014 and will run until end of January 2016.

TM/SG: *Who are the members of AFE-INNOVNET network and how is this thematic network evolving?*

ASP/JW: The AFE-INNOVNET thematic network project is run by a consortium of 29 partners from 16 EU member states, and is coordinated by AGE. The consortium includes 13 cities, 6 regions, 5 large EU networks active in the field of aging or representing large numbers of local authorities or seniors' organizations, 4 research centers specialized in ICT and aging policies and a communication agency. Most of the consortium partners are actively involved in the EIP AHA D4 action group on AFE. To ensure timely and effective implementation of the project, a SG has been nominated. It is composed of all Work Package leaders (AGE, PAU Education—the communication agency, Council of European Municipalities and Regions, Polibienestar research centre/University of Valencia, and TNO, a Dutch research centre) and of Louth County Council. The SG co-leads the project and is responsible for ensuring that progress is made toward the project's objectives and deliverables are developed within the agreed deadline. The SG members have daily e-mail exchanges, teleconferences every two weeks, and face-to-face meetings three times a year.

The overarching goal of AFE-INNOVNET is to set up a large EU wide community of local and regional authorities and other relevant stakeholders across the EU who wants to work together to find smart and innovative evidence-based solutions to support active and healthy aging and develop AFE across the EU. To achieve this objective, the project activities seek to:

- mobilize a wide range of local and regional authorities and other stakeholders—industries, including SMEs, research center/universities, civil society organizations—to link up, benefit from each other's experience, and work together to promote initiatives on AFE across the EU;

- develop methodologies to help local and regional authorities assess the socio-economic impact of AFE and the benefits of involving older people in the coproduction of AFE solutions;
- develop a repository of notable and replicable practices in innovative ICT and services solutions with associated socioeconomic evidence;
- facilitate pilot projects clustering to stimulate local/regional investment in ICT and services innovation and thus local/regional economic activity (with a focus on SMEs);
- facilitate public access to research/expertise and generic methodologies in the field of AFE;
- ensure convergence and strong coordination with WHO Age-Friendly Cities and Healthy Cities initiatives;
- facilitate information sharing between all interested stakeholders in the field of AFE;
- frame and launch an EU Covenant on Demographic Change to create the necessary political and technical framework to bring together in a more formal and long-term structure local and regional authorities—and other stakeholders—across the EU who want to cooperate and implement smart and innovative evidence-based AFE solutions.

As soon as its website was set up and running, the AFE-INNOVNET thematic network was open to all interested parties. End of August 2015, more than 300 stakeholders from the EU and a few from outside Europe have joined, and the list continues to grow.

Most information is posted on the public part of the project website and can be translated online in some 60 languages to give at least an overview of the project and of published good practices to a wider range of visitors at grass root level. In addition among the SG members, there is several multilingual staff and we can cope with almost all EU languages if needed.

TM/SG: *What will be the added value of the proposed Covenant on Demographic Change in relation to the existing WHO GNAFCC*?

ASP/JW: There is at present no EU structure to enable all stakeholders interested in promoting and supporting AFE to link up, benefit from each other's experience, and work together on shaping the EU agenda on active and healthy aging. For this reason, an increasingly large and diverse group of stakeholders have joined forces with AGE and support the launch of an EU Covenant on Demographic Change to create such framework for cooperation and to populate an EU repository of good practices. Local or even national authorities do not have the capacity to link up with a wide range of other local/regional/national stakeholders and to build an EU-wide evidence-based repository for age-friendly innovation. This can only be done at EU level.

While the WHO GNAFCC focuses on local authorities and covers the whole world, the Covenant on demographic change will connect the different levels of governance (local, regional, national and EU) and the different categories of interested stakeholders (public authorities, civil society organizations, researchers, industry), and its main focus will remain the EU, eventhough non-EU stakeholders will be able to join.

Our objective is to ensure that the Covenant can be used to recruit more cities and communities in the EU to join the age-friendly movement. We see both the GNAFCC and the Covenant networks developing as two circles which will partly overlap where EU cities and communities are concerned.

We are working in close cooperation with WHO to ensure that the application, monitoring and reporting for cities and communities located in the EU can be done through the Covenant and be valid for the GNAFCC as well without extra burden for the applicants. In a nutshell, the Covenant on demographic change will among objectives become the GNAFCC affiliate program for the EU to allow EU cities and communities to benefit from their membership in both networks within one simple application and reporting procedure process managed through the Covenant. This will ensure a strong bridge between the two networks that will benefit members of both communities.

Several AFE-INNOVNET SG members are also actively involved in the scientific steering board and advisory board of the WHO-Europe AFEE (Age-Friendly Environments for Europe) project funded by the EU. Through this project WHO-Europe is developing tools and guidance inspired by the WHO GNAFCC tools and methodology on age-friendly cities and communities and is adapting them to the EU context, so that they can be used by the upcoming Covenant.

TM/SG: *What will be the incentive for cities and communities to join the Covenant rather than apply directly to the GNAFCC? Will it become compulsory for EU cities and communities to apply through the Covenant?*

ASP/JW: We do not want to force anyone. Cities and communities based in the EU will still have the choice to join the GNAFCC directly if so they wish. However, they will be encouraged to join through the Covenant and when doing so, they will automatically become eligible for the GNAFCC if they wish to join the global network as well. There will be advantages for them to be in both networks. By joining the Covenant on Demographic Change they will join a network of actively committed EU stakeholders that include other levels of governance and other types of actors with whom they can share their experience and from whom they can learn a lot. They will also have direct access to useful EU-related information such as the outcomes of the EIP AHA other Action Groups and the EIP on Smart Cities and Communities,[3] the EU work on Design-for-All and interoperability standards, relevant EU projects, EU funding opportunities. They will get support to find

[3]AGE, CEMR, and Polibienestar have submitted a commitment under the EIP on Smart Cities and Communities to promote age-friendly environments and build synergy with the AFE-INNOVNET project.

potential partners if they wish to join the EU project application and will follow what the Knowledge for Innovation Community (KIC) for healthy living and active aging that has been selected by the European Institute for Innovation and Technology (EIT) in December 2014 is doing, etc.

By joining the GNAFCC as well, they will be able to benefit from WHO expertise at global level, learn from other GNAFCC members in other parts of the world, without any extra work on their side. We are coordinating with WHO GNAFCC to ensure that they will only need to fill in one application form, send data once when their application is accepted, and report through one single procedure through the Covenant. Their data will be transferred automatically to the GNAFCC webplatform by the Covenant.

TM/SG: *Can you tell us a bit more about the Covenant*?

ASP/JW: For the Covenant we are trying to build as much as possible on what already exists. For example, the first AFE-INNOVNET workshop was organized on purpose in Dundalk (Louth County, Ireland). During the final session, we discussed how we could build synergy with the Dublin Declaration on Age-Friendly Cities and Communities in Europe. As you know this, Declaration was signed by some 60 mayors and senior political representatives of European cities, municipalities, and communities during the EU Summit on Active and Healthy Aging in Dublin in June 2013. By signing this Declaration they committed to promote WHO GNAFCC model and to take action to make their community more age-friendly.

Since this is also the objective of the AFE-INNOVNET project, we have decided to use the Dublin Declaration as vision statement for the proposed Covenant on Demographic Change and to invite all its signatories to join the Covenant when it will be launched end of 2015. There is no need to reinvent the wheel. The Declaration states clearly the vision we want the Covenant to promote, so let's use it and valorize it as a key document of the future covenant.

TM/SG: *Anything else you would like to add to conclude*?

ASP/JW: We hope that the planned Covenant will continue to be supported by the European Commission both politically and financially, because we are convinced that it can play an important role in reducing geographical inequalities in Europe, an indispensable target for reaching the EIP AHA headline objective of improving by two years the Healthy Life expectancy, and for helping local and regional authorities which meet their demographic challenge in innovative and sustainable ways.

We also believe that given the huge potential of ICT and social innovation in support to AFE, mobilizing local and regional stakeholders who can pick up such solutions, and deploy them across the EU will support healthy competition and a higher competitiveness among "silver economy" industries.

Since the proposed Covenant on demographic change will work in close cooperation with the WHO GNAFCC and will support its objectives, it should also bring progress in other parts of the world and will hopefully create new opportunities for partnership and cooperation around AFE that will benefit aging populations around the globe. There is so much we need to do and so much we can learn from each other!

Reference

European Commission. (2012). *The 2012 ageing report. Economic and budgetary projections for the 27 EU Member States (2010–60)*. Luxembourg: Publications office of the European Union.

Part III
Challenges From and For Age-Friendly Cities and Communities

Chapter 15
Age Friendliness, Childhood, and Dementia: Toward Generationally Intelligent Environments

Simon Biggs and Ashley Carr

15.1 Introduction

It is possible to see a city as a concretion of certain channels of social relationship and as a complex set of arrangements; both reflect and continue to shape the way people behave and interact over time and in space. This shaping of public space has spatial and temporal consequences as it is continually remade in response to changing priorities, a process that both determines the possibilities for social engagement and makes a statement about how people of different ages and circumstances are made to live with each other. Citizens are, of course, not equal in their ability to affect the shape a city or neighborhood takes, and the degree to which this complexity is expressed reflects the interests of different groups with different degrees of power. Here, the age-friendly narrative is discussed, both in terms of a focus on active aging and claims to be 'for all ages.' It is compared to initiatives around child and dementia friendliness which hopefully both specify the focus of age-friendly cities and environments (AFE) and examine the commonalities and differences between these groups when it comes to their access to public space. Children and people living with dementia are not chosen because they might have life course or experiential similarities, however, but in order to draw out how different initiatives have conceptualized the problem of public space and its ability to

S. Biggs (✉)
Brotherhood of St Laurence Chair of Gerontology and Social Policy,
School of Social and Political Sciences, Melbourne University,
John Medley Building, Parkville, VIC 3010, Australia
e-mail: biggss@unimelb.edu.au

A. Carr
School of Social and Political Sciences, Melbourne University, Melbourne, VIC, Australia
e-mail: ashley.carr@unimelb.edu.au

A. Carr
Brotherhood of St Laurence, Fitzroy, VIC, Australia

© Springer International Publishing Switzerland 2016

T. Moulaert and S. Garon (eds.), *Age-Friendly Cities and Communities in International Comparison*, International Perspectives on Aging 14,
DOI 10.1007/978-3-319-24031-2_15

accommodate beyond dominant uses, such as to pursue the demands of working life. In many ways, the third group of 'active agers' follows the same channels as those laid down by working life, but from the perspective of consumption rather than production. It may therefore be less difficult to address the needs of the active ager than to look at the predicament of children and especially people living with dementia.

As power relations are pertinent to interaction between age groups, we examine how active agers, children, and people living with dementia are reflected in policy on, what turns out to be mostly urban environments. We interrogate public policy as a way in which perspectives on age and generation are made visible and legitimize certain forms of relationship. If urban spaces are designed with certain groups in mind, principally adults of working age, either as producers or as consumers, then the question arises, how far are other groups taken into account, in terms of inclusion or separation and their importance to the main project of public space. Age and generation are interesting in this regard because they add a longitudinal dimension to the use of urban space which reflects personal life course time, rather than, on the one hand, time-of-day use, and on the other historical time. One is arguably too short, though repetitious; the other, while a source of spatial determination, is too long a time frame to catch the impact on and of human development across a lifetime. The extent to which urban environments are planned for future use, to affect future functioning, as well as intergenerational relations in the here and now are also important factors that require critical assessment when examining the way in which intergenerational relations are conceived. Sometimes, it is important to look at the periphery to determine the underlying assumptions on which the center is built. This is particularly true when examining the two 'ends' of the life course: age and childhood, and increasingly, dementia.

At the time of writing, the 'age-friendly' debate is in a period of rapid development. In terms of nomenclature, this is reflected in a move from age-friendly cities, to age-friendly environments, municipalities, ruralities, and second homelands. In this chapter, we have restricted our inquiry to the discussion of public space. This is used as a way of engaging with the question of the intergenerational use of space, primarily as it is expressed in policy statements. We therefore will not be considering private spaces such as personal homes or housing in any detail. By public space, we mean the social interaction that occurs in public areas and their interconnection, such as on streets, in neighborhoods, public buildings and facilities, parks, and squares. The argument is principally about narratives (from international organizations and from Australia) rather than specific examples, which would be the subject of a much larger enquiry.

15.2 Age Friendliness and Spaces for All Ages

A growing interest in the city as "age-friendly" and "friendly for all ages" raises a number of questions for the future design of the built and social environments, the generational nature of discourse on urban design, and the degree to which diverse and intergenerational spaces are possible and indeed desirable. Behind these questions, it is also possible to discern an anxiety about the management of demographic change and aging populations which, as the work of the World Health Organization (WHO), the United Nations (UN) and the World Economic Forum (WEF) indicates, has global implications. Anxiety is perhaps heightened further by the emergence of dementia as a global, public health priority (WHO 2012).

According to the UN (2014), about 3.9 billion people currently live in urban areas, with this figure expected to grow to about 6.3 billion by 2050 and account for approximately two-thirds of the world's total population. Australia is at the higher end of the scale, with approximately 90 % of the population currently living in urban areas (UN 2014). Incidence of dementia is set to increase from 44 million world-wide in 2013 to 135 million by 2050 (Alzheimer's Disease International 2013), with Australian figures rising from 320,000 to 891,000 in the same time period (Australian Institute of Health and Welfare 2012). For the United Nations Population Fund (2007), global urban growth is occurring unevenly, both geographically and demographically, with a 'youth bulge' expected to occur in the demographic profile of some developing countries, primarily among low-income groups. However, the general global trend is one of the population aging. The WHO (2007) claims that between 2006 and 2050 the world's population of older adults will double from 11 to 22 %, with the developing world expected to age at a faster rate than developed countries. Making cities and communities age-friendly has emerged as an attractive policy response to these twin challenges of population aging and urban growth. Like the United Nations Children's Fund (UNICEF)'s Child-Friendly Cities model that pre-dated it, WHO's Age-Friendly Cities project examined the hazards and opportunities of urban living to develop an optimal public environment for these growing numbers of older people.

The dominant narrative arising from the WHO initiative has been one of the 'active ageing.' Walker (2002) traces the emergence of active aging from a concentration on activity theory in the 1960s, to a modern incarnation which combines elements of productive aging with an emphasis on health and quality of life. During the 1990s, the concept gained considerable ground among international organizations, individual states, and in the academy and represented a non-dependency-based response to demographic change, one that was able to incorporate new and developing concepts of aging and operate at different levels of society (Walker 2009). Olshansky et al. (2012), for example, have argued that a goal for public policy should be to increase healthy active aging thus creating a 'longevity dividend' that boosts the productivity of older adults.

By 1999, WHO was making the claim that "active ageing makes a difference" and the idea of active aging formed the underlying rationale of the age-friendly project when launched in 2005. It is possible to read 'Age friendly Cities' as one answer to the question, articulated in the WHO document 'Active Ageing' (2002), of how to provide older adults with health, participation, and security, in urban environments.[1]

---

[1]Note from the editors: for detailed discussions of 'active ageing' and 'age friendly cities,' see the chapters of Walker, of Kalache and of Moulaert and his colleagues.

An attractive feature of age friendliness was that it proposed a 'holistic vision of personal development,' which consisted of eight inter-related factors of urban living: the built environment, transport, housing, social participation, respect and social inclusion, civic participation and employment, communication, and community support and health services. In 2010, the WHO launched the 'Global Network of Age-friendly Cities' with 47 individual cities that, in Australia, included Melville (WA) and Melbourne (Vic). Currently, Boorondara (Vic), Canberra (ACT), Clarence (Tas), Lane Cove (NSW) Melville (WA), Rockingham (WA), Unley (SA), and Warnambool (Vic) are listed as active members of the network. However, all except Canberra are local councils, suggesting that while popular at the local government level, the model has not been adopted by larger state or governmental authorities. State governments have tended to elaborate aging-well policies, which incorporate aspects of age friendliness. However, while 'active ageing' has been progressively restricted to reflect a longer working life (Moulaert and Biggs 2013), age friendliness has remained associated with a broader notion of activity.[2] Indeed, as active agers, engaging with their urban environments and public spaces, it is possible to see elements of what Gilleard and Higgs (2005) refer to as a habitus or consumption-dependent lifestyle emerging. The age-friendly city would thereby adjust both to the needs of adult workers, including older workers and active older adults who consume leisure, volunteering and health-related activities. On a wall adjacent to the Melbourne Cemetery, there are graffiti saying "work, buy, consume and die." Clearly, someone intends to critique the current use of this, the most commonly voted "world's most liveable city" (The Economist Intelligence Unit Limited 2012). Age friendliness, however, has not yet developed into a wider critique of public space; it has, rather, emphasized the modifications that may need to be made to make urban environments more inclusive within the twin priorities of consumption and production.

15.3 Age and Elder Friendliness Plus Cities 'For All Ages'

The most common justification for the inclusiveness of age-friendly initiatives, made on a number of Global Network websites and in the State of Victoria's *Inquiry into opportunities for participation of Victorianseniors* (Family and Community Development Committee 2012), is that it creates an environment for all ages. The WHO supports this trend, stating "An age-friendly city emphasizes enablement rather than disablement; it is friendly for all ages and not just 'elder-friendly'" (WHO 2007, p. 72). In this section, we explore the different ways which 'friendliness' has been labeled, both in relation to categories for later life and as a link between age groups.

[2]Note from the editors: in their chapter, Brasher and Winterton present an empirically based critical discussion on the Australian case which completes the present chapter.

While 'age-friendly' community initiatives have been introduced in localities throughout the world, and at various governmental levels (WHO 2014), there is currently considerable confusion as to what 'age-friendliness' might actually mean. Lui et al. (2009) note the different terminologies used to describe what makes an environment age-friendly, suggesting that these differences are not problems in themselves, but indicate different emphases in practice and research. Terms, such as age-friendly (WHO 2007), aging-friendly (Scharlach and Lehning 2013), and elder-friendly (Lui et al. 2009), may, however, denote different approaches in the design and development of age-friendly communities as well as different understandings of age friendliness. What they all share in common, according to Scharlach and Lehning (2013, pp. 112–113) drawing on examples from North America, is the "goal of modifying the physical and social environment to improve the health and well-being of older residents and increase elders' social participation," Central here is the capacity of the urban environment for modification to meet specific generational needs. Thus the WHO Age-Friendly Guide claims that "It should be normal in an age-friendly city for the natural and built environment to anticipate users with different capacities instead of designing for the mythical 'average' (i.e. young) person" (WHO 2007, p. 72). A problem for this approach is that, in terms of intergenerational dynamics, it focuses attention on a relatively fixed group, identified by a particular cohort or adult life stage. 'Aging' friendly attempts to resolve this limitation by focusing on the process rather than the state of later life and thereby opens the door to other parts of the life course.

A second, and perhaps intergenerationally more interesting shift, has been where the age-friendly program has gravitated toward the notion that age friendliness is inherently friendly for all. As Scharlach (2009, p. 8) states, "an ageing-friendly community promotes the physical and psychosocial well-being of community members throughout the life cycle" and is essentially a society for all ages. Indeed, the notion that 'age-friendliness benefits all ages' comprises one of the arguments to support investment in urban modification, and in North America a number of organizations and government agencies have built a strong case on the mutual benefits of age-friendly design. The focus is largely on the physical aspects of urban design: sidewalks, parks and recreation facilities, and transport services, with the modeling for 'community for all ages' (Miller and Annesley 2011; Arizona State University Herberger Center for Design Excellence 2005; National Association of Area Agencies on Aging 2007), according to the idea "that by accommodating the needs of older people, it is possible to better accommodate the needs of all groups" (Miller and Annesley 2011, p. 23). Similar developments have taken place in the UK (Department for Work and Pensions 2009), Australia (Commonwealth of Australia 2006), and in Canada (see for example District of Saanich 2014). As suggested by van Vliet (2011, p. 351), such approaches may apply across generations, as "intergenerational integration of urban liveability initiatives will result in more efficient use of physical facilities and funding sources." The notion of 'a design for all ages' has been closely associated with principles of universal design, which arising from inclusive design for disability also maintains that "Design for the young and you exclude the old; design for the old and you include the young" (cited in Noyes 2001, p. 72). It also resonates with the Madrid International Plan of Action on Ageing in 2002 (UN 2002, p. 19) which states that "A society for all ages encompasses the goal of providing older persons with the opportunity to continue contributing to society." The Australian Government, which claimed in 2005 to have been the first nation to adopt the society for all ages approach to population aging, launched a speakers series, *A Community for All Ages—Building the Future*. While short-lived, the aim was to raise awareness and stimulate debate among planners, designers, and policy-makers on the need to plan and build better communities in preparation for the changes an aging population will bring (Commonwealth of Australia 2006). It, however, made little mention of other age groups and its focus was primarily on issues of physical design, particularly housing.

What appears to be taking shape is a narrative of shared principles between adult (including older adult) and child friendliness. Surprisingly, there has been little research on the convergence of such principles. When Carr et al. (2013) undertook a review of municipal activity on both Child and Age friendliness in Australia, it was found that while policy initiatives had been undertaken, often in parallel to each other, there was little empirical evidence to demonstrate either the implementation of shared generational strategies, or their positive effects.

A closer reading of a narrative shift toward environments for all ages reveals a tension in the way that age friendliness has been adopted on the ground. On the one hand, 'all ages' emerge as a trope, to advance the cause of design that takes specifically older adults into account while hitching it to the wagon of a universal good. As such it parallels early thinking on the value of productive aging as a means to increased social inclusion and reduced ageism (Caro et al. 1993). 'For all ages' leaves the group of older persons targeted vague and open to wide interpretation and, as a tactic to address demographic change in a predominantly ageist society, may bear fruit. The growth in the WHO network and the enthusiasm with which specifically 'age friendly' (urban) environments have been taken up in policy indicates a parallel dynamic at work. It is precisely because 'age friendliness' offers 'something to do' (or at least write) in the face of population aging that the program has grown so quickly. As such it is unclear how far, and in which direction, it has been threatened by what Biggs (2004, p. 103) has identified as "the colonisation of the goals, aims, priorities and agendas of one age group onto and into the lives of other age groups." A shift in narrative from age, to aging, through the 'all ages' claim, runs the danger of eclipsing the specific needs of a particular age group and reinvents a 'universal urbanite,' while the age-friendly narrative gives a method by which policy-makers can adjust to changing demographics. If the analysis above is accepted, then age-friendly narratives may themselves be a trope for a path of least difficulty. On the one hand, they include active agers and older workers into a predominantly urban project; on the other hand, they are suitably vague about the actual benefits to other age-identified out-groups.

15.4 Child Friendliness in Policy Settings

This is not to say that no other age-based initiatives exist. At least one other age group has its champions in public space. UNICEFs Child-Friendly Cities (CFC), established as part of the UN Conference on Human Settlements (Habitat II) in 1996, has been founded by the 1989 UN Declaration on the Rights of the Child. As such a "Child-Friendly City is committed to the fullest implementation of the Convention on the Rights of the Child" (UNICEF 2004). The notion of rights has proven a subject of debate, with disagreement on how far children's rights extend, or indeed for some, whether specific children's rights are valid at all (see for example Freeman 2007). The development of children's rights from the 1950s to the present reflects some important, but subtle shifts since the UNs first declaration on the rights of the child in 1959, with a growing emphasis on children as valued and capable participants (Tranter and Sharpe 2007). UNICEFs child-friendly cities program picks up on the development of positive rights and children's participation as "the very essence of the process of building a Child-Friendly City" (UNICEF 2004). From this basis, the child-friendly cities program has worked to enhance features of the built and natural environment that can advance children's rights, including parks, play spaces, and children services. The children's rights agenda remains strong primarily because it is linked to a wider human development agenda. It remains difficult, however, to find instances where the interests of children are not subsumed by wider economic and political concerns.

Child-friendly cities function to ensure that local governments and authorities translate commitments made at the global, national, and state levels into specific urban environments (Riggio 2002). In Australia, for example, children's rights form an important part of national childhood and child development policies. The Council of Australian Government's (COAG) National Childhood Development Strategy affirmed a commitment to children's rights, but evidences a drift toward seeing children as principally future workers, rather than as citizens and active participants in their own right, as intended by Article 12 of the original Convention. Thus COAG states

> Children are also important for their future contribution to society – as the next generation of leaders, workers, parents, consumers and members of communities. Their ability to participate fully in society as adults will be largely shaped by their childhood experiences. Children who have a good start in life are more likely to develop the capabilities that will better equip Australia to compete in a global society. This will be increasingly important as our workforce shrinks due to population ageing and low fertility rates (COAG 2009, p. 7).

This policy narrative claims that an investment in childhood development has long-term benefits for the individual, society, and the economy.

Unlike for children, there is no international covenant on the rights of older adults (Fredvang and Biggs 2012); indeed, the possibility of a convention is contested in so far as it might infantilize an otherwise active adult citizenship status, neither has a rights-based approach featured significantly in the age-friendly cities model. Children often cannot speak for themselves and therefore can have their rights championed by others, while active elders can speak and are concerned not to be identified as an out-group. Both approaches make claims that they benefit all ages, but do not to date specify exactly how that benefit takes place. Some common links appear in an emphasis on productivity. Children, at least in Australia, are portrayed as valuable future workers, while active agers are encouraged, almost universally, to continue to work (Moulaert and Biggs 2013). If aged friendliness principally addresses active aging, and active aging has been colonized by productive aging, then both children and older citizens conform to the needs of cities as designed for working life and for the consumption of its products.

15.5 Risks and Opportunities in Urban Environments

There is a further way in which children and older adults share a narrative link, and that is in terms of risk and vulnerability. The risks of urban living are often the most salient aspects associated with the invisibility of children and older adults in the built environment. Historically, the planning and development of modern cities functioned largely to support productive activities and overlooked the needs of young and old (van Vliet 2011). In response proponents for each age group make similar claims: Smith (2009, p. 1) argued that the needs of older adults have come after those of "children, young people and adults of working age," while Woolcock and colleagues (2010, p. 178), championing children and young people, state that "other population groups have claimed the centre stage of public debate since the 1970s." Both reflect a sense that public debate and policy have neglected their respective age groups and that not only should the urban environment function to meet specific age-based needs, but also should position children and older adults at the center of development and growth.

For older adults, the risk of social isolation has emerged as a principal concern, as reflected in the WHO's (2007) emphasis on the interconnection of transport, services, and other civic amenities and of social participation. Gleeson (2001) points to the fact that urban environments may present physical and institutional barriers that can exacerbate social exclusion, isolating particular social, and age groups from mainstream society. Evidence suggesting that urban environments and neighborhoods work to exclude older adults is persuasive (see for example Scharlach and Lehning 2013; Scharf et al. 2002) and examples of unsafe, insecure, and unwelcoming urban settings for older adults have been documented (Smith 2009). Scharlach and Lehning (2013) have proposed the age-friendly city as one that fosters social inclusion, addressing the dangers of economic vulnerability, disabling environments, and ageist social norms. And these aspirations are widely reflected in the policy documents within the age-friendly network, suggesting ways in which vulnerabilities can be reduced.

Similar evidence highlights the dangers of urban living for children, including unsafe surroundings, traffic congestion, poor transport, and in some areas lack of services and amenities (Bartlett 2002). According to Bartlett (2002, p. 7), what particularly worries children "is the extent to which they feel threatened by and excluded from their urban surroundings." In much the same way, children's vulnerabilities are at the center of child-friendly initiatives. Like older adults, they are particularly sensitive to changes in their immediate environment, and display the same connection to home and neighborhood. The children's rights approach presupposes a more substantial critique of urban environments. The rights agenda centers heavily on poverty and disadvantage, and its prevalence among urban populations. With increasing urban growth, children's rights become more important, because as UNICEF's (2010, p. 28) report card on inequality among children states "children are not to be held responsible for the circumstances in which they are born." However, a particular characteristic of current urban development links both initiatives: the unplanned, unmediated form of urban growth threatens to leave both children and older adults behind, and in Brecht's (cited in Davies 2004) words, to create countless 'victims of the Metropolis.'

Cities and urban environments also provide resources and opportunities to address people's needs. As the United Nations Population Fund report states in *Unleashing the Potential of Urban Growth* (2007, p. 1): "Cities concentrate poverty, but they also represent the best hope of escaping it." The WHO (2007, p. 4) states "Cities are the centre of cultural, social and political activity, they are a hothouse for new ideas, products and services that influence other communities and therefore the world." For UNICEF (2004, p. 1), the "importance of cities within national and political systems" underscores their potential to support child-friendly features and children's rights. Urban environments are argued to be best able to protect and advance children's rights and provide older adults with the best opportunities for health, independence, and autonomy in later life. A strong connection to locality is observed among children (see Ward 1978; Christensen and O'Brien 2002) and both groups rely heavily on their immediate environment for assistance and support (Buffel et al. 2012; Wahl and Oswald 2010). According to this argument—neighborhood and place attachment, access to a wide range of networks and availability of social and cultural resources—local space provides the most beneficial opportunities in terms of positive public space.

While there is tension between the perils and promises of urban environments, creating age and child-friendly communities appears as a task of transforming risks into opportunities. Again the dangers and threats of urban life are the potential resources to meet age-based and other social needs. Here, the proximity to amenities and services, the attachment to particular localities and social networks, provides the best opportunities from which optimal environments can be created (Macintyre and Ellaway 2000; Richard et al. 2009).

15.6 Toward Dementia-Friendly Spaces

However, there is one group that has not been invited to the party. And this is particularly true of public space. People living with dementia are not children, nor are they generally a group with whom other older adults wish to be included. In addition, they are rarely considered potential workers. Designing for dementia is cast along the lines of future-proofing or safeguarding, largely due to an expected increase in the overall numbers of people living with it. Dementia raises critical questions about later life and the assumptions upon which active, successful, and positive aging has been built. The loss of multiple cognitive functions and progressive cognitive impairment (see Claire 2003) is thought to undermine an essential component of successful aging, namely the presence and operation of a strong sense of self (Chapman 2005) with Alzheimer's Disease, the most common type of dementia, being largely regarded as a loss of self and personal identity (Herskovits 1995). In fact, dementia is one of the most feared conditions of older age (Alzheimer's Disease International 2012).

Baars and Phillipson (2013, p. 17) suggest that "interdependence and mutual recognition" are important values of the aging experience. Both are challenged by dementia itself but both are also present in the literature on creating social environments for people living with dementia and protecting personal identity in the face of mental and physical decline (Kitwood 1997).

It is possible to see the history of dementia in public space as one that has moved from a medieval perception of 'holy fools' in the community, through a period of progressive incarceration, and most recently into new forms of gated community. The next step would, according to this trajectory, be the return to a visible presence in the public sphere.

In Australia, dementia-friendly design was first developed within hospitals and residential facilities (see Fleming et al. 2003; Alzheimer's Australia 2004). The focus of much early work was on creating closed therapeutic environments (Day et al. 2000), and while this remains important (see for example Alzheimer's Australia 2014), from here it has expanded, as with much aging and health policy, outward to the wider community (Alzheimer's Australia 2013). Here, the narrative changes from dementia-containing institutions to dementia-friendly communities or societies, with the principle aim being to raise public understanding and reduce stigma. Alzheimer's Australia (2013) uses the following definition, developed by Davis et al. (2009, p. 187) to describe dementia-friendly care as a

> Cohesive system of support that recognises the experiences of the person with dementia and best provides assistance for the person to remain engaged in everyday life in a meaningful way.

The most recent example of this kind of environment is the dementia village, examples of which are given below. The dementia village exemplifies the progressive expansion of dementia-friendly environments, back into the public domain. Dementia Villages can take a number of forms, with different approaches to the creation of meaningful community living.

De Hogeweyk in Weesp, the Netherlands, takes the form of a self-contained village within which 150 people live with varying stages of dementia. The design of the village and living areas/units provides people with the ability to continue living the lifestyle they had prior to their need for care with social and retail facilities within the village itself. However, the community is gated and access is monitored. Inside, the village contains seven neighborhoods, reflecting different social class, cultural, and faith-based identities. In addition,

> The reality of an individual's life is that not every aspect of life takes place inside the home. Normalization, in its richest form, means allowing the residents to go out for grocery shopping, go out to have lunch {…}go out to the theatre {…} [and] leaving the home to see the doctor {…}so many people in aged-care settings do not experience these normal activities (Anderzhon et al. 2012, p. 149).

A project labeled Dementiaville, (mainly by its critics) exists in the Swiss village of Wiedlisbach near Bern. A mock-1950s village is planned specifically for people with dementia and Alzheimer's disease. Like De Hogeweyk the notion of maintaining normalcy is paramount, but plans are made to disguise carers as gardeners, hairdressers, and shop assistants. According to UK's *Daily Mail* (Hall 2012), the project will create an illusory world of 1950s houses and gardens, with the aim of fostering a sense of familiarity and security. Not everyone agrees with such an approach. Director of Switzerland's Sonnweid home, Michael Schmieder, stated "The very notion is an attempt to fake the normality that people with dementia don't have" (Paterson 2012).

Hammondville in NSW Australia co-locates a number of different care services run by the charity HammondCare. It is designed as an environment that promotes independence, while compensating for physical disability and cognitive impairment; small, cottage-like living units that are domestic and familiar, good visual access; and environment cueing to promote independence and safety (Anderzhon et al. 2012). The design principles reflect those developed by Fleming et al. (2003), stressing safety and security, familiarity, links to the community, domesticity, and the reduction of unwanted stimuli. Environmental cues, such as the use of color and signage, enhance way-finding and safety, while compensating for physical and psychological deficits.

In the city of Bruges, Belgium, a dementia-friendly community project has adopted the symbol of a knotted handkerchief to denote a dementia-friendly place or business wherein people with dementia will be empathically welcomed (see Alzheimer's Australia 2013). In each example, the material environment provides physical props that facilitate certain forms of supportive social relations.

15.7 Props to Identity and the Design of Shared Public Space

If what we are witnessing is the progressive re-emergence of people living with dementia into public space, to equal groups such as active agers and children, it raises questions, not simply about the limits of current thinking on age friendliness, but also the relationship between public space and supports to identity. In terms of policies that focus on age-friendly environments, examining child and dementia friendliness causes the inclusion of voices from the periphery to enhance and inform the center.

Thinkers such as Debord (1955) and Lefebvre (1968, 1996) have examined the ways in which built space determines certain forms of action, with Lefebvre claiming that urban space is 'organized ideologically.' The 'right to the city' he argued needs to be reclaimed from an emphasis on public control and the needs of production. He was particularly concerned with the ghettoization of disempowered parts of the population and their exclusion from certain areas of public space. In the 1960s, Henri Lefebvre and the Urban Situationist Guy Debord were both interested in the everyday experience of the City, the latter even invited Lefebvre (1968) to walk together to explore his ideas on 'psychogeography.' Whereas Lefebvre tended to emphasize the interaction of functions in city life (work, leisure, privacy and family life), Debord became interested in the ways in which a pre-existing urban environment could freeze the fluidity of social interaction. This fixing of social expectations, which he called 'the spectacle,' forced people to perform pre-determined roles and pathologized civic interaction around the needs of commerce and work discipline. Debord was particularly interested in the notion of 'drifting' and 'play' as behaviors that subverted conventional uses of space. A channeling of urban behavior by the determining nature of public space may help to explain why the consumer/producer active elder has more easily been assimilated into public space centerd on work, than children and people living with dementia. Further, it may be the case that handled correctly, the 'spectacle' held in place by public space may be negatively related to self-expression for working groups, but has the potential to support groups who are either intimidated by or made invisible by public space.

For the current discussion, Debord supplies an interesting connection between the material opportunities that urban environments supply and the possibilities for identity management. If, as Biggs (1999a, b) has argued elsewhere, age often becomes a performance, then the use of props supplied by suitably designed public space might extend the life of such performances, with a progressive transfer of dependence on personal masquerade to reliance of props embedded in the built environment itself. For people with advanced dementia, this might flow in the direction of increasingly engineered environments, while at the opposite end of the life course, children may require these props to be progressively withdrawn. The examples of dementia-friendly villages described above can be seen, from this perspective, as diverse experiments, providing the props to meaningful engagement for people living with dementia.

The notion of play, as what you do when freed from the discipline of work, and from being a spectacle to others and oneself, also gives meaning to the peripheral. The 'purposelessness' of advanced dementia can be seen as play in this sense, which appears to those without empathy as meaningless, but rather is a form of drift, freed from the bonds of forced direction. Similarly, the 'play' exhibited by children is unstructured, non–work determined, and increasingly contained within fixed physical and institutional boundaries. Here, Debord (1967) was particularly interested in the segregation of lived activity as determined by pre-existing material spatial arrangements. However, rather than attempting to unify these different spheres, as was the tendency in Lefebvre, he emphasized moving beyond them, and as such may hold greater promise as a means of exploring the new challenges presented by changing demographic relations. As a group that largely stands outside the material empathic creation of urban space, people with dementia and children both tell us how it works, and point to ways to make it more humane.

15.8 Toward Generationally Intelligent Space: The Material and Emotional in the Here and Now

Our public space is currently generationally configured, institutionally (Kohli 1985) and in terms of segregated geography (Laws 1995). However, the use of cities and their different sections changes over time in age-determined ways. In the UK, the role of London's suburbs is an example of space designed for families with young children which is now largely utilized by active older adults (Office of the Deputy Prime Minister 2004). Melbourne has seen a series of reports that attempt to engage with the challenge of demographic shifts including an *Inquiry into opportunities for participation of Victorian seniors* (Family and Community Development Committee 2012). The New York Active Design initiative argues that in the twenty-first century, designers can again play a crucial role in combating the most rapidly growing public health epidemics of our time: obesity and its impact on related chronic diseases such as diabetes, heart disease, and some cancers (City of New York 2010). The authors show that physical activity and a healthy diet can be increased by changing established architectural and urban design strategies. The redesign of neighborhoods, buildings, streets, and urban spaces, so that they encourage walking, bicycling, active transportation, and recreation, has been a key element in incorporating the principles of preventive public health into the urban environment.

What these examples show is not only that public space can be modified over time, but that it can be used by and to influence the well-being of specific generational groups. It would also suggest that notions like design or age friendliness 'for all ages' require qualification. And what can be learned from an excursion into the worlds of children and with the very different groups of people living with dementia is a need to focus on the material and emotional in the here and now.

Here, at least two answers emerge from the literature. The first would be to follow Gehl's (2010) promotion of the urban form according to a human scale and person-based perspective. He argues for incremental improvements that re-establish the livability of urban and public space, such as evidenced by the transformation of Copenhagen into a walkable city. It is an approach based on changing the material environment that determines public space and the forms of interaction that take place there. Second, Biggs et al. (2012) suggest that generationally intelligent spaces, ones that allow different generational groups to meet, interact and include ways of negotiating the shared use of their environment, would be a key element in intergenerationally sustainable living. Here, the authors focus on the ambivalence that arises as part of intergenerational interaction and the establishment of common goals based on an emotional understanding of the life priorities of each age group. In our case, this would require empathic understanding of the requirements of groups as diverse as children, working age adults, active agers, and people living with dementia. Both of these approaches would shift the terms of the debate toward connecting what might be called the emotional-material rhythms of public space and life. In terms of intergenerational design that balances specific with common needs, these emotional-material rhythms suggest the following points of departure:

- First, recognizing that each generational group will have life projects, arising from the point they have reached in their life course, which give rise to distinctive requirements of public space.
- Second, those intelligent generational spaces are both the result of and should enhance the ability to negotiate the sustainable use of those spaces in generational terms. The uses made of these spaces, by specific age groups separately and through shared activity, should be able to stand the test of time and produce mutually advantageous use value.
- Third, those environments should be designed so that they enhance the cooperative and emotionally empathic capacity to share space. This will include recognition of the need for generational privacy as well as of interaction and solidarity.

The aim of intergenerational urban space should be to enhance social and emotional understanding between age groups, increase harmony, and reduce generational ambivalence. An intergenerational understanding of space implies that cities are more than simply rat runs between centers of work, consumption, and closed door domesticity. They are also more than places for the active aged to relax and socialize with each other and more than a setting for healthy exercise regimes. While the complexity of urban life will invest cities with all of these elements, we must also ask "What makes urban space creative, public and delightful and what is the role of intergenerational living in all this?" In beginning to answer this question, we suggest the following: a functioning and sustainable urban space entails taking shared and distinctive generational requirements into account, negotiating diverse and possibly contradictory uses, and designing structures that can stand the test of generational time.

References

Alzheimer's Australia. (2004). *Dementia care and the built environment.* Position Paper 3, June (Online). Available: https://fightdementia.org.au/sites/default/files/20040600_Nat_NP_3DemCareBuiltEnv.pdf. (Accessed November 21, 2014).

Alzheimer's Australia. (2013). *Dementia friendly societies: The way forward. A report for Alzheimer's Australia.* Paper 31, May (Online). Available: https://fightdementia.org.au/sites/default/files/20130604_NAT_PUB_Paper31DementiaFriendlySocieties.pdf. (Accessed November 21, 2014).

Alzheimer's Australia. (2014). *Dementia care in the acute hospital setting: Issues and strategies. A report for Alzheimer's Australia.* Paper 40, June (Online). Available: https://fightdementia.org.au/sites/default/files/Alzheimers_Australia_Numbered_Publication_40.PDF. (Accessed November 21, 2014).

Alzheimer's Disease International. (2012). *World Alzheimer Report 2012: Overcoming the stigma of dementia.* ADI, London (Online). Available: http://www.alz.org/documents_custom/world_report_2012_final.pdf. (Accessed November 21, 2014).

Alzheimer's Disease International. (2013). *Policy Brief for Heads of Government: The Global Impact of Dementia 2013–2050.* ADI, London (Online). Available: http://www.alz.co.uk/research/GlobalImpactDementia2013.pdf. (Accessed November 21, 2014).

Anderzhon, J. W., Hughes, D., Judd, S., Kiyota, E., & Wijnties, M. (2012). *Design for aging: International case studies of building and program.* Hoboken, New Jersey: Wiley.

Arizona State University Herberger Center for Design Excellence. (2005). *Livable communities: An evaluation guide.* Washington, DC: AARP Public Policy Institute.

Australian Institute of Health and Welfare. (2012). *Dementia in Australia.* Cat. no. AGE 70. Canberra: AIHW (Online). Available: http://www.aihw.gov.au/WorkArea/DownloadAsset.aspx?id=10737422943. (Accessed November 21, 2014).

Baars, J., & Phillipson, C. (2013). Connecting meaning with social structure: Theoretical foundations. In J. Baars, J. Dohmen, A. Grenier, & C. Phillipson (Eds.), *Ageing, meaning and social structure: Connecting critical and humanistic gerontology* (pp. 11–30). Bristol, UK: Policy Press.

Bartlett, S. (2002). Building better cities with children and youth. *Environment and Urbanization, 14*(2), 3–10.

Biggs, S. (1999a). The 'Blurring' of the Lifecourse: Narrative, memory and the question of authenticity. *Journal of Ageing and Identity, 4*(4), 209–221.

Biggs, S. (1999b). *The mature imagination.* Buckingham: Open University Press.

Biggs, S. (2004). Age, gender, narratives and masquerades. *Journal of Aging Studies, 18*(1), 45–58.

Biggs, S., Carstensen, L., & Hogan, P. (2012). Social capital, lifelong learning and social innovation. In J. R. Beard, S. Biggs, D. E. Bloom, L. P. Fried, P. Hogan, A. Kalache, & S. J. Olshansky (Eds.), *Global Population Ageing: Peril or Promise* (pp. 39–41). Geneva, Switzerland: World Economic Forum, 2011.

Buffel, T., Phillipson, C., & Scharf, T. (2012). Ageing in urban environments: Developing 'age-friendly' cities. *Critical Social Policy, 32*(4), 597–617.

Caro, F. G., Bass, S. A., & Chen, Y.-P. (1993). *Achieving a productive aging society.* Westport, Conn: Auburn House.

Carr, A., Kimberley, H., & Biggs, S. (2013). *Child-Friendly Cities and Age-Friendly Cities: Discussion paper prepared for Hobsons Bay City Council.* Working Paper (in press). Melbourne, VIC: Brotherhood of St Laurence.

Chapman, S. A. (2005). Theorizing about aging well: Constructing a narrative. *Canadian Journal of Aging, 24*(1), 9–18.

Christensen, P., & O'Brien, M. (2002). *Children in the city: Home, neighbourhood and community.* London: Routledge.

City of New York. (2010). *Active design guidelines.* New York: City of New York.

Commonwealth of Australia. (2006). *A community for all ages—building the future*. Canberra, ACT: Department of Health and Ageing, Commonwealth of Australia.

Council of Australian Governments (COAG). (2009). *Investing in the early years—a national early childhood development strategy*. Barton, ACT: COAG, Commonwealth of Australia.

Claire, L. (2003). Managing threats to self: Awareness in early stage Alzheimer's disease. *Social Science and Medicine, 57*(6), 1017–1029.

Davies, M. (2004). Planet of slums. *New Left Review, 26*, 5–34.

Davis, S., Byers, S., Nay, R., & Koch, S. (2009). Guiding Design of dementia friendly environments in residential care settings: Considering the lived experience. *Dementia, 8*(2), 185–203.

Day, K., Carreon, D., & Stump, C. (2000). The therapeutic design of environments for people with dementia: A review of the empirical research. *The Gerontologist, 40*(4), 397–416.

Debord, G. (1955). *Introduction to a critique of urban geography* (Online). Available: http://library.nothingness.org/articles/SI/en/display/2. (Accessed November 21, 2014).

Debord, G. (1967). *The Society of the Spectacle*, translation (1970) by F. Perlman & J. Supak. New York: Black & Red.

Department for Work and Pensions. (2009). *Building a society for all ages*. Norwich, UK: Department for Work and Pensions, TSO.

District of Saanich. (2014). *Age Friendly Cities and the Municipality of Saanich* (Online). Available: http://www.saanich.ca/parkrec/community/agefriendly.html. (Accessed November 21, 2014).

Family and Community Development Committee. (2012). *Inquiry into opportunities for participation of victorian seniors*. East Melbourne, VIC: Parliament House.

Fleming, R., Forbes, I., & Bennett, K. (2003). *Adapting the ward for people with dementia*. Sydney: NSW Department of Health.

Fredvang, M., & Biggs, S. (2012). The rights of older persons: protection and gaps under human rights law. Social Policy Working Paper no. 16, Brotherhood of St Laurence & Centre for Public Policy, Melbourne, VIC: University of Melbourne.

Freeman, M. (2007). Why it remains important to take children's rights seriously. *International Journal of Children's Rights, 15*, 5–23.

Gehl, J. (2010). *Cities for people*. Washington, DC: Island Press.

Gilleard, C., & Higgs, O. (2005). *Contexts of ageing: Class, cohort and community*. Cambridge: Polity Press.

Gleeson, B. (2001). Disability and the open city. *Urban Studies, 38*(2), 251–265.

Hall, A. (2012). Welcome to Dementiaville: £17 million village in Switzerland will keep sufferers in fake reality. *Daily Mail*, January 31 (Online). Available: http://www.dailymail.co.uk/news/article-2094155/Dementiaville–17m-village-Switzerland-sufferers-fake-reality.html. (Accessed November 21, 2014).

Herskovits, E. (1995). Struggling over subjectivity: Debates about the 'Self' and Alzheimer's Disease. *Medical Anthropology Quarterly, 9*(2), 146–164.

Kitwood, T. (1997). *Dementia reconsidered: The person comes first*. Buckingham, UK: Open University Press.

Kohli, M. (1985). Die Institutionalisierung des Lebenslaufs: Historische Befunde und theoretischeArgumente [The institutionalization of the life course: Historical findings and theoretical arguments]. *Kölner Zeitschrift für Soziologie und Sozialpsychologie, 37*, 1–29.

Laws, G. (1995). Embodiment and emplacement. *International Journal of Aging and Human Development, 40*(4), 253–280.

Lefebvre, H. (1968). *The production of space*, translation (1991) by D. Nicholson-Smith. Oxford, UK: Blackwell.

Lefebvre, H. (1996). *Writings of cities,* selection, translation and introduction by E. Kofman & E. Lebas. Oxford, UK: Blackwell.

Lui, C. W., Everingham, J.-A., Warburton, J., Cuthill, M., & Bartlett, H. (2009). What makes a community age-friendly: A review of international literature. *Australasian Journal on Ageing, 28*(3), 116–121.

Macintyre, S., & Ellaway, A. (2000). Ecological approaches: Rediscovering the role of the physical and social environment. In L. Berkman & I. Kawachi (Eds.), *Social epidemiology* (pp. 332–348). Oxford: Oxford University Press.

Miller, G., & Annesley, A. (2011). *Re-positioning age friendly communities: Opportunities to take AFC mainstream*. Toronto, Ontario: Canadian Urban Institute.

Moulaert, T., & Biggs, S. (2013). International and European policy on work and retirement: Reinventing critical perspectives on active ageing and mature subjectivity. *Human Relations, 66*(1), 23–43.

National Association of Area Agencies on Aging. (2007). *A blueprint for action: Developing a livable community for all ages*. Washington DC: National Association of Area Agencies on Aging and Partners for Livable Communities.

Noyes, J. (2001). *Designing for humans*. East Sussex, UK: Taylor and Francis Group.

Office of the Deputy Prime Minister. (2004). *Sustainable cities and the ageing society*. London: Brook Lyndhurst Ltd.

Olshansky, S. J., Beard, J. R., & Börsch-Supan, A. (2012). The longevity dividend: Health as an investment. In J. R. Beard, S. Biggs, D. E. Bloom, L. P. Fried, P. Hogan, A. Kalache, & S. J. Olshansky (Eds.), *Global population ageing: Peril or promise* (pp. 57–60). Geneva, Switzerland: World Economic Forum, 2011.

Paterson, T. (2012). Switzerland's 'Dementiaville' designed to mirror the past. *The Independent*, January 24 (Online). Available: http://www.independent.co.uk/news/world/europe/switzerlands-dementiaville-designed-to-mirror-the-past-6293712.html. (Accessed November 21, 2014).

Richard, L., Gauvin, L., Gosselin, C., & Laforest, S. (2009). Staying connected: Neighbourhood correlates of social participation among older adults living in an urban environment in Montréal, Québec. *Health Promotion International, 24*(1), 46–57.

Riggio, E. (2002). Child friendly cities: Good governance in the best interests of the child. *Environment and Urbanization, 14*(2), 45–58.

Scharf, T., Phillipson, C., Smith, A., & Kingston, P. (2002). *Growing older in socially deprived areas: Social exclusion in later life*. London: Help the Aged.

Scharlach, A. E. (2009). Creating ageing-friendly communities. *Generations, 33*(2), 5–11.

Scharlach, A. E., & Lehning, A. J. (2013). Ageing-friendly communities and social inclusion in the United States of America. *Ageing and Society, 33*(1), 110–136.

Smith, A. E. (2009). *Ageing in urban environments: Place attachment and social exclusion*. Bristol, UK: The Policy Press.

The Economist Intelligence Unit Limited. (2012). *A Summary of the Liveability Ranking and Overview August 2013*. The Economist Intelligence Unit Limited (Online). Available: http://www.eiu.com/public/topical_report.aspx?campaignid=Liveability2013. (Accessed November 21, 2014).

Tranter, P., & Sharpe, S. (2007). Children and peak oil: An opportunity in crisis. *International Journal of Children's Rights, 15*, 181–197.

UNICEF. (2004). *Building Child Friendly Cities: A Framework for Action.*UNICEF, International Secretariat for Child Friendly Cities, Florence: Innocenti Research Centre.

UNICEF. (2010). *The Children Left Behind: A league table of inequality in child well-being in the world's rich countries*. Innocenti Report Card 9, UNICEF, Florence: Innocenti Research Centre.

UN. (2002). *Political Declaration and Madrid International Plan of Action on Ageing*. Second World Assemble on Ageing, Madrid, Spain. New York: UN (Online). Available: http://www.un.org/en/events/pastevents/pdfs/Madrid_plan.pdf. (Accessed November 21, 2014).

UN. (2014). *World Urbanization Prospects: The 2014 Revision, Highlights (ST/ESA/SER.A/352)*. Department of Economic and Social Affairs, Population Division (Online). Available: http://esa.un.org/unpd/wup/Highlights/WUP2014-Highlights.pdf. (Accessed November 21, 2014).

United Nations Population Fund. (2007). *The state of world population 2007*. New York: United Nations Population Fund.

van Vliet, W. (2011). Intergenerational cities: A framework for policies and programs. *Journal of Intergenerational Relations, 9*(4), 348–365.

Wahl, H.-W., & Oswald, F. (2010). Environmental perspectives on ageing. In D. Dannefer & C. Phillipson (Eds.), *The SAGE handbook of social gerontology* (pp. 111–124). London: SAGE Publications.

Walker, A. (2002). A strategy for active ageing. *International Social Security Review, 55*(1), 121–139.

Walker, A. (2009). Commentary: The emergence and application of active aging in Europe. *Journal of Aging and Social Policy, 21*(1), 75–93.

Ward, C. (1978). *The child in the city*. London: The Architectural Press.

Woolcock, G., Gleeson, B., & Randolph, B. (2010). Urban research and child-friendly cities: A new Australian outline. *Children's Geographies, 8*(2), 177–192.

WHO. (2002). *Active ageing, a policy framework.* Geneva: WHO.

WHO. (2007). *Global age-friendly cities: A guide*. Geneva: WHO.

WHO. (2012). *Dementia: a public health priority*. Geneva: WHO and ADI (Online). Available: http://www.who.int/mental_health/publications/dementia_report_2012/en/. (Accessed November 21, 2014).

WHO. (2014). *WHO Age friendly Environments Programme* (Online). Available: http://www.who.int/ageing/age_friendly_cities/en/. (Accessed November 21, 2014).

Chapter 16
Active and Healthy Ageing: Blended Models and Common Challenges in Supporting Age-Friendly Cities and Communities

Thibauld Moulaert, Kim Boudiny and Mario Paris

16.1 Introduction

At the birth of the twenty-first century, population ageing has become a central issue for governments all over the world. Taking its impact on human development and world economy into account, different international organizations have proposed their vision of ageing since the first United Nations (UN) World Assembly on Ageing in Vienna in 1982. Researchers have joined policymakers and advocates in supporting one or another perspective. While the case of the World Bank and the

T. Moulaert (✉)
REIACTIS, Dijon, France
e-mail: tmoulaert@gmail.com

T. Moulaert
School of Public Health, Université Catholique de Louvain, Louvain-la-Neuve, Belgium

T. Moulaert
Human and Social Sciences Institute, Université de Liège, Liège, Belgium

T. Moulaert
School of Social Work, University of Sherbrooke, Sherbrooke, QC, Canada

K. Boudiny
PhD FWO-Flanders, Faculty of Political and Social Sciences, CELLO (Research Center for Longitudinal and Life Course Studies), University of Antwerp, Antwerp, Belgium
e-mail: kim.boudiny@uantwerpen.be

M. Paris
Research Centre on Aging, University of Sherbrooke, Sherbrooke, QC, Canada
e-mail: mario.paris@usherbrooke.ca

© Springer International Publishing Switzerland 2016
T. Moulaert and S. Garon (eds.), *Age-Friendly Cities and Communities in International Comparison*, International Perspectives on Aging 14,
DOI 10.1007/978-3-319-24031-2_16

Organization for Economic Co-operation and Development (OECD) as supporters of a neoliberal perspective on ageing has already been explored[1] (Phillipson 2003), recent discussions focus more specifically on the World Health Organization (WHO), its promotion of "active ageing" (WHO 2002) as a more comprehensive vision of ageing, and the application to cities via "Age-Friendly Cities" (AFC) (WHO 2007) and "Age-Friendly Cities and Communities" (AFCC) since the 2010 WHO "Global Network of Age-Friendly Cities and Communities" (GNAFCC).

AFCC is a multilevel project at the crossroads of environmental gerontology (Phillipson 2004) or ecological theory (Menec et al. 2011), public policies, civil society, and the private sector. The environment is conceived as both physical (pavement, access to public areas, etc.) and social (ageist stereotypes, social relations, etc.). A variety of practices aimed at making communities more "age-friendly" have been observed in the United States (Scharlach and Lehning 2013) and around the world (Lui et al. 2009). Efforts have been made to go beyond the original WHO initiative (WHO 2007) to better address rural issues (Menec et al. 2011; Menec and Nowicki 2014), the diversity of ageing processes (Keating et al. 2013; Liddle et al. 2014; Menec et al. 2011), and the balance between bottom-up *and* top-down perspectives (Buffel et al. 2012; Garon et al. 2014; Lui et al. 2009), as the WHO initiative, and its original checklist, has often—and wrongly according to Plouffe et al. *in this book*—been described only as a top-down perspective (Buffel et al. 2012, 2013; Lui et al. 2009).

When the reference to the WHO programme is mentioned, few authors identify in depth the historical link with the WHO version of "active ageing" (WHO 2002), for example while Buffel et al. (2012) explicitly make the link by recalling the WHO's role during the 2000s in promoting a comprehensive definition of the "active" term and in consequently supporting a wide range of policy programmes, the notion of "active ageing" is only generally mentioned by Liddle et al. (2014) and Lui et al. (2009) and occasionally does not even appear in studies on Age-Friendly Cities or Communities (e.g. Greenfield et al. 2015; Warner and Morken 2013).[2] Similarly, in studies focusing on "active ageing" and acknowledging the WHO in this regard, a connection to AFC(C) is rarely made (e.g. Boudiny 2013; Clarke and Warren 2007; Walker 2009) and if it is, it tends to be rather limited (e.g. the study of Lassen and Moreira (2014) in which the relation is limited to the support of physical activity and its correlative, a "lifestyle focus", but

[1]According to Phillipson, these organizations support a neoliberal vision of ageing at a global level, characterized by a reduction of pension systems and State intervention; an extension of working lives without necessarily adapting workplaces; a global concurrency between workers in an extended free market.

[2]Although AFC(C) studies tend to refer to the notion of active ageing in only a general fashion, this is not to say that programmes offering a comprehensive perspective on ageing issues and using similar principles as those gathered in AFC(C) do not exist; for example, the *Community for a Lifetime* from Florida State existing since 1999 (Florida Department for Elder Affairs 2007) presents a close relation with the WHO perspective of supporting local communities in ageing policies.

in which the inclusion of such activity within community development support—as explored in AFCC—is largely neglected). Quebec is probably a unique case study where the links between "active ageing" (WHO 2002) and the WHO's original AFC perspective (WHO 2007) are strongly embedded in social research and policy practice (Garon et al. 2014; Paris et al. 2013b; Plouffe 2011).

In this chapter, we consider the importance of identifying the origins of active and healthy ageing behind AFCC; in doing so, this chapter further contributes to efforts to refine the original WHO AFC perspective (for a detailed discussion on the WHO methodology and implementation, see the chapter of Louise Plouffe and colleagues). To this end, we (re)introduce empowerment as a key element of "active and healthy ageing". Indeed, empowerment is not entirely absent from some active-ageing approaches (e.g. Walker 2002); however, in our view, it has not been fully described in its different meanings for ageing. More interestingly, it can be applied to AFCC. Similarly to "active ageing", empowerment is also not entirely absent from definitions of environmental gerontology. For example, Liddle et al. (2014, p. 1606) argue that "an overarching principle is that 'community' is an inclusive concept that prioritizes the participation and empowerment of its members, especially older people, in its creation and functioning". In their classification of "age-friendly communities", Lui et al. (2009) contrast top-down approaches with bottom-up empowering ones that strengthen the capacity of older people to enhance their neighbourhood and community, the latter being illustrated by the Quebec case (Garon and Veil 2011). However, authors rarely discuss various definitions of empowerment, mostly referring to an "ideal" vision without elaborating how important aspects (e.g. *multilevel perspective*, *insider's view*: infra) can be realized and integrated in practice.

Inspired by previous works on active ageing (Boudiny and Mortelmans 2011; Boudiny 2013; Moulaert and Biggs 2013; Moulaert and Paris 2013; Moulaert and Viriot Durandal 2013) and its application within AFCC programmes (Garon et al. 2014; Moulaert and Houioux 2013; Paris et al. 2012, 2013a), this chapter aims to answer a fundamental question: Where do we go from here? The first two sections present the notions of active and healthy ageing within their historical context. The third section addresses three critics applicable to both of them. Finally, the last section proposes one step forward in giving a deeper importance to empowerment. While this element has not been absent from some definitions (e.g. Walker 2002), we propose placing it at the centre of the argument. In doing so, we have the opportunity to take initial steps towards an integrative framework on active and healthy ageing that could be applied in AFCC programmes.

16.2 Active and Healthy Ageing: The Original Blend

As we look at where the WHO comes from in terms of its active and healthy-ageing notions, it is remarkable to see where it is heading with AFCC practices. Indeed, active ageing has not always been interpreted in terms of social ecology. Benefits of physical activity on health, initially resumed into the notion of "healthy ageing" (WHO 1993), have been a starting point for the WHO to promote a first vision on active ageing. For the 1999 UN International Year of Older Persons, WHO launched a project titled "Active ageing makes the difference" in which active ageing was defined individualistically as "{…}the capacity of the people, as they grow older to lead productive and healthy lives in their families, societies and economies" (WHO 1999), a perspective that is still central for some researchers (Lassen and Moreira 2014).

Although there is an important relationship between health and activity, it is important—as we will argue further on—to bear in mind the differences in nuance between the concepts of active and healthy ageing. Unfortunately, both scholars and policymakers often fail to take these differences into full consideration (Boudiny 2013). The confusion between active and healthy ageing might be the clearest in the very first official appearance of active ageing in the WHO lexicon. Indeed, in 1997, WHO titles its review "Active Ageing" (Fig. 16.1). Nevertheless, once one turns to the front page of the magazine, the table of contents largely describes healthy ageing (Fig. 16.2). Since then, the active-ageing definition used by the WHO has further evolved, but the health component continues to take a central place as expressed in its now fixed definition: "Active ageing is the process of optimizing opportunities for health, participation and security in order to enhance quality of life as people age. Active ageing applies to both individuals and population groups" (WHO 2002, p. 12).

Fig. 16.1 Cover page—active ageing. *Source* WHO (1997). Reproduced, with the permission of the publisher, from WHO (1997)

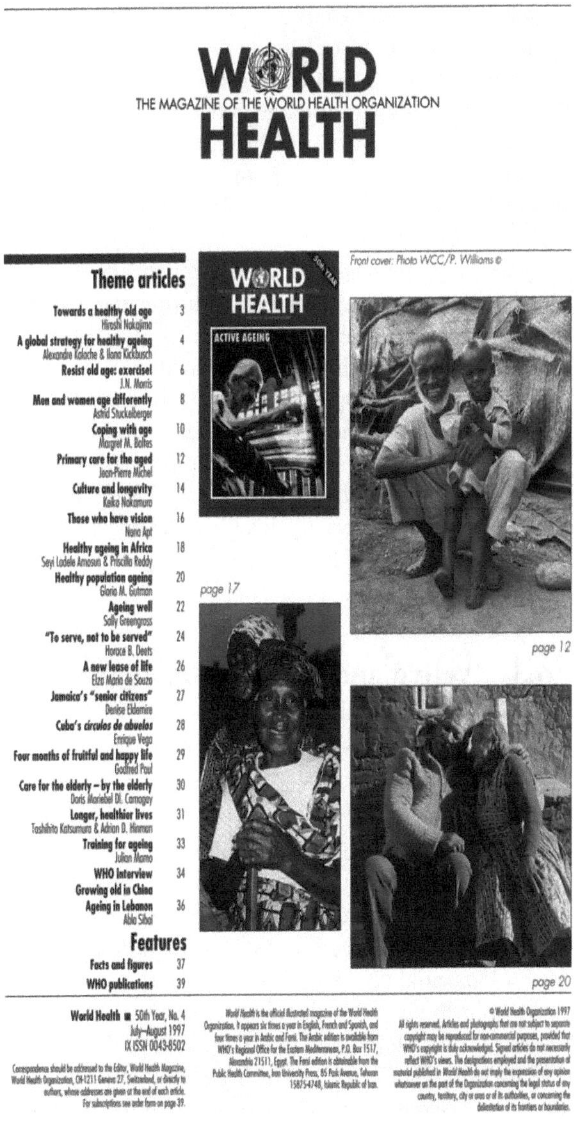

WORLD HEALTH
THE MAGAZINE OF THE WORLD HEALTH ORGANIZATION

Theme articles

Features

World Health ■ 50th Year, No. 4
July–August 1997
IX ISSN 0043-8502

Correspondence should be addressed to the Editor, World Health Magazine, World Health Organization, CH-1211 Geneva 27, Switzerland, or directly to authors, whose addresses are given at the end of each article. For subscriptions see order form on page 39.

World Health is the official illustrated magazine of the World Health Organization. It appears six times a year in English, French and Spanish, and four times a year in Arabic and Farsi. The Arabic edition is available from WHO's Regional Office for the Eastern Mediterranean, P.O. Box 1517, Alexandria 21511, Egypt. The Farsi edition is obtainable from the Public Health Committee, Iran University Press, 85 Park Avenue, Tehran 15875-4748, Islamic Republic of Iran.

© World Health Organization 1997
All rights reserved. Articles and photographs that are not subject to separate copyright may be reproduced for non-commercial purposes, provided that WHO's copyright is duly acknowledged. Signed articles do not necessarily reflect WHO's views. The designations employed and the presentation of material published in *World Health* do not imply the expression of any opinion whatsoever on the part of the Organization concerning the legal status of any country, territory, city or area or of its authorities, or concerning the delimitation of its frontiers or boundaries.

Fig. 16.2 Table of contents—healthy ageing. *Source* WHO (1997). Reproduced, with the permission of the publisher, from WHO (1997)

Figures 16.1 and 16.2. The ambiguity of active *versus* healthy ageing.[3]

The fact that the concepts of active and healthy ageing are often used interchangeably and have been blended in early WHO documents is most likely related to the historical links shared by both concepts. Both of them received increasing attention as part of a transformation of public discourses. Traditionally, older

[3]For instance, the first two articles are "Towards a Healthy Old Age" and "A Global Strategy for Healthy Ageing".

people's limitations were emphasized in the socio-gerontological literature and the societal debate. In a society characterized by three stages throughout the life course (learning, working, and resting (i.e. retirement)), old age was negatively depicted. This negative view of old age generally shifted towards more positive models in the late 1980s and especially throughout the 1990s. Fueled by anxieties about the economic implications of population ageing, such notions as productive ageing, healthy ageing, and active ageing came to the forefront (Bowling 2005; Jacobs 2004; Walker 2002).

As active ageing has received much attention from international organizations and researchers since it emerged during the 1990s (Moulaert and Viriot Durandal 2013), it has produced some recurring (though not universally applied) definitions and/or strategies, either from international bodies (WHO 2002) or from academy (Walker 2002). It appears that even more ambiguity exists on healthy-ageing definitions. From a review of 18 studies that describe and quantify the outcome of healthy ageing, Peel and her colleagues (2004, p. 115) refer to the Health Canada's definition of healthy ageing as "a lifelong process optimizing opportunities for improving and preserving health and physical, social and mental wellness, independence, quality of life and enhancing successful life-course transitions".

16.3 Active and Healthy Ageing: General Trends and Definitions

16.3.1 Active Ageing

In his chapter, Alan Walker explored active ageing as emerging at the junction of international organizations and social gerontology. There exists great variety in definitions on active ageing, both among scientific scholars and international organizations. D'Souza's (1993)[4] definition has been identified by Moulaert and Viriot Durandal (2013) as the oldest one in the literature. Since then, other scholars have contributed to active-ageing discussions. Also different international organizations like OECD, International Labour Organization (ILO), European Commission (EC), or even Non-Governmental Organizations (NGO) like AGE Platform have used a variety of definitions of "active ageing" since the mid-1990s. As described in Boudiny (2013), three general types of definitions can be discerned as commonly used in active-ageing discourses. All of them tend to stimulate the ongoing participation of older adults in society—in one way or the other—instead of equating the oldest phase of life with rest as has been traditionally done before (Jacobs 2004).

[4]"The concept of Active Aging refers to the social reconstruction of the situation of the aging in modern societies, which would provide the aging with developmental opportunities besides humanitarian care, benefiting both the aging and the society" (D'Souza 1993, p. 333).

A first group of perspectives are the *unidimensional approaches*. Given the increasing costs for pensions and health care associated with population ageing, it is not surprising that several studies focus either on fostering labour market participation or on increasing physical activity (and thus health) of older adults (Boudiny 2013). The interpretation of active ageing in an exclusively economic framework is especially prevalent—both among researchers and policymakers (Boudiny 2013; Clarke and Warren 2007; Moulaert and Biggs 2013). This was, however, not always the case. While the WHO is often presented as the only large-scale organization with a comprehensive vision of active ageing, it has been shown that the OECD was actually the first organization with a more holistic version (Moulaert and Viriot Durandal 2013). In 1998, active ageing was described as "the capacity of people, as they grow older, to lead productive lives in the society and economy. Active ageing implies a high degree of flexibility in how individuals and families choose to spend their time over life—in work, in learning, in leisure and in care-giving" (OECD 1998, p. 14). The OECD, however, abandoned this notion only two years later by considering that the active-ageing idea

> is useful in formulating longer-term policy alternatives. It can be a useful framework for developing supports to policy such as data development. However, it is far too inclusive to be a manageable theme for international monitoring and sharing of experiences. It encompasses the transition from school to work, as well as from work to retirement. It encompasses policies that range from lifelong learning to healthy ageing (OECD 2000, p. 138).

The forgoing of a holistic perspective on active ageing for the benefit of a reduced version that focuses mainly on the goal of extending careers is particularly clear in Europe[5] (Moulaert 2012). The EC stated in 1999 that active-ageing practices include lifelong learning, working longer, retiring later, and more gradually, being active after retirement and engaging in capacity-enhancing and health-sustaining activities (European Commission 1999); however—in practice—the all-encompassing potential of active ageing was largely overlooked after this 1999 document as the EU began to focus overwhelmingly on one specific dimension of it: employment (Walker 2009). As a result, Walker (2009, p. 84) was forced to admit that "nowhere do we find the comprehensive strategy set out in the 1999 EC policy document and hinted at in the 2006 one". Originally conceived as a way of redefining what it means to grow old and of transforming the identity of older people, the story of active ageing has turned into a discourse exploited by the European Employment Strategy. The latter acts on the labour supply and may encourage older workers to accept more flexible and less remunerative jobs:

> The older worker needs to be activated for the purpose of economic growth and to avoid a waste of human resources while an individualisation of the responsibilities of the activated older worker is discursively promoted. S/he is expected to be employed for longer in an

[5]In Québec, the interpretation of active ageing in a work context received little attention. Moulaert and Paris (2013) hypothesized that Quebec may be associated with a liberal market where the position of older workers in the job market is less regulated.

increasingly insecure labour market and to save for retirement in a rather insecure financial investment market, where s/he will bear the risk of any investment (Carmel et al. 2007, p. 397).

As such, the dominant focus on work in many current active-ageing approaches is potentially fueled by such broader societal trends as the trend of individual responsibility introduced by neoliberal policies. The neoliberal perspective, in which the capitalist system determines values and norms by promoting responsibility and individualism, transposes the contribution of older people into a utilitarian discourse. In this context, the concept of active ageing is defined by neoliberal ideas about the need for older people to contribute economically to society. The transition from welfare to "active societies" (promoted by OECD) has most likely played a role as well. Both welfare and "active societies" focus on work, but the former is based on an employment standard (full-time jobs for male breadwinners), while the latter expands employment to include other groups (youth, women, older people, etc.). Just as the feminist movement has rightly criticized the model of the "male breadwinner" and, at the same time, served the cause of the active society, we can apply this argument to the notion of "active societies" and gerontology. While gerontological researchers in political economy have rightly seen forms of exclusion in retirement (Guillemard 2010; Walker 1981), we can paraphrase Walters (1997, p. 226) to observe that "what the logic of [active ageing] represents, in many respects, is a turning of these partial victories against the groups which won them". That is, by criticizing retirement or early retirement, these authors provided, in spite of themselves, implicit support to the primacy of paid work. While there is thus no evidence that "working longer" is no longer the dominant format of "active ageing" in Europe (Lassen and Moreira 2014), interesting developments around "age-friendly environments" with the European Innovation Partnership on Active and Healthy Ageing (EIP-AHA) and other initiatives from the European Year 2012 for Active Ageing and Intergenerational Solidarity should be carefully followed (for more details, see the chapter by Parent and Wadoux from AGE Platform).

Notwithstanding the common reduction of active ageing to a discourse mainly focusing on labour market participation, some authors explicitly draw attention to the various ways in which one can age actively. In the so-called *multidimensional approaches* (Boudiny 2013, p. 1082), active ageing is envisioned as a concept referring to the continuous participation of older adults in several domains of life. Houben et al. (2004), for example, consider activities that require physical and/or mental effort and that occur largely outdoors (i.e. social activities). Building upon these criteria, the authors developed an active-ageing index consisting of five indicators: paid labour, care, voluntary work, sports and active recreation outdoors (e.g. participation in club life).

At last, there are the *definitions that transcend the mere behavioural level* by including such aspects as health and economic circumstances. The inclusion of health and independence is particularly prevalent in such definitions (Boudiny 2013). Bowling (2005, p. 230), for instance, conceptualizes active ageing as "continuing physical, psychological, social health, participation, independence, autonomy, control for the enhancement of quality of life". Also the WHO considers health as one of the three principal dimensions of active ageing, which is not that surprising given its mandate. This emphasis on health and independence is—not surprisingly—also present in definitions of healthy ageing, discussed below.

16.3.2 Healthy Ageing

The WHO website defines health as "a state of complete physical, mental, and social well-being and not merely the absence of disease or infirmity" (WHO 2014). However, even if many scholars and organizations recognize the WHO's definition of health, there is far less consensus about healthy-ageing operationalizations and definitions (Peel et al. 2004).

Nevertheless, we can say that healthy ageing relies on three main principles. First, like the WHO's definition of health, definitions of healthy ageing tend to offer a global approach with multiple dimensions. For instance, Peel and her colleagues refer to the 2001 Health Canada definition that states: "Healthy ageing is described as a lifelong process [of] optimising opportunities for improving and preserving health and physical, social and mental wellness, independence, quality of life and enhancing successful life-course transitions" (Health Canada 2001, underlining added). Even if healthy ageing relies on a global approach, the latter focuses mainly on the individual level of health and well-being and often does not intend to grasp the macro-level like communities. Bryant et al.'s (2001) study in which they confront healthy-ageing factors as identified in the literature (perceptions of health; well-being; valued abilities, activities, and relationships; social support; control; sense of coherence; and personal outlook) with older adults' perceptions on healthy ageing is a good example of this emphasis on the individual level. These authors have developed a model based on a specific dynamic called "going and doing": "healthy aging meant going and doing something meaningful—older people who do something meaningful feel healthier than those who do not" (Bryant et al. 2001, p. 932). With this model, they have underlined the role of subjectivity and signification in healthy ageing by four components: (1) having something worthwhile and desirable to do; (2) possessing the required abilities to meet perceived challenges; (3) obtaining the necessary resources; and (4) having the will to go and do.

Second, authors tend to underline the role of social interaction in healthy ageing, for instance by—as done by Sirven and Debrand (2013)—examining and underscoring the influence of social participation (i.e. voluntary/charity work, training course, sport/social club, religious organization, political/community organization) on self-reported health. In doing so, healthy ageing considers the individual in his social environment. To that end, Clarke and Nieuwenhuijsen (2009) use a specific framework to define healthy ageing: the International Classification of Functioning, Disability and Health (ICF). Embedded in an environmental approach, ICF "is based on the concept that health and social functioning are influenced by complex interactions between environmental factors and body functions and structures (e.g. hemiparesis due to stroke), as well as activities and participation (e.g. mobility)" (Clarke and Nieuwenhuijsen 2009, p. 15). ICF addresses three levels of functionality: the body, the activities and the social participation. It gives relevancy to environmental factors in the ageing process but, again, with a micro-level point of view: the changes this model is trying to introduce are more easily made at the level of the individual (e.g. lifestyle modifications). Instead of a concentric vision that includes environment by referring it in the end to the individual, we will argue later on (Sect. 16.5) for a parallel vision: on one side the policy level, on the other side the individual and at their crossroad actions and actors.

Third, healthy-ageing definitions often focus on the general process of ageing, while also explicitly referring to old age. In doing so, it is not surprising to observe such definitions concerned with the quantity and quality of life of people in general, with a focus on the maintenance of health through strategies of prevention and promotion. Many (quantitative) studies focus on levels of physiological, cognitive, and/or functional abilities or observed changes (e.g. deteriorations) in this regard (Bryant et al. 2001). However, in line with studies on older adults' own interpretations of active ageing (see infra), some—mainly qualitative—studies emphasize that "healthy aging does *not* mean the absence of limitations but a level of health and adaptation[6] to the aging process acceptable to the individual" (Bryant et al. 2001, p. 928).

16.4 Limitations

For a detailed overview of the criticisms for each of the three definitions of active ageing, we refer to previous work (Boudiny 2013; Moulaert and Biggs 2013; Moulaert and Paris 2013). The current section focuses on some *overall* criticisms on both the active-ageing and healthy-ageing discourse, which will foster our discussion later on regarding the way forward using empowerment and its potential for AFCC.

16.4.1 Absence of Decline

The stories of active and healthy ageing—as unfolded in many scientific and public policy discourses—have rarely been able to fully integrate the notion of decline into their play. As mentioned by Biggs et al. (2006), in the past few years several

[6]We will come back to this idea of adaptation later in the chapter: in addition to making adaptations in the sense of accepting altered levels of functioning, particular attention will be paid to alternative ways of maintaining engagement with life in face of these altered abilities.

authors have welcomed the transformation of public policy discourses because old age has never been presented in such a positive light. However, this emphasis on active, healthy, successful, or productive ageing can paradoxically undermine the value of or reject certain individual experiences associated with old age, such as disease, infirmity, impotence, frailty or vulnerability (Cardona 2008). By promoting health, physical fitness, productivity and employability, active and healthy ageing establish a normative ageing between a normal and a pathological life course. According to Holstein and Minkler (2003, p. 791), this concept of normative ageing is important because "{…} we are situated selves, embedded in society and culture and resonating with what is valued in the environment". For a number of researchers (e.g. Biggs et al. 2006), old age has lost its symbolic and cultural references, whereas youth has made corresponding gains: "Older citizens are encouraged not just to dress 'young' and look youthful, but to exercise, have sex, take holidays, socialize in ways indistinguishable from those of their children's generation" (Blaikie 1999, p. 104). Others (Boudiny 2013; Boudiny and Mortelmans 2011; Bowling 2008; Ranzijn 2010) argue that the purpose of active ageing to provide a healthy, independent and "active" old age is far from reality and fails to take diversity within the group of older adults into account—at least in common interpretations of the word "active" in which active activities are rather narrowly defined according to the interpretative criteria of researchers and policy-makers rather than older adults' own interpretations. Following from the ambiguous distinction between active and passive activities, the activities that are typically included in definitions of active ageing are—as argued in Boudiny (2013)—mainly suitable for the young-old while they appeal relatively less to the day-to-day reality of many of the old-old. See for instance the emphasis on outdoor social activities included in Houben and colleagues' (2004) multidimensional definition discussed previously: by a priori excluding solitary leisure activities, several sources of home-based leisure (e.g. reading, crossword puzzling, watching news programmes) are not taken into account while they may actually account for a substantial amount of the older old's time allocation and be perceived—by these older adults themselves—as a means to stimulate their cognitive abilities and skills and to remain engaged with life (Boudiny 2013; Boudiny and Mortelmans 2011; Klumb 2004; Pettigrew and Roberts 2008; Verbrugge et al. 1996).

Critics thus objected to the active-ageing play on the grounds of its failure to allow for dependency and senescence in old age. Biggs et al. (2006, p. 242) ironically commented: "Nobody, it seems—in the sense of dependency, withdrawal from society, plus a limited ability to both produce and consume—is old anymore". Although Walker's story embraces frailty to a certain extent, the plays of active ageing as enacted through public policies and practices do not seem to follow the "play script" (Moulaert and Paris 2013). One reason for this could be the gerontological tradition from which the active-ageing discourse stems. Driven by activity theories that stressed the importance of an active lifestyle in old age for personal life satisfaction in the 1940s–1950s (Lynott and Lynott 1996), current—overly ambitious—conceptions of active ageing seem to have fallen in the same pitfall by becoming too idealistic once again[7] (Boudiny 2013). By doing so, they (perhaps unconsciously) distinguish themselves from the other major social gerontological theory—i.e. the disengagement theory—that was proposed by Cumming and Henry (1961) and which looked at the ageing process in a fundamentally different way by focusing on the mutual withdrawal between ageing persons and society. By lapsing into one of the two, the possibilities that (still) exist for the more vulnerable older adults to remain engaged with life tend to be ignored. To push the active-ageing discourse forward, more nuanced perspectives are needed that strive for the golden mean between both theories by allowing for the diversity in patterns of ageing and explicitly acknowledging the fact that older adults' disengagement and limitations in certain areas do not necessarily preclude their engagement and possibilities in others.

16.4.2 Individualization of Responsibility

While discourses such as active and healthy ageing appear to reflect a positive view of ageing, it has been argued that they are underpinned by an economic imperative to reduce the 'burden' on society (Powell 2001; Ranzijn 2010). By arguing that older people's engagement in such activities as prolonged paid employment or voluntary work helps to bear the costs related to population ageing, the active-ageing discourse holds the implication that "if it is possible for older people to work they should do so, and that this is the responsible thing to do" (Ranzijn 2010, p. 717). Also Carmel et al. (2007) hypothesized that the European reforms based on the active-ageing story were designed to make individuals more accountable. Some political discourses clearly identify this responsibility of individuals in the management of population ageing:

> Preparing for longer, more active and better lives, working longer, retiring more gradually and seizing opportunities for active contributions after retirement are the best ways to secure the maximum degree of self-reliance and self-determination throughout old age. This is true even in the face of fading faculties and growing dependency (European Commission 1999, p. 22).

As mentioned earlier, active-ageing stories can thus lead to a great emphasis on responsibility for individuals from an economic standpoint. This individualization trend is, however, not strictly limited to older adults' economic contributions. Both active and healthy-ageing discourses tend to provide homogenized models of ageing that overly emphasize individual responsibility while not taking adequately into account the "structural inequities that affect people's lifecourse experiences and their likelihood of fitting into these glorified ways of being old" (Martinson 2006, p. 62). To this limitation, similarly to the constraints identified in a parallel story of

---

[7]The activity theory was criticized as overly idealistic because of its emphasis on the maintenance of activity patterns typical of middle age (Walker 2002).

healthy ageing, that is "successful ageing" (Rozanova 2010), Cardona (2008, p. 481) adds the following:

> The allocation of ageing as a risk category becomes a call for middle aged and older individuals to act 'responsibly' to reduce these risks by self-intervention in the name of health. The ethical obligation to exercise responsibility to minimize risks becomes an enactment of citizenship obligations - our moral duty to the community.

Such an individualistic lens may entail a type of victim blaming towards those not ageing healthy or actively in the traditional sense: since active and healthy-ageing policies, if backed up by the provision of health-promoting physical activities (e.g. exercise classes) in a supportive social environment, presumably lead to an improved and prolonged health, there is no excuse for not being active or even productive in old age (Boudiny 2013; Martinson 2006; Ranzijn 2010). As Holstein and Minkler (2003, p. 794) indicate, however: "By suggesting that the great majority of those elders in wheelchairs could indeed have been on cross-country skis had they but made the right choices and practiced the right behaviors can burden rather than liberate older people". After all, the ability to make wise choices and maintain them is partially dependent upon environmental conditions and perceived options (Fahlberg and Fahlberg 1996). We can thus ask how responsible older people really are for the success or failure of their old age—a concern raised by several critics (e.g. Laliberte Rudman 2006; Moulaert 2012). To make sure that we do not further marginalize the already marginalized, both individual and contextual factors need to be taken into account, thereby paying explicit attention to a broader analysis of why some subgroups are disadvantaged (Holstein and Minkler 2003).

16.4.3 Confusion of Terms

The transformation of public discourses has brought a range of positive notions (e.g. productive ageing, active ageing, healthy ageing) into being, many of which have been used interchangeably by various authors (e.g. Fernández-Ballesteros 2008; Lamb 2014) as Ranzijn (2010) already noted. For the purpose of the current chapter, the confounding of the terms active ageing and healthy ageing—which becomes most apparent in the third type of active-ageing definitions discussed before—is addressed.[8] Those active-ageing definitions, which exceed the mere behavioural level and treat health and independence as ultimate goals, cross the boundaries with healthy ageing (Boudiny 2013), as the latter notion implies a focus on the maintenance of health (Davey 2002). In doing so, the specific potential of the active-ageing discourse is negated. As argued by Boudiny (2013, p. 1085):

[8]For a detailed overview on overlaps between various terms, see Boudiny (2013).

> Healthy ageing and active ageing emphasise different aspects in the interrelationship between health and activity. Some physical, mental and social activities are beneficial to personal health. Good health, in turn, may positively influence personal activity levels. The central focus of active ageing is on an active, involved lifestyle. The maintenance of health is a potentially beneficial consequence. In contrast, healthy ageing stresses the identification of programmes to enhance older adults' health. Successful policies in this area may increase labour-market participation and the performance of other activities (Oxley 2009).

While this difference may seem negligible at first sight, Boudiny (2013, p. 1086) continues that

> it is precisely because of the shift in emphasis away from health and independence {…} that the active-ageing discourse allows for a crucial question: How can the active-ageing ideal be realised under circumstances of declining health? How can active ageing be fostered under circumstances of dependence?

By considering health as a determinant of active ageing rather than a constituent—i.e. by conceiving health as a means to active ageing, not the end—the common criticism on the active-ageing discourse "for offering no alternative to less able or seriously ill persons" (Perek-Białas et al. 2006, p. 568) can be tackled. New ways to remain engaged with life, in spite of health-related limitations, can be discovered (Boudiny 2013). In this regard, it is interesting to note that the AFC Guide (WHO 2007) already materializes a more distant relation with health relative to the WHO active-ageing framework (2002): as Paris et al. (2013a) showed, health quotations decreased, while social participation quotations, on the opposite, rose (Fig. 16.3). If social participation is broadly interpreted (see Sect. 5.2, the "human factor"), then such a perspective can offer opportunities also for the oldest old.

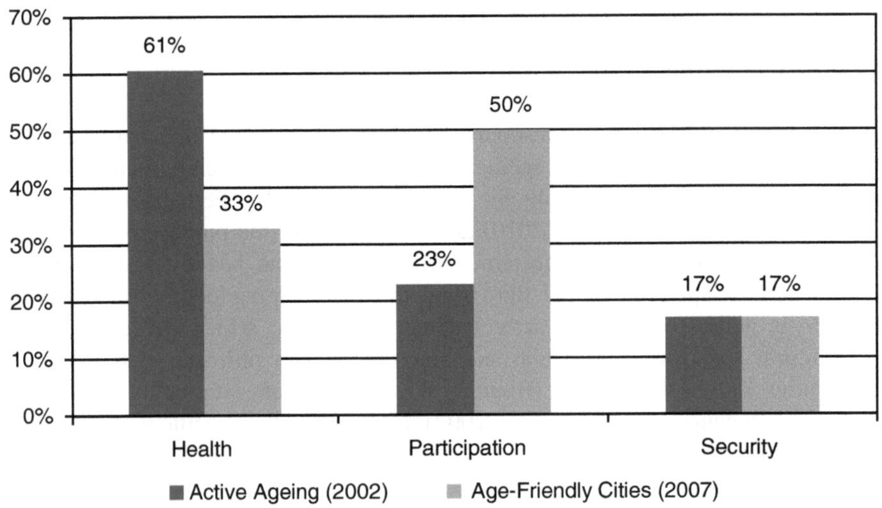

Fig. 16.3 Qualitative content analysis of WHO's Active ageing and AFC guide. *Source* Paris et al. (2013a)

In the next section, we propose to go one step further. Taking into account the limits of active and healthy ageing, we suggest forging a stronger perspective on ageing to be able to achieve an integrative framework covering the whole of the life course. In doing so, we refer to empowerment as a paradigm. This should be a strong proposition to link together active and healthy ageing and to better understand potentials for AFCC.

16.5 Empowerment: The Way Forward

Empowerment is neither entirely absent from definitions of active or healthy ageing, nor from definitions of "age-friendly" environments. Used by scholars, it is considered a tool to promote self-responsibility of midlife people for their "healthy active ageing" through the provision of information (Secker et al. 2005) or it may become one piece of the puzzle as it conceptualizes active ageing in a "combination of top-down policy action to enable and motivate activity with, also, opportunities for citizens to take action from the bottom up" (Walker 2002, p. 125). Used by international organizations such as the UN or the WHO, the term refers to a general framework to promote autonomy and individual capacity against inequality, the UN being particularly active in such a promotion, either in the domain of ageing as demonstrated by Moulaert and Viriot Durandal (2013) or in gender mainstreaming as presented by Bacqué and Biewener (2013). Inspired by the WHO perspective, local programmes like Quebec's *Municipalités Amies des Aînés* consider empowerment of older people a key element (Garon et al. 2014; Garon and Veil 2011), while environmental gerontologists similarly consider it a fundamental element for a comprehensive definition of what an "age-friendly" community should be (Liddle et al. 2014).

When it gives priority to autonomy in the service of competition and self-regulation, in the same vein as the search for "active citizenship" (Goul Andersen et al. 2005), active ageing may consequently be considered a particular form of behavioural management, with a potential negative aspect: through an emphasis on individuals' "rights and duties", their "obligations", active-ageing discourses can produce a form of "blaming the victim" criticism (cf. supra). We should, however, not fall into the other extreme of "blaming the system" either. In AFCC, such a critique is addressed to all "top-down" programmes, with a particular focus on the WHO guide (WHO 2007) and its checklist, considered as the search for an "ideal city" (Buffel et al. 2012). In order to avoid both limits ("blaming the victim"—"blaming the system"), we suggest adopting the position of Van Regenmortel (2009) by considering empowerment as a paradigm that emphasizes the role of shared responsibility in empowering processes. We further reinforce her vision with a well-informed history of empowerment calling for caution about its plural meaning throughout the world since its American origin in the 1970s (Bacqué and Biewener 2013).

Even if it is based on a variety of personal references to empowerment,[9] we justify the need for a greater centration on empowerment in the conceptualization of active ageing and its translation into AFCC applications with three arguments.

First, both active/healthy ageing and empowerment followed a similar historical track in promoting proactive action and an emancipative position. Besides the fact that they both started heavily expanding throughout different parts of the world in the 1990s, they both tend to provide a framework for a *different*, more positive story: just like active ageing distances itself from the classical deficit model (which focused on older adults' limitations) through an emphasis on older adults' competencies (Boudiny and Mortelmans 2011; Jacobs 2004), the empowerment paradigm can also be called "challenging" because it provokes the classical way of thinking and acting by focusing on health, well-being, strengths and environmental influences instead of on problems, risk factors, and the individualization of problems with the risk of blaming the victim (Van Regenmortel 2009). Indeed, similar to Walker's (2002) strategy for active ageing, Van Regenmortel's (2009, p. 29; all citations from Van Regenmortel are our translations) vision of empowerment radically challenges our common way of seeing and doing as "the empowerment paradigm gives specific "glasses" to look at socially vulnerable groups" (e.g. persons belonging to minority groups, persons in situations of dependency) with consequences for approach and policy. Through a fundamental acknowledgment of vulnerabilities, the focus of empowerment is on potentials and strengths. Since the empowerment paradigm, as envisioned by Van Regenmortel, takes *both* the strengths and vulnerabilities of people into account, the paradigm can be particularly important for older people as their strengths have tended to be overemphasized in many current approaches of active and healthy ageing (see dominant focus on the young-old, discussed earlier). Such a perspective can also be valuable for the practice of AFC(C). Here too, the criticism of a poor diversity in ageing profiles holds: Menec and Nowicki (2014) criticized AFC for only focusing on urban areas. By taking rural areas more seriously, a higher diversity of ageing profiles can be reached, including people with poor resources.

Second, both the notions of active/healthy ageing and empowerment are exposed to the same risk, expressed in terms of "empty rhetoric" for active ageing (Boudiny 2013) and in terms of "a container concept with erosion of content" for empowerment (Van Regenmortel 2009). Although positive notions such as active and healthy ageing are intuitively appealing and have become increasingly popular during the last two decades, their use is often little more than empty rhetoric, considering the prevalent focus on labour market participation (and other activities traditionally included) without truly acknowledging the multiple pathways through

[9]The different cultural and national origins of the authors of this chapter have been a strong driver for discussions; while we all work with the notion of empowerment, we build here a common knowledge and present how empowerment holds potential for AFCC.

which older adults can age actively. Similarly, while Van Regenmortel is pleased with the increased attention for empowerment and its diverse applications to target groups, themes, etc., she warns that the broad overarching framework of empowerment threatens to fade into the background. In practical applications such as AFCC, it is therefore important to examine how the overall ideas behind the notions of active/healthy ageing and empowerment can be translated to a concrete context.

Third, both active/healthy ageing and empowerment are potential strong drivers for a more complex and adaptive social policy taking into account both personal and social responsibilities and both personal and social resources and powers; they both have the potential of re-evaluating the issue of responsibility by balancing power between the individual and society. While traditional approaches to active and healthy ageing currently tend to put too much emphasis on individual responsibilities, AFCC offers a possible application for a more balanced perspective. While the social is central through amongst other things the "community" perspective, the individual is not absent: the "socially participating" older people integrated into Age-Friendly programmes are asked to be involved (and are sometimes skilled) into the local process. For example, the WHO (2007, p. 64) simultaneously calls for "a personal and collective responsibility" in exchanging information (getting informed on one side, producing reliable and accessible information on the other). All in all, this section aims to apply an "empowerment-way of thinking" to active ageing and, extensively, to AFCC.

16.5.1 Empowerment as a Paradigm

According to Van Regenmortel, empowerment is

> a process of strengthening in the course of which individuals, organizations and communities get a grip on their own situation and their environment and this by obtaining control, fostering a critical mind and stimulating participation (Van Regenmortel 2002, p. 76).

She states

> Empowerment implies strengthening and connecting people, organizations and groups in society. The paradigm prioritizes vulnerable individuals and groups, focusing on their strengths as well as recognizing their vulnerabilities and respecting their experiences. Empowerment implies strengths-oriented care which is based on the two basic premises of a positive basic attitude and suitable participation. Empowerment can be put into practice by means of appropriate methodologies (Van Regenmortel 2009, pp. 22–23).

This conception has connections with the US origin of empowerment and the development of community psychology driven by Julian Rappaport. Rappaport and followers are described as a "radical" family of empowerment by Bacqué and Biewener (2013), with some links to Paulo Freire's theory of liberation and to radical feminism; this perspective is focused on self-determination, social redistribution, political rights, social change and bottom-up power. Next to this first family, Bacqué and Biewener present the liberal (or social-liberal) family that articulates social cohesion, equal opportunities, governance and autonomy, this model being less interested in a radical discussion of social inequalities; the third and last family (called the neoliberal family) puts the market on the forefront in terms of values and practices; the State does not disappear, but it supports the "self-entrepreneur"; here, empowerment is a tool against poverty via the market and the subject is a rational actor. Even if it has different historical pathways, we argue that empowerment should essentially be understood as a *multilevel* construction. According to Van Regenmortel (2009, p. 25), at least two levels are distinguished: an individual-psychological level and a more collective sociopolitical level.

> Rappaport and Zimmerman distinguish three levels: the individual or psychological level, the level of the organization, and the community level. It is important that these levels are connected to each other. This points to the interactional nature of empowerment. Social problems such as poverty, discrimination, domestic violence arise and are perpetuated by a combination of factors on the micro-, meso- and macro level (Van Regenmortel 2009, p. 25).

By avoiding a naive faith in self-care (as well as the other extreme of paternalistic meddling by professionals), empowerment can challenge a radically liberal conception of autonomy that disregards social connections (Van Regenmortel 2009). This perspective should permit practical and political action under the heading of empowerment, even if the political level has (to date) often disappeared from social intervention according to Bacqué and Biewener (2013). For Scharlach and Lehning (2013), a similar critique (i.e. a sociopolitical level that is not sufficiently taken into account) can be addressed to the notion of "social capital" by Navarro, because of which they prefer the notion of "social inclusion". To achieve results in practice, the articulation of levels of intervention (individual, collective/community, political) is of paramount importance. Even if it comes with complex intervention, it should be oriented by a clear end. Simply put, in Ninacs' (2008) words, empowerment is the capacity to (1) choose, (2) decide and (3) act.

The diversity of local and national experiences of AFCC exposed in the second section of this book is of particular interest under such perspectives. Here, we will apply our reflection on empowerment to active and healthy ageing and discuss potential avenues for AFCC. Contrary to a main focus on "active ageing" explored elsewhere (Boudiny 2013; Walker 2002) where the primary objective was to clarify one term and explore its opportunities (a "comprehensive" vision), the current target is slightly different. We make an attempt to "integrate" the merits of both active and healthy ageing by focusing on the interplay between them. We prefer the term "integrative" as it supposes a mutual exchange between different parts of a situation and a process to produce their "coming-togetherness". We include examples to present the challenges for AFCC.

16.5.2 An Integrative Life Course Perspective on Active and Healthy Ageing Through Empowerment

Recent authors (e.g. Jacobs 2004; Moulaert 2012; Walker 2002) have critically challenged active ageing and attempts have been made to extend it to different subgroups (e.g. old-old by Boudiny 2013 or minority groups by Ranzijn 2010). For AFCC, the tension lies on a balance between "age-integrated" and "age-specific" initiatives (Scharlach 2009), which project participants in the WHO's original AFC study argued as follows:

> it should be normal in an age-friendly city for the natural and built environment to anticipate users with different capacities instead of designing for the mythical 'average' (i.e. young) person. An age-friendly city emphasizes enablement rather than disablement; it is friendly for all ages and not just 'elder-friendly' (WHO 2007, p. 72).

Building upon these contributions and inspired by the empowerment paradigm, we propose an integrative perspective on the life course ("before old age"; "young-old"; "older old") based on the forces of both active and healthy ageing. In doing so, we particularly insist on two central aspects of empowerment: the *insider's perspective* (i.e. valuing experience-knowledge and expertise, which is not only a strong source of empowerment for the (older) adults involved, but which also improves the quality of the initiatives and services developed within the context of active ageing, healthy ageing and AFCC) and the *multilevel* sites of action (i.e. personal support should come with environmental and political action at all levels) (Van Regenmortel 2009). The first element is central for a critical perspective on AFCC that focuses on "experience" instead of "ideal" (Buffel et al. 2012) and helps,[10] together with the second element, to (re)insist on social inequalities (Buffel et al. 2013).

Before old age. Although Walker (2002) underscores the importance of active ageing to be a preventive concept covering the whole of the life course, in practice scientists and policymakers concerned with active ageing have often focused on older adults (e.g. centrality of prolonging labour market participation). Moreover, from an *insider's view*, employment can only be one important element next to other dimensions of life (family, friends, social and civic participation), as recalled by Scharlach and Lehning (2013). The limited attention given to the period before old age somewhat reoccurs in the AFCC perspective, which builds on the WHO's active-ageing framework. Interestingly, the French translation of AFC is *Villes amies des aînés* (Elder-friendly cities). This gives the feeling that actions do not concern other generations (adults, young people) and it may deflect attention from potentially useful preventive campaigns while continuity theories suggest the importance of encouraging participation in potentially empowering activities early

---

[10] Van Regenmortel (2009) discusses research on socially vulnerable groups, which shows that the *insider's perspective* helps so-called outsiders to gain more insight into the various meso- and macro-level factors that attribute to a vulnerable situation besides people's own share of factors.

on. Even though the WHO sought to consider ageing as a process in formulating its general AFC approach and other generations are not denied—e.g. they are solicited when activities support or target intergenerational relations and solidarity, even as an ideal (see Fig. 16.4 at the end of the chapter). In practice, further attention could be paid towards such a public, not only as "young vs. old" but also for themselves, as "future cohorts of older people". In contrast, current healthy-ageing discourses probably already hold a clearer preventive perspective that does not focus specifically on old age: encouraging healthy lifestyles from childhood on is of particular importance as it is an important predictor for later health. For our integrative framework, this means sufficient attention must be given to younger generations (both with regard to health and empowering participation). Such an experience is being tested by the city of Dijon through the creation of its *Observatoire de l'âge* where older people constitute only one kind of citizen.

Young-old. While references to autonomy or freedom of choice occur in the active and healthy-ageing literature, it has to date not been adequately realized in practice. First, such freedom of choice is limited through the existence of age-related structural barriers that limit older adults' ability to engage in certain

Fig. 16.4 Cover WHO (2007). *Source* WHO (2007). Reproduced, with the permission of the publisher, from WHO (2007)

activities, or more generally, in life (Boudiny 2013). Mandatory retirement ages[11] that continue to exist in some industrialized countries offer one example of structural barriers related to the young-old (Boudiny 2013; Walker 2002). Removing such barriers requires a *multilevel perspective* that not only focuses on individual factors but also takes into account structural forces. Additional attention needs to be paid to other forms of inequality. For instance, inequality can resurface between genders (with, e.g. less power for those responsible for older parents), between cultures (Lamb 2014) and, in AFCC, between urban settings and rural areas (Keating et al. 2013; Menec and Nowicki 2014). From a multilevel perspective in which authorities at the meso- and macro-levels make efforts to facilitate individuals' active- (and healthy-) ageing experiences, Quebec's case is interesting as it has proposed actions at the local level (or at the level of the lived neighbourhood for big cities, just as the New York AFC programme developed) through a standardized training of an NGO active on the whole territory; this has been further complemented by the support of a positive vision of ageing at the macro-level (provincial level). Second, older adults' freedom of choice has been further restricted through a generally rather limited interpretation of the word "active"[12] (cf. supra). When, according to the *insider's view*, older adults' own interpretations are taken into account, supporting older adults' engagement in several of the activities traditionally emphasized in definitions of active ageing can be meaningful and empowering for the healthy young-old as long as it is not too narrowly focused on productive activities, and there is—as such—still some room for open interpretation (Boudiny 2013). For instance, Clarke and Warren (2007) find that many older adults re-engage with life precisely when they are freed from certain responsibilities, such as work or caring for family members, and are able to pursue other activities (e.g. travel, further education). In this regard, such AFC practices as the one in Quebec and Wallonia, which support a better local information system (via the city journal or website, via local press, via social services and home cares, etc.) to promote the importance of social, cultural and physical activities against isolation could be encouraged and extended. In addition to improving older adults' freedom of choice in the context of active ageing, a final point deserves attention. For the young-old, it may be particularly interesting for the concept of healthy ageing to complement the "activity" focus of active ageing. The implementation of programmes (e.g. nutrition campaigns, healthy lifestyle campaigns) to maintain good health or improve health where (still) possible, is useful. Since such health-improving initiatives could, in turn, exert a positive influence on the young-old's participation in several kinds of activities, a "circular" logic arises in which the notions of active and healthy ageing (and their respective merits) are truly integrated.

---

[11]Recently, there is an observable trend towards either the abolishing of mandatory retirement or an increase in age limits (OECD 2013).

[12]Instead alternative ways of ageing actively should be acknowledged, hereby presenting the older person with a wider choice of active-ageing activities (Boudiny 2013).

Older old. Active ageing is often dismissed for the older old, the umbrella notion of successful ageing (including healthy ageing) being attacked of being even a "counterproductive model" (Lamb 2014, p. 51). The strongest conceptual step forward is probably situated here. When health starts to decline rapidly, or when we focus on the old–old, the concept of active ageing becomes particularly relevant by focusing on older adults' engagement with life rather than primarily focusing on their health or reducing the active-ageing concept to economic engagement or engagement in highly physical or highly social ("youthful") activities (Boudiny 2013). In this regard, Boudiny (2013) proposed three principles that can shape active-ageing policies throughout various phases of life (including situations of declining health and care dependency) and which were—although not explicitly stated at the time—built against an empowerment background. From a *multilevel perspective*, the power of adaptability principle mainly responds to the importance of *individual* strengthening as it aims to help older people accept and integrate age-related changes into their lives. For the old-old, supporting a refocusing process towards alternative ways of ageing actively may be one of the useful practices in this regard, as—in line with the *insider's perspective*—qualitative research shows that many of them consider "ordinary" activities (e.g. reading, crossword puzzling, gardening) as a more important indicator of their involvement with life than the typically included "youthful" activities (e.g. Clarke and Warren 2007). The second guideline, the human factor, acknowledges the social nature of human beings as even for the old-old, who often experience a shift from large social networks towards emotionally close relationships, engagement in social life (though altered) remains important. Measures directed at the *collective* level can facilitate older adults' use of power sources in their own environments (Jacobs et al. 2005), e.g. social support (and participation) from family and other long-term care residents in caregiving settings. In AFCC, this refers to attempts to better include the image of older people (including those with disabilities) in the neighbourhood. In this regard, Walloon initiatives of embedding such practices as playing cards in rural settings are interesting to follow. At last, the primacy of agency over age-related structural barriers, relates to the *broader political–societal level* and the importance of tackling (as much as possible) ageist practices and patterns of inequality that are (structurally) embedded in societies (Jacobs et al. 2005; Thompson and Thompson 2001). Not only the young-old's agentic capacities are at risk of being limited (e.g. mandatory retirement ages, supra), also the old-old's potential for maintaining engagement with life may be limited by barriers exclusively based on age or dependency (for instance, structural forces in caregiving settings that limit older care receivers' contribution to decision-making processes, see Boudiny 2013). Since the Protocol of Vancouver, the WHO is conscious of the need to include the older old into consultations (by listening to their own voices or the one of their caregivers) to avoid an ageist bias in favour of the young-old. However, it is unclear how far this broader level has impacted local and national programmes, through the

GNAFCC. The challenge of including the voices of the older old remains huge for AFCC: they should not only be listened to, but also be enacted upon. To do so, the importance of a well-considered vision on empowerment appears obvious.

16.6 Conclusion: Empowerment and Its Potential for AFCC

This chapter aims to examine active- and healthy-ageing perspectives behind AFCC and to better integrate both concepts through a nuanced proposition structured by a twofold application of empowerment, the first taking account of the *insiders' perspective*, the latter promoting a *multilevel perspective*. This nuanced proposition, illustrated by some applications in AFCC, is also an attempt to go beyond the limits of active and healthy ageing by better addressing the diversity of ageing profiles.

Interestingly in this regard are the ambiguous title pages of the first use of active ageing by the WHO (recall Figs. 16.1 and 16.2). While the image on the cover page focuses on labour and the table of contents is highly concerned with health-related issues and healthy ageing, the accompanying images tell a somewhat different story. By presenting images of (very) older people, from less developed countries, the WHO (perhaps) unconsciously suggested a first more nuanced perspective. All elements were present at that time (the term active ageing; the content of images also referring to apparently very old people; intergenerational relations; work and non-work activities; and smiles on the faces), but they were not yet sufficiently integrated. Through an empowerment perspective, that insists on an *insider's view* and a *multilevel perspective* for action, we believe that a more integrative perspective on ageing can be applied, here to AFCC. This can be illustrated by another WHO image, i.e. the cover of the AFC guide (Fig. 16.4) (WHO 2007).

This image deals with crucial elements of a more nuanced perspective on active ageing and AFCC. First, in line with an *insider's view*, other aspects of life are represented than (merely) labour or health-promoting physical activities. Second, because of the use of silhouette-like images, the persons represented are somewhat "stripped of their age". Apart from the children, it is hard to tell which generation some of the persons come from. This could relate to the WHO's ideal of a city that is friendly for all ages, not just "elder-friendly" (cf. life course perspective). Finally, next to the people, a *multilevel view* can be applied to the "environment". At first sight, the bench promotes the "human factor" as it permits older people to continue to get involved in their neighbourhood. Interestingly, at a second level, the "city" appears as a background that acts as a source of "social support" (Scharlach and Lehning 2013). If the bench reveals a potential empowerment process, it also corresponds to the search for the "person-environment fit" (Keating et al. 2013). From the "person" perspective, ageing deals with its own history to find appropriate ways to continue to be engaged in its personal, familial and neighbourhood life.

From an "environment" perspective, ageing deals with the physical and social barriers to enhance social participation in local significant places.

In conclusion, in this chapter, we tried to work towards an integrative life course perspective on active and healthy ageing through empowerment. Future research could build upon this line of investigation and further explore potentials of active and healthy ageing for AFCC. In order to do so, it could be useful to gather more knowledge on how AFCC practices around the world rely in their routine on active and/or healthy ageing.[13] In addition, a deeper knowledge of healthy ageing and its links with "Healthy cities" could reveal some important similarities and/or differences with AFCC. With these suggestions in mind, we hope that AFCC may contribute to a more adapted and inclusive society, not only for older people, but for all ages.

References

Bacqué, M.-H., & Biewener, C. (2013). *L'empowerment, une pratique émancipatrice*. Paris: La Découverte.

Biggs, S., Phillipson, C., Money, A.-M., & Leach, R. (2006). The age-shift: Observations on social policy, ageism and the dynamics of the adult lifecourse. *Journal of Social Work Practice, 20* (3), 239–250.

Blaikie, A. (1999). *Ageing and popular culture*. Cambridge: Cambridge University Press.

Boudiny, K. (2013). "Active ageing": From empty rhetoric to effective policy tool. *Ageing and Society, 33*(06), 1077–1098.

Boudiny, K., & Mortelmans, D. (2011). A critical perspective: Towards a broader understanding of "active ageing". *Electronic Journal of Applied Psychology, 7*(1), 8–14.

[13]Back to the introduction of this chapter, it might help to build more bridges worldwide, including with American gerontologists who produce comprehensive and innovative thinking on "Age-friendly community initiatives" (see the recent publication of Greenfield et al. 2015) without noticing the potential relations with the European thinkers of "active ageing" (e.g. Boudiny 2013; Walker 2009). With their attention to the prevention of problems and 'consumers' sharing in the power of decision-making, the work of Greenfield and colleagues (2015, p. 192) shows some common grounds (that could be explored in future research) with ideas expressed in this chapter. The same holds for US researchers linking child- and age-friendly communities based on the idea that both population groups have similar needs (Warner and Morken 2013): by asking for planning tools such as affordable housing and accessible parks that can be used to promote *both* a child- and age-friendly community, the authors hope to design communities that meet the needs of all residents *across the entire life course*. While no attention is given to the notions of active and healthy ageing, the incorporation of these terms could be useful, because of the life course perspective that is inherently embedded in these terms (at least in the way they have been conceptualized throughout this chapter). By doing so, the authors' main focus on the implementation of tools and services that are beneficial to various age groups, could be complemented by attention for the enhancement of active and healthy ageing across the life course 'within individuals' (e.g. developing healthy lifestyles from childhood on)—and this not solely driven by an economic imperative (i.e. not just to build "the workforce of the future", Warner and Morken, p. 55), as it is also presented in the chapter by Biggs and Carr. Hopefully this book may contribute to the enhancement of such (worldwide) connections.

Bowling, A. (2005). *Ageing well: Quality of life in old age*. Maidenhead, UK: Open University Press.
Bowling, A. (2008). Enhancing later life: How older people perceive active ageing? *Aging and Mental Health, 12*(3), 293–301.
Bryant, L. L., Corbett, K. K., & Kutner, J. S. (2001). In their own words: A model of healthy aging. *Social Science and Medicine (1982), 53*(7), 927–941.
Buffel, T., Phillipson, C., & Scharf, T. (2012). Ageing in urban environments: Developing "age-friendly" cities. *Critical Social Policy, 32*(4), 597–617.
Buffel, T., Phillipson, C., & Scharf, T. (2013). Experiences of neighbourhood exclusion and inclusion among older people living in deprived inner-city areas in Belgium and England. *Ageing and Society, 33*(1), 89–109.
Cardona, B. (2008). "Healthy ageing" policies and anti-ageing ideologies and practices: On the exercise of responsibility. *Medicine, Health Care and Philosophy, 11*(4), 475–483.
Carmel, E., Hamblin, K., & Papadopoulos, T. (2007). Governing the activation of older workers in the European Union. The construction of the "actived retiree". *International Journal of Sociology and Social policy, 27*(9–10), 387–400.
Clarke, A., & Warren, L. (2007). Hopes, fears and expectations about the future: What do older people's stories tell us about active ageing? *Ageing and Society, 27*(4), 465–488.
Clarke, P., & Nieuwenhuijsen, E. R. (2009). Environments for healthy ageing: A critical review. *Maturitas, 64*(1), 14–19.
Cumming, E., & Henry, W. E. (1961). *Growing old. The process of disengagement*. New York: Basic Books.
Davey, J. A. (2002). Active Ageing and education in mid and later life. *Ageing and Society, 22*(1), 95–113.
D'Souza, V. (1993). The concept of active aging. *Indian Journal of Social Work, 54*(3), 333–344.
European Commission. (1999). *Towards a Europe for all ages. Promoting prosperity and intergenerational solidarity*. Brussels: European Commission.
Fahlberg, L. L., & Fahlberg, L. A. (1996). Exercise programmes and the promotion of health. In J. Kerr, A. Griffiths, & T. Cox (Eds.), *Workplace Health—employee fitness and exercise* (pp. 101–115). London: Taylor & Francis.
Fernández-Ballesteros, R. (2008). *Active aging. The contribution of psychology*. Cambridge, MA: Hogrefe & Huber Publisher.
Garon, S., Paris, M., Beaulieu, M., Veil, A., & Laliberté, A. (2014). Collaborative partnership in age-friendly cities: Two case studies from Quebec, Canada. *Journal of Aging and Social Policy, 26*(1–2), 73–87.
Garon, S., & Veil, A. (2011). Les Villes amies des aînés au Québec : un mouvement de changement à large échelle en faveur des aînés. *Vie et Vieillissement, 9*(1), 6–12.
Goul Andersen, J., Guillemard, A.-M., Jensen, P. H., & Pfau-Effinger, B. (2005). *The changing face of welfare: Consequences and outcomes from a citizenship perspective*. Bristol: Policy Press.
Greenfield, E. A., Oberlink, M., Scharlach, A. E., Neal, M. B., & Stafford, P. B. (2015). Age-friendly community initiatives: Conceptual issues and key questions. *The Gerontologist, 55*(2), 191–198.
Guillemard, A.-M. (2010). *Les défis du vieillissement. Age, emploi, retraite. Perspectives internationales*. Paris: Armand Colin.
Health Canada. (2001). *Workshop on Healthy Aging*. (Online). Available fromhttp://www.hc-sc.gc.ca/seniors-aines/pubs/healthy_aging/intro_e.htm. Accessed May 23, 2002 by Peel et al. (2004).
Holstein, M., & Minkler, M. (2003). Self, society, and the "new gerontology". *Gerontologist, 43*, 787–796.
Houben, M., Audenaert, V., & Mortelmans, D. (2004). Vrije tijd en tijdsbesteding. In T. Jacobs, L. Vanderleyden, & L. Vanden Boer (Eds.), *Op latere leeftijd: de leefsituatie van 55- plussers in Vlaanderen* (pp. 225–252). Antwerpen: Garant.

Jacobs, G., Braakman, M., & Houweling, J. (2005). *Op eigen kracht naar gezond leven. Empowerment in de gezondheidsbevordering: concepten, werkwijzen en onderzoeksmethoden*. Utrecht: Universiteit voor Humanistiek.
Jacobs, T. (2004). "Actief ouder worden" en "autonomie":twee kernbegrippen voor het ouderenbeleid? In T. Jacobs, L. Vanderleyden, & L. Vanden Boer (Eds.), *Op latere leeftijd: de leefsituatie van 55- plussers in Vlaanderen* (pp. 329–342). Antwerpen: Garant.
Keating, N., Eales, J., & Phillips, J. E. (2013). Age-friendly rural communities: Conceptualizing "Best-Fit". *Canadian Journal on Aging, 32*(4), 319–332.
Klumb, P. L. (2004). Benefits from productive and consumptive activities: Results from the Berlin aging study. *Social Indicators Research, 67*(1–2), 107–127.
Laliberte Rudman, D. (2006). Shaping the active, autonomous and responsible modern retiree: An analysis of discursive technologies and their links with neo-liberal political rationality. *Ageing and Society, 26*, 181–201.
Lamb, S. (2014). Permanent personhood or meaningful decline? Toward a critical anthropology of successful aging. *Journal of Aging Studies, 29*, 41–52.
Lassen, A. J., & Moreira, T. (2014). Unmaking old age: Political and cognitive formats of active ageing. *Journal of Aging Studies, 30*, 33–46.
Liddle, J., Scharf, T., Bartlam, B., Bernard, M., & Sim, J. (2014). Exploring the age-friendliness of purpose-built retirement communities: Evidence from England. *Ageing and Society, 34*(09), 1601–1629.
Lui, C.-W., Everingham, J.-A., Warburton, J., Cuthill, M., & Bartlett, H. (2009). What makes a community age-friendly: A review of international literature. *Australasian Journal on Ageing, 28*(3), 116–121.
Lynott, R. J., & Lynott, P. P. (1996). Tracing the course of theoretical development in the sociology of aging. *The Gerontologist, 36*(6), 749–760.
Martinson, M. (2006). Opportunities or obligations? Civic engagement and older adults. *Generations, 30*(4), 59–65.
Menec, V. H., Means, R., Keating, N., Parkhurst, G., & Eales, J. (2011). Conceptualizing age-friendly communities. *Canadian Journal on Aging, 30*(3), 479–493.
Menec, V. H., & Nowicki, S. (2014). Examining the relationship between communities' "age-friendliness" and life satisfaction and self-perceived health in rural Manitoba, Canada. *Rural and Remote Health, 14*, 14.
Moulaert, T. (2012). *Gouverner les fins de carrière à distance. Outplacement et vieillissement actif en emploi*. Bruxelles: Peter Lang.
Moulaert, T., & Biggs, S. (2013). International and European policy on work and retirement: Reinventing critical perspectives on active ageing and mature subjectivity. *Human Relations, 66*(1), 23–43.
Moulaert, T., & Houioux, G. (2013). *Les projets « villes-amies des ainés » wallons en pratique. « Pour et par les aînés ». Enjeux de la participation sociale et du politique. Rapport de recherche final*. Namur: Ministre de la Santé, de l'Action sociale et de l'Égalité des Chances.
Moulaert, T., & Paris, M. (2013). Social policy on ageing: The case of "Active Ageing" as a theatrical metaphor. *International Journal of Social Science Studies, 1*(2), 113–123.
Moulaert, T., & Viriot Durandal, J.-P. (2013). De la notion au référentiel international de politique publique : Le savant, l'expert et le politique dans la construction du vieillissement actif. *Recherches Sociologiques et Anthropologiques, 44*(1), 11–31.
Ninacs, W. A. (2008). *Empowerment et intervention : Développement de la capacité d'agir et de la solidarité*. Québec: Les Presses de l'Université Laval.
OECD. (1998). *Maintaining prosperity in an ageing society*. Paris: OECD.
OECD. (2000). *Reforms for an ageing society*. Paris: OECD.
OECD. (2013). *Ageing and Employment Policies: Norway 2013: Working better with Age*. Paris: OECD Publishing.
Oxley, H. (2009). *Policies for healthy ageing: An overview*. OECD Health Working Papers No. 42. Paris: OECD Publishing.

Paris, M., Beaulieu, M., Garon, S., & Bigonnesse, C. (2013a). «Selon le temps, la manière». Ou en quoi «Villes-amies des aînés» se distingue de quatre documents internationaux sur le vieillissement. *Canadian Review of Social Policy, 2*(68–69), 10–23.

Paris, M., Garon, S., & Beaulieu, M. (2012). Le projet "Villes-amies des aînés au Québec". *Les Politiques Sociales, 72*(1–2), 91–100.

Paris, M., Garon, S., Chapon, P.-M., Laliberté, A., & Veil, A. (2013b). Le programme «Municipalités amies des aînés au Québec», un modèle pour la France? *Retraite et société, 1* (64), 181–189.

Peel, N., Bartlett, H., & McClure, R. (2004). Healthy ageing: How is it defined and measured? *Australasian Journal on Ageing, 23*(3), 115–119.

Perek-Białas, J., Ruzik, A., & Vidovićová, L. (2006). Active ageing policies in the Czech Republic and Poland. *International Social Science Journal, 58*(190), 559–570.

Pettigrew, S., & Roberts, M. (2008). Addressing loneliness in later life. *Aging and Mental Health, 12*(3), 302–309.

Phillipson, C. (2003). Globalization and the reconstruction of old age: New challenges for critical gerontology. In S. Biggs, A. Lowenstein, & J. Hendricks (Eds.), *The need for theory: Critical approaches to social gerontology* (pp. 163–179). Amityville, NY: Baywood Publishing.

Phillipson, C. (2004). Urbanisation and ageing: Towards a new environmental gerontology. *Ageing and Society, 24*, 963–972.

Plouffe, L. (2011). Villes amies des aînés: de ses origines à son déploiement international. *Vie et Vieillissement, 9*(1), 14–18.

Powell, J. (2001). Theorizing gerontology: The case of old age, professional power, and social policy in the United Kingdom. *Journal of Aging and Identity, 6*(3), 117–135.

Ranzijn, R. (2010). Active ageing-another way to oppress marginalized and disadvantaged elders? Aboriginal elders as a case study. *Journal of Health Psychology, 15*(5), 716–723.

Rozanova, J. (2010). Discourse of successful aging in The Globe & Mail: Insights from critical gerontology. *Journal of Aging Studies, 24*(4), 213–222.

Scharlach, A. E. (2009). Frameworks for fostering aging-friendly community change. *Generations, 33*(2), 71–73.

Scharlach, A. E., & Lehning, A. J. (2013). Age-friendly communities and social inclusion in the United States of America. *Ageing and Society, 33*(1), 110–136.

Secker, J., Bowers, H., Webb, D., & Llanes, M. (2005). Theories of change: What works in improving health in mid-life? *Health Education Research, 20*(4), 392–401.

Sirven, N., & Debrand, T. (2013). La participation sociale des retraités en Europe: instrument du bien vieillir ou facteur d'inégalités sociales et de santé?*Retraite et Société, 65*(2), 59–80.

Thompson, N., & Thompson, S. (2001). Empowering older people beyond the care model. *Journal of Social Work, 1*(1), 61–76.

Van Regenmortel, T. (2002). Empowerment en Maatzorg. Een krachtgerichte psychologische kijk op armoede. In J. Vranken, K. de Boyser, D. Geldof, & G. Van Menxel (Eds.), *Armoede en Sociale Uitsluiting* (pp. 71–84). Leuven/Leusden: Acco.

Van Regenmortel, T. (2009). Empowerment als uitdagend kader voor sociale inclusie en moderne zorg. *Journal of Social Intervention: Theory and Practice, 18*(4), 22–42.

Verbrugge, L. M., Gruber-Baldini, A. L., & Fozard, J. L. (1996). Age differences and age changes in activities: Baltimore longitudinal study of aging. *The Journals of Gerontology Series B: Psychological Sciences and Social Sciences, 51B*(1), S30–S41.

Walker, A. (1981). Towards a political economy of old age. *Ageing and Society, 1*(1), 73–94.

Walker, A. (2002). A strategy for active ageing. *International Social Security Review, 55*(1), 121–139.

Walker, A. (2009). Commentary: The emergence and application of active aging in Europe. *Journal of Aging and Social Policy, 21*(1), 75–93.

Walters, W. (1997). The "active society": New designs for social policy. *Policy and Politics, 25* (3), 221–234.

Warner, M. E., & Morken, L. (2013). Building child and age friendly communities in tight fiscal times. *The Municipal Year Book*, 47–56.

WHO. (1993). *Health for All.* Copenhagen: WHO.

WHO. (1997). Active Ageing. *World Health, WHO Magazine*, 50th Year (4).
WHO. (1999). *Global movement for active ageing*. Geneva: WHO.
WHO. (2002). *Active ageing*.A contribution of the World Health Organization to the Second United Nations World Assembly on Ageing. Madrid: WHO.
WHO. (2007). *Global Age-friendly Cities: A Guide*. Geneva: WHO.
WHO. (2014). *Health*. (Online). Available from http://www.who.int/trade/glossary/story046/en/. Accessed May 20, 2014.

Chapter 17
Age-Friendly Cities: For Whom? By Whom? For What Purpose?

Andrew E. Scharlach

17.1 Introduction

This chapter addresses complexities and controversies regarding the conceptualization and creation of age-friendly cities (AFC). Many of these reflect unresolved issues concerning the very nature of AFC, including what they are, whom they are for, how they can be created, and indeed whether or not AFC are even desirable or possible. The Chapter first examines overarching perspectives and underlying conceptualizations of AFC, then explores the characteristics that are indicative of age friendliness, follows with a critical examination of approaches to helping cities become more age-friendly, and closes with a number of issues that require further attention if we truly are to create livable communities for all ages.

17.2 What Does It Mean to Be an "Age-Friendly City"?

Despite the growing worldwide attention to creating AFC, it is rather remarkable how little attention has been given to the basic question of what is an age-friendly city, or for that matter the tacit assumptions and conceptualizations that undergird existing models of age friendliness. For the most part, there seems to be the sense that "you know an age-friendly city when you see one" and, furthermore, that the same definitions of age friendliness can (or should) be equally salient for all cities and all populations.

That underlying conceptualizations might be contested is suggested by the variations in terminology that are used. Even a quick survey of existing initiatives

A.E. Scharlach (✉)
The Center for the Advanced Study of Aging Services, School of Social Welfare, University of California, Berkeley, Berkeley, USA
e-mail: scharlach@berkeley.edu

© Springer International Publishing Switzerland 2016

T. Moulaert and S. Garon (eds.), *Age-Friendly Cities and Communities in International Comparison*, International Perspectives on Aging 14,
DOI 10.1007/978-3-319-24031-2_17

and reports reveals recurrent use of the following terms: age-friendly, elder-friendly, aging-friendly, community for all ages, healthy city, livable city, livable community for all ages, etc. The World Health Organization (WHO) focuses on creating "AFC and communities." American Association of Retired Persons (AARP), an organization whose entire mission concerns promoting the well-being of older Americans, on the other hand, has focused its efforts on creating "livable communities"; while the National Association of Area Agencies on Aging (N4A), charged by the Older Americans Act to plan and provide for the needs of persons age 60 and older, takes this idea one step further by advocating for "livable communities for all ages."

Some argue that these terms all mean the same thing, that the basic idea is a community that is good for older adults and for everyone, and that differences are purely semantic. While such a position may make sense politically, in order to smooth over differences as the AFC movement strives to grow by being as inclusive as possible, such claims blur critical distinctions that need to be addressed, for practical as well as intellectual reasons.

I suggest that we would be better served by deconstructing the notion of "age friendliness" in order to clarify what it is and what it is not. In an effort to move forward our collective thinking in this regard, I propose that AFC are evidenced by three related, but conceptually and empirically distinct, constructs: (1) Livability (i.e., a city whose physical and social environments promote the health and well-being of all of its human inhabitants); (2) Elder Friendliness (i.e., supporting the well-being of elderly residents); and (3) Aging Friendliness (i.e., supporting constructive developmental processes, especially but not exclusively in the latter part of the life cycle). In this section, I also consider the importance of adopting a perspective on age friendliness that is communal as well as transactional.

17.2.1 Livability

Livability is perhaps the most basic condition of age friendliness, transcending age, and other individual or group characteristics. It is difficult to see how a city can be age-friendly if it is not a reasonably decent place for human beings to live. The WHO definition of "age-friendly community," for example, includes the following criteria: "optimiz[es] opportunities for health, participation and security" (WHO 2007, p. 1). While health, participation, and security seem like intuitively desirable goals for older adults, one would be hard put to find any particular population for whom such goals would not be relevant. AARP defines a "livable community" as one that is safe and secure, has affordable and appropriate housing and transportation options, and has supportive community features and services (AARP 2013). These descriptions echo the general characteristics of a "healthy city" identified by WHO: basic sanitation and hygiene, access to health care, a health-supportive environment, and the opportunity to achieve a good quality of life (http://www.who.int/healthy_settings/types/cities/en/).

17.2.2 Elder Friendliness

An age-friendly city certainly must, at a minimum, support the well-being of its elderly residents, including their needs and their capabilities. Such efforts vary in terms of their relative focus on compensatory goals and enabling goals. Compensatory perspectives on age friendliness tend to focus on age-related deficits, and the potential supportive roles of physical and social environments in meeting older residents' needs for safety and security in the face of declining abilities. Lawton's concept of person–environment (P×E) fit, for example, reflects the ability of physical and social environments adapting to the declining functional capacities of individuals as they age.

Much of the attention given to compensatory needs is focused on helping older community members to live in their own homes and apartments rather than in care facilities, a priority of individual elders for psychological and social reasons and a priority of governments for economic reasons. This focus is captured in the term "ageing in place," which is defined by the Centers for Disease Control and Prevention (CDC) as "the ability to live in one's own home and community safely, independently, and comfortably, regardless of age, income, or ability level" (http://www.cdc.gov/healthplaces/terminology.htm). CDC's Healthy Aging Research Network asserts that healthy aging is likely to be achieved when "physical environments and communities are safe, and support the adoption and maintenance by individuals of attitudes and behaviors known to promote health and well-being, and by the effective use of health services and community programs to prevent or minimize the impact of acute and chronic disease on function" (Healthy Aging Research Network Writing Group 2006, p. 3).

Elder friendliness is similar to, but distinct from, disability friendliness, which focuses primarily on accessibility. The United States (US) Americans with Disability Act (ADA), for example, focuses primarily on increasing accessibility for individuals with observable disabilities, especially mobility limitations, and particularly the needs of young people using wheelchairs (e.g., curb cuts, wide doorways, etc.). Less attention is given to environmental modifications for individuals with usual age-related limitations, such as decreased physical and cognitive stamina, reduced resilience, slower functioning physically and cognitively, and lack of familiarity with recent technological innovations. Some "disability-friendly" accommodations actually serve as barriers to older adults, such as the removal of benches, poles, and other "mobility barriers" (for wheelchairs), which tend to serve as "mobility supports" for unstable and tired older adults. Similarly, the cul-de-sac design so common among suburban housing developments in the post-World War II period may have provided safety for children years ago, but it may now contribute to isolation and limit the ability to get to shopping and other destinations when driving becomes problematic.

Enabling perspectives tend to focus on optimizing physical and social functioning, including opportunities for older residents to participate more fully in valued social roles in areas such as employment, social participation, and independent living. "Active ageing," for example, has been a primary focus of recent European efforts, as noted by Alan Walker in his chapter. AARP sees "livable communities" as helping older adults to "enhance personal independence; allow residents to age in place; and foster residents' engagement in the community's civic, economic, and social life" (AARP 2013).

Recent conceptualizations of age friendliness have attempted to include both compensatory and enabling goals. Since 2010, the WHO's Age-Friendly Environments Programme, for example, defines an age-friendly city as one that "encourages active ageing by optimizing opportunities for health, participation and security in order to enhance quality of life as people age." The Milken Institute's *Best Cities for Successful Ageing* index, similarly, includes general indicators that are apt to be relevant regardless of age (e.g., cost of living, crime and safety, overall economic prosperity, and weather), as well as specific indicators of particular relevance to older persons (e.g., high-quality health care and wellness programs, the availability of specialized housing and living arrangements, financial factors, transportation and convenience, continuing education and job training programs, and community engagement).

17.2.3 Aging Friendliness

Finally, an age-friendly city supports constructive aging processes, across the life cycle. Aging neither begins nor ends at a certain chronological age. Rather, aging can be understood as an ongoing developmental process, which has certain underlying themes which vary in prominence and salience at different points in life and under different conditions. This lifespan developmental approach builds upon the contributions of Eric and Joan Erikson, among others, in identifying psychosocial developmental themes that manifest themselves to various degrees as individuals transition through their lives.

I have elsewhere identified six developmental themes that are especially prominent in later life: continuity, compensation, control, connection, contribution, and challenge. *Continuity* refers to the ability of individuals to maintain lifelong interests and activities even as they experience normal aging. From this perspective, an aging-friendly community is one that removes age-based barriers that might otherwise make it difficult for older individuals to continue personally and societally meaningful life patterns as they age. *Compensation* refers to the availability of adequate supports and accommodations to assure that the basic health and social needs of individuals with age-related disabilities are met. *Control* refers to the actual and perceived ability to achieve one's aims, including goal achievement as well as distress reduction, whether through active modifications of external environments or internal adaptation. *Connection* refers to the increasing salience of emotionally salient interpersonal relationships as we age and move closer to the end of life. *Contribution* refers to the lifelong need to have an impact on one's social environment, especially in age-graded societies in which aging brings fewer opportunities for generativity and socially valued activities, undermining self-worth and a sense of personal meaning. *Challenge* refers to the lifelong need for stimulation in order to maintain psychological and organismic fitness, including opportunities to develop new sources of fulfillment, productive engagement, and social interaction.

An aging-friendly community, therefore, is one that enables individuals to continue to engage in lifelong interests and activities as they age, retain a sense of control over their lives, maintain fulfilling social and emotional connections, contribute in meaningful ways to the well-being of other persons and the community at large, enjoy opportunities to develop new interests and sources of fulfillment, and receive necessary supports and accommodations to help them accomplish those tasks and meet their basic needs (Scharlach 2009). These characteristics of aging-friendly communities echo to some degree Lawton's three principles of person–environment fit (maintenance, support, stimulation) (Lawton 1983), as well as the notion of vital involvement posited by the Eriksons and by Kivnick and Wells (2014).

17.2.4 Communal

Age friendliness has focused primarily on enhancing the well-being of individuals as they age, apparently reflecting Western values of individualism and separateness. Considerably less attention has been given to communal values and norms, including the social contexts of which the individual is but a part. Many cultures understand aging primarily in terms of relational goals, such as familism, harmonious relationships, intergenerational transmission of values, or simply playing well one's appointed role in communal life. From this perspective, community age friendliness is not simply a reflection of the well-being of individuals and their physical and social environments, but it also reflects the nature of the community as a whole. A community is greater than the sum of its parts. Communities are holistic entities that transcend the characteristics of any single group of individuals or community programs or institutions. An age-friendly community therefore includes the very essence of what it means to be a community—properties such as sense of community, social capital, sense of security, and positive group identification.

17.2.5 Transactional

Scholarship on age-friendly communities, and environmental gerontology more generally, has tended to focus more on environments than on the individuals that inhabit them. Aging is conceived as decline, and person–environment fit is achieved by adapting physical and social environments to be more supportive and ameliorative of functional deficits. However, just as environments impact individuals, so the individual members of those communities affect their environments in a multitude of ways. For example, people select where they live, although obviously constrained by a variety of structural and individual characteristics. Individuals together create informal social structures that are often at least as important as formal structures created by their representatives in government. Moreover, individuals not only adapt to their environments but also adapt their environments to meet their needs.

A transactional perspective takes the concept of person–environment fit one step further, conceptualizing fit not simply at one point in time but understanding that person–environment interactions change over time as individuals and their environments inevitably change over time. For example, we certainly understand that our bodies change over time, with implications for physical and social functioning that require changes in our physical environment in order to optimize PxE fit, which is the primary focus of most age-friendly approaches.

However, it is equally the case that environments change over time as well, necessitating adaptation by individuals. For example, not surprisingly, older adults are apt to live in older homes, which succumb to the same forces of entropy that lead to decreased capacity in aging bodies. Communities, and their inhabitants, furthermore, have a history, which provides an implicit foundation upon which the future is built (Keating et al. 2013). This suggests a transactional perspective that understands individuals and their environments as continually adapting together over time. From this perspective, age-friendly communities are ones where individuals and their environments adapt in constructive ways to support positive aging. Furthermore, an aging-friendly community is not simply one that is age-friendly at one point in time, but which is actively adapting overtime to become more aging-friendly.

This dynamic perspective is reflected in WHO's definition of a healthy city: "a healthy city is one that is continually creating and improving those physical and social environments and expanding those community resources which enable people to mutually support each other in performing all the functions of life and developing to their maximum potential" (Nutbeam 1998). Partners for Livable Communities (PLC), furthermore, suggests that livable communities are resilient communities, which PLC defines as the capacity of a community to persevere through environmental, economic, and social hardships (Partners for Livable Communities 2013).

17.3 Aspects of Age Friendliness

17.3.1 Age-Friendly Characteristics

Most scholarship on age-friendly communities has focused on the specific characteristics that are considered to be indicators of a city's "age friendliness." The Visiting Nurse Service of New York's *AdvantAge Initiative*, for example, identified

33 characteristics arrayed in four overarching domains: basic needs; physical health and well-being; independence for the frail and disabled; and social and civic engagement.

Building upon this model, the WHO identified eight domains of age friendliness, which currently are being expanded and refined: housing; outdoor spaces and buildings; transportation; social participation; civic participation and employment; community support and health services; communication and information; and respect and social inclusion (WHO 2007).

AARP's "livable community" construct features four domains: safety, housing, transportation, and supportive services. Similarly, Stanford Center on Longevity and the MetLife Mature Market Institute identified three "livable community indicators for aging in place": housing options, transportation/accessibility, and community services and supports.

Age-friendly characteristics often are considered to reflect features either of the built environment (i.e., physical infrastructure) or the social environment (i.e., social infrastructure). Physical characteristics typically include housing, transportation, and land use. Social characteristics typically include employment and volunteer opportunities, social activities, and an array of community programs and supportive services.

Another social factor, albeit one that has received less attention, is social capital, defined as "social networks and norms of reciprocity" and the benefits derived therefrom (Putnam 2004, p. 668). Three types of social capital have been proposed: bonding (e.g., potential support from peers); bridging (e.g., potential support from other groups); and linking (e.g., potential support from external sources). A related concept is social inclusion, reflecting the extent to which the perspectives and contributions of all community residents and population sectors are valued. Together, social capital and social inclusion contribute to an overall sense of community, "a feeling that members have of belonging and being important to each other, a shared faith that members' needs will be met by the commitment to be together" (Chavis et al. 1986, p. 25).

17.3.2 Interrelated Nature of Age-Friendly Domains

Some have suggested that the entire range of age-friendly characteristics can be collapsed into a single dimension. Lui et al. (2009), for example, developed a two-dimensional model of age-friendly community initiatives, with domains displayed on a single continuum ranging from physical infrastructure (outdoor spaces, transportation, housing, information, services) to social infrastructure (community planning, employment, civic participation, social inclusion, safety). However, such efforts to create unidimensional representations are limited by the interrelated nature of physical and social infrastructure components, and age-friendly domains in general. For example, a community is simultaneously a physical space (e.g., neighborhood), a social space (e.g., social boundaries and norms), and a set of social bonds (e.g., a sense of community). Furthermore, physical and social components interact and interpenetrate to imbue spaces with meaning, thereby creating a sense of place.

As one example, transportation and mobility needs (i.e., the ability to get where one wants/needs to go) can be met in multiple ways. Certainly, an adequate physical infrastructure (buses, cars, roads) is needed, but so are drivers and companions. The same mobility need can be met by public fixed-route or para-transit services, by a taxicab, by a companion who accompanies a senior in his or her own car and provides cueing, by a volunteer who drives the senior, or by a companion who accompanies the senior on a public bus. Moreover, social factors such as the density of interpersonal social networks and the amount of social participation are highly dependent upon housing factors contributing to residential stability (Chaskin 1997).

Housing, furthermore, is important to the extent that it provides a safe and secure base for meeting one's physical and social needs, while helping to preserve a relatively stable sense of personal identity, continuity, and meaning. From this perspective, adequate housing involves not simply a physical structure, but considerations of location, affordability, customization, maintenance, ambient assistive technologies, supportive services, and a variety of alternative housing options within the community.

Clearly, physical infrastructure is only part of the story, and efforts to differentiate aging-friendly components as either physical or social infrastructure are apt to be problematic. Lui et al. (2009), for example, argues that the ideal age-friendly community has an integrated physical and social environment, where older residents are actively involved in all aspects of community life and governance, within a climate of inclusiveness and respect.

Age-friendly community environments, therefore, can be expected to include the following:

1. adequate general physical and social infrastructures that promote health and well-being for the entire community;
2. minimal age-related barriers faced by older community members in trying to access that infrastructure;
3. compensatory and enabling features that respond to the particular age-related needs and sensibilities of older community members;
4. mechanisms for engaging older adults as valued contributors to community life.

Intersecting with these overarching domains are a number of cross-cutting considerations, including access, disparities, and scope. Access refers to issues such as availability, affordability, physical and social accessibilities, appropriateness, and acceptability. Disparities include variations based on economic, geographic, linguistic, cultural, and ability factors. Scope refers to the space under consideration, whether a neighborhood or municipality, urban or rural, and a geographic/spatial community or a virtual/consensual community.

17.3.3 Identification of Age-Friendly Characteristics

The characteristics that serve as indicators of age friendliness have generally been identified through focus groups with select groups of older adults or the judgments of community planners and other experts. The eight domains of age friendliness utilized by the WHO's *Global Age Friendly Cities and Communities Project*, for example, are based primarily on characteristics identified through focus groups of older adults in four US cities conducted by the Visiting Nurse Service of New York's *AdvantAge Initiative* and 14 focus groups with older adults and caregivers conducted by AARP's *Livable Communities* project.

Other efforts, such as the National Association of Area Agencies on Aging's (N4A) *Aging in Place Initiative* and the US Environmental Protection Agency's (EPA) *Building Healthy Communities for Active Aging* award program, have been based primarily on the expert opinions of key stakeholders and planners. Dawn Alley and her colleagues, similarly, used a Delphi technique with 15 national experts to develop a set of "elder-friendly community characteristics" (Alley et al. 2007).

In one of the few empirically based efforts, Smith and colleagues utilized an Exploratory Factor Analysis (EFA) of Detroit area data to examine characteristics suggested by the EPA's four domains of physical health, environmental safety, mobility, and social participation (Smith et al. 2013). EFA yielded six factors with eigenvalues greater than 1 and individual items having face validity and factor loadings greater than or equal to 0.4: (a) access to business and leisure, (b) social interaction, (c) access to health care, (d) neighborhood problems, (e) social support, and (f) community engagement.

17.3.4 Connecting Characteristics with Goals

Implicit in the identification of age-friendly characteristics is an assumption that their presence will result in desired goals, typically including community members' physical, psychological, and/or social well-being. AARP's definition of a livable community, for example, suggests that intended goals (e.g., ageing in place, personal independence, community engagement) are an inevitable result of safety, affordable housing, adequate transportation, and other supportive community features and services (Farber and Shinkle 2011).

For the most part, the characteristics of "age friendliness," and the various measures developed to assess them, have been developed with little formal attention to conceptual models or empirical evidence regarding environmental factors associated with elder well-being. In this regard, the consistency among the various domains and characteristics that have been identified is notable. To some degree, this is a result of the fact that many of these efforts can be traced back to the early work by AARP and the Visiting Nurse Service of New York, but it also speaks to the fact that many "age friendly" characteristics seem intuitively obvious. For example, it seems obvious that elders need to be able to get from place to place, in order to meet their needs and simply to avoid becoming prisoners in their own dwellings. However, uncritical adoption of some of these assumptions runs the risk of leading planners and even researchers down alleyways that do not always lead to desired destinations.

Frameworks for describing or assessing aging-friendly communities have focused primarily on domains reflecting existing programs or structures, often from the perspective of planners or policy-makers, rather than the function(s) that those programs or structures are intended to serve. As just one example, age-friendly city efforts typically include a substantial focus on public transportation (e.g., fixed-route buses and trams), even though public transportation reflects only a tiny percentage of all trips taken by older persons, at least in the United States context where 51 % of older adults live in suburban settings. While fixed-route and flexible-route public transportation are important components of any transportation system, the larger issue for older adults is one of mobility, i.e., how to get where they want and need to go when they want and need to get there. “Optimal mobility,” for example, has been defined as “relative ease and freedom of movement in all of its forms” (Satariano et al. 2012, p. 1508).

Focusing on mobility rather than transportation expands the discussion to include consideration of a wide range of issues, including walkability, road design, vehicle design and customization, volunteer drivers, companions, and personal assistance. It also allows researchers and planners to examine older adults’ own perspectives and considerations as they attempt to solve mobility problems, rather than imposing preconceived “solutions.” For example, as driving becomes more difficult, so does public transportation use, leading older individuals to either continue driving themselves or become passengers in someone else’s vehicle. This suggests a range of potential solutions, beginning with cars that are safer to drive. Having a companion would seem to be a potentially effective low-cost midrange solution, to help with way finding, point out potential hazards, and serve as a reminder of safety practices. Similarly, some public transportation providers have senior ambassador programs, where volunteers are available to accompany older adults and others who may need assistance using public transportation, to help them learn how best to use the system and to get to their desired destinations safely and easily. Recent developments in web-based as-needed volunteer and private sector transportation systems, often combining transportation with companionship (e.g., ITN, Uber, Lyft), may be especially promising.

Further research also is needed regarding barriers to driving, particularly from a person-in-environment perspective that understands driving limitations as involving gaps in the interface between individuals, automobiles, and roads. Most of the attention on older drivers has focused on the limitations of the older drivers themselves, rather than on the adaptive capacity of the environment. As examples, seats that move up and down and forward and back are common features on nearly every automobile, as a way to improve driving comfort and safety for individuals who are shorter or taller than average. It does not seem like too much of a stretch to assume that automobiles might also be designed so as to accommodate variations based on age and ability. While some limitations in driving capacity are virtually inevitable with age (e.g., reduced peripheral vision, decreased neck flexibility, sensitivity to glare, decreased ability to attend to multiple sensory inputs), most of these changes can be accommodated using available technologies (e.g., mirrors that increase the visual field, sensors that let drivers know when there are cars or other obstacles outside of an individual's peripheral vision). Of course, self-driving cars are likely to obviate many of the barriers to safe driving among older adults.

17.3.5 *Differential Needs and Priorities of Various Groups and Stakeholders*

By adopting a single set of age-friendly characteristics to be applied to communities in various settings throughout the world, what often is missed is the opportunity to construct conceptualizations and perspectives on age friendliness that reflect the priorities, meanings, and considerations of particular groups of older adults in their own social, cultural, economic, and political contexts. It seems likely, for example, that "age friendliness" will have different meanings for individuals from different contexts, reflecting their particular goals and situations. Even geography and weather can affect what is considered age-friendly, and what is needed to promote age friendliness. In some communities, the greatest need is for snow and ice removal; in others, getting trash cans out to the street is a problem.

Different priorities and perspectives lead to different ideas about what community attributes are most important or relevant. Community characteristics ranked highest by respondents to AARP Public Policy Institute's national survey of more than 4500 adults age 50 and older (Harrell et al. 2014), for example, varied in accordance with the needs and interests of particular groups. Seniors with disabilities or with family caregivers, for example, wanted better transportation for people with limitations. They were less interested than those without disabilities in improved parks, pedestrian friendly streets, or better buses or subways. Yet, these seniors seldom are able to attend community town hall meetings or participate in the kinds of focus groups convened by community planners and researchers, which typically recruit members from active senior center participants.

Among respondents with the lowest incomes, the top three priorities were increased police presence, transportation services, and increased affordable housing, reflecting the realities of having fewer resources and being more vulnerable, resulting in a greater need for basic services. We also should note that increasing police presence consistently was ranked first or second, even though personal safety from crime is seldom included in AFC models. Not surprisingly, personal safety was found to be more of a concern for non-drivers, people with disabilities, and those with lower incomes. Having more public parks was a priority only for the highest income group, most likely because they have the discretionary time to make use of such amenities.

One final example of how characteristics considered to be age-friendly reflect the priorities of particular stakeholders is the US government, which has approached livability primarily in terms of environmental and economic sustainability. For example, the US Government's Partnership for Sustainable Communities (a partnership of Department of Housing and Urban Development, Department of Transportation and EPA) considers economic competitiveness as a key indicator of community livability, in addition to transportation and housing (Partnership for Sustainable Communities n.d.). Similarly, providing easy access to employment opportunities and enhancing economic and social well-being are key goals of the US Department of Transportation's Livability Initiative, along with enhancing transportation options.

17.3.6 Assessing a City's Age Friendliness

Indicator systems. Age-friendly domains and their identifying characteristics are used as the basis for efforts to assess the "age friendliness" of individual communities. Cities and communities implementing the WHO *Global AFC* methodology, for example, typically convene focus groups of older adults, family members, and service providers to discuss the age friendliness of their community in each of the eight domains, and to identify areas for improvement. AARP's *Livable Communities Evaluation Guide* includes a toolkit that local community volunteers can use to assess the age friendliness of their community on eight different domains. The *AdvantAge Initiative* created a survey instrument they has been administered to older adults in communities throughout the United States, assessing respondents' personal well-being and their perspectives on community well-being regarding the 33 characteristics of age friendliness.

The Milken Institute has developed a particularly comprehensive framework, the *Best Cities for Successful Ageing* index. The index examines 78 factors thought to affect seniors' quality of life, using publicly available data from the US Census and other sources. The resulting index generates a composite ranking for each metropolitan area, which is used to rank small and large cities in terms of their potential for "successful aging." This approach has been criticized because of its reliance on Census data that imperfectly reflect meaningful aging-friendly characteristics and are available only for metropolitan areas.

The AARP Public Policy Institute has developed a web-based tool, a "Livability Index," that incorporates mapping technology, preference survey results, quantitative measures, and public policies to assess community livability (https://livabilityindex.aarp.org/). The tool uses nationally available data to measure several essential attributes of livability, engage stakeholders, and draw attention to livability issues specific to older adults.

Information sources. In considering efforts to characterize age friendliness, it is important to be cognizant of variations based on the methodologies that are employed. Some needs assessment efforts (e.g., N4A's *Maturing of America* surveys) rely on the perspectives of planners or other local experts, who typically are local government employees and seldom trained in gerontology. Other efforts (e.g., WHO's *Global AFC*) engage seniors and/or persons knowledgeable about seniors, typically through focus groups or interviews with non-representative groups recruited from senior centers and similar sources. Community needs assessments (e.g., *AdvantAge Initiative*) include surveys of older community members regarding their needs, characteristics, and perspectives; such efforts tend to be the most inclusive, but still underrepresent persons who are less able to participate in mailed, online, or telephone surveys (e.g., persons with sensory or cognitive disabilities, persons with inconsistent access to communication technologies, or persons in residential care settings).

Public opinion surveys (e.g., AARP's "What Is Livable? Community Preferences of Older Adults") provide a mechanism for obtaining the perspectives of older adults and other community members regarding their attitudes and preferences about desirable community characteristics. Such efforts frequently are national in scope and include individuals from a wide variety of locations and age groups; while highly representative, these surveys do not allow detailed information about specific communities and their characteristics, and tend to underrepresent persons who are less able to participate in surveys. Finally, some recent efforts (e.g., *Best Cities for Successful Ageing*) utilize publicly available information regarding the characteristics of particular communities (e.g., census data, police records regarding crime rates, real estate records on housing costs, and other online sources), which tend to be rather imperfect indicators of age friendliness.

17.4 Strategies for Making Cities More Age-Friendly

While the characteristics of AFC have received substantial attention, there has been comparatively less scholarly attention to community change processes and their implementation. Existing reports are mostly descriptive, with little critical analysis of what does and does not work, on behalf of what goals, and under what conditions. Age-friendly community initiatives seldom are based on an explicit theory of change, nor linked to the rich body of conceptual and empirical scholarship regarding community change processes in general (e.g., Weil and Gamble 1995). It is not clear, therefore, whether the process of making communities more age-friendly differs from other community change processes, and if so in what ways.

Based on previous conceptualizations, I postulate three basic types of aging-friendly community change initiatives: community-wide planning, interorganizational collaboration, and community development. Community-wide planning efforts, such as the WHO's Global AFC Program, typically include government-initiated community-wide needs assessment and strategic planning efforts, employing social planning methods such as individual and community assessment, intervention planning, and community education. Interorganizational collaborations, such as the Robert Wood Johnson Foundation's Community Partnerships for Older

Adults (CPFOA) and the Naturally Occurring Retirement Communities Supportive Services Program (NORC-SSP), involve interorganizational collaborations across existing service delivery sectors in order to enhance existing programs and services for older adults. Finally, community development efforts, such as the "Village" model, engage community residents in consumer-driven mutual assistance and advocacy efforts.

17.4.1 Community Planning Approaches

Community planning approaches mostly are "top-down" efforts sponsored by local governments, typically relying on experts to solicit input from key stakeholders, based on pre-existing typologies of age friendliness. These mostly are local initiatives, rather than central government efforts. The primary role of central governments is, at most, to provide a vision, or set an agenda, and sometimes provide modest support for local or regional efforts. While the WHO's Global Age-Friendly Cities initiative might never have happened without the support of the Public Health Agency of Canada, the Canadian government's subsequent role has been rather limited, leaving responsibility to individual provinces, cities, and towns. To some extent, this reflects efforts by central governments in some countries to devolve responsibility for services, and often financial responsibility as well, to local governmental and non-governmental organizations.

In the United States, the 1996 reauthorization of the Older Americans Act included funding for a Community Innovations for Aging in Place (CIAIP) demonstration program, which provided three years of support for 14 communities in the United States to implement local initiatives. However, nearly all of these initiatives were focused on individual programs, or cross-sector collaborations, rather than community-wide system change.

While it can be argued that age-friendly initiatives are by their very definition local in nature, and need to reflect local interests and conditions, the lack of centralized leadership and support has prevented widespread adoption of aging-friendly policies and programs. In the United States, for example, transportation systems are subject to federal highway safety and disability access regulations, which have some potential for increasing the age friendliness of transportation locally. On the other hand, housing and urban design are largely subject to local planning efforts and priorities, with the federal role mostly limited to preventing discrimination and subsidizing or requiring a certain level of affordability. Most published reports about AFC initiatives have described efforts in large urban settings, with relatively little attention to rural settings or suburbs (where the majority of older adults in the US now live) (Lui et al. 2009).

Community planning approaches might be understood as helping communities to become more age-friendly from the outside in. Their success ultimately is dependent upon buy-in and engagement from those who ultimately are most likely to be affected. Because of their non-representative and non-inclusive nature, many such initiatives result in plans that are only partially implemented. While such initiatives often receive substantial attention at first, interest frequently peters out as political forces and resources shift (e.g., US Administration on Aging's CIAIP, California's Long-Range Strategic Plan on Aging). Success requires maintaining sufficient momentum and public attention for long enough to create ripple effects that can survive long after political forces shift and that engage a variety of human and economic resources. Such efforts are apt to work best when a central initiative is used as the basis for local, sustainable efforts that engage multiple local stakeholders, such as Age-Friendly New York City's adoption of Aging Improvement Districts (AID). AIDs' accomplishments, however, have been relatively modest (e.g., seniors-only swimming pool hours) and are highly dependent upon support provided by the New York Academy of Medicine; they face an uncertain future following the election of a new mayor of New York City, given the extent to which the entire initiative was linked to former Mayor Bloomberg (for a more detailed description of recent New York City development, see the chapter of Goldman and her colleagues).

While the focus of age-friendly community initiatives is appropriately on community-wide system change, the reality is that most initiatives only succeed, if they succeed at all, in improving the situation for relatively small segments of local communities. Even when there are major changes in community-wide systems, those changes affect community residents differentially, as we have seen. Some changes, such as improvements in the public transportation system, are likely to be experienced only by those older adults who use those systems; and in the case of public transportation, that is most likely to be individuals who are sufficiently functionally able to take public transportation, but do not have the economic or social resources to drive a car or be driven.

17.4.2 Interorganizational Collaborations

Collaborations across existing service delivery sectors represent an effort to transcend existing programmatic and funding silos in order to improve community-based systems of services and supports for older adults. While not typically thought of as enhancing the age friendliness of communities, any intervention that improves person–environment fit for aging community members could be thought of as an age-friendly intervention. This might include efforts to enhance the functioning of individuals or groups, enhance service access, improve service coordination, or work together to modify physical environments or public policies that support well-being in later life. Better coordination of supportive services, for example, can enhance service efficacy, increase service satisfaction, reduce caregiver strain, shorten lengths of inpatient hospitalization, and decrease inappropriate use of emergency rooms (Rizzo and Rowe 2006).

Increasing reliance on a mixed economy of care appears to be a global phenomenon, at least among highly developed countries, as a result of worldwide economic, demographic, social, and political pressures (Burau et al. 2007). Yet, the private sector has been largely ignored in most cross-sector aging-friendly community initiatives. While some community planning efforts intentionally try to involve private sector representatives, their role tends to be relatively minor. Aging-Friendly New York City strives to include local businesses in the Aging Improvement District planning councils, and other initiatives sometimes include representatives from Chambers of Commerce or other business interests. However, businesses are seldom the main driver of efforts; nor are they seen as the primary source of age-friendly solutions. Yet, at least in the United States context, individuals typically look to the marketplace for products and services that will enable them to live more safely and easily in their homes and apartments, get to where they want and need to go, or otherwise enhance their ability to live well as they age. Even in societies that traditionally have relied primarily on families or public services to meet the needs of older adults, and have generally eschewed private sector solutions, there has been dramatic privatization in the areas of home care, transportation, and even healthcare. Moreover, technological solutions, which perhaps hold the greatest promise for helping individuals to age well and reducing the gap between the needs and desires of older adults, on the one hand, and the age friendliness of the external environment, on the other, have been developed almost entirely within the private sector.

Of course, private sector solutions are not universally available, and can tend to exacerbate existing health and economic disparities, if not adequately monitored. Moreover, the most vulnerable older adults are likely to be especially subject to market failure, because of their inability to compete for the most affordable and high-quality services, the difficulty of estimating the risks involved, the limited options available, and the constraints on consumers' ability to locate and take advantage of the most desirable and cost-effective alternative source of care.

While some cross-sector collaborations have been highly successful, partnerships among disparate organizations and interests generally are notoriously difficult to sustain over long periods of time. NORC-SSPs, for example, are partnerships among housing providers, health and social service organizations, and a variety of other governmental and non-governmental service provision and funding entities, who collaborate to enhance services and supports for older adults living in geographically defined areas with relatively high densities of older adults, but which were not specifically designed to serve aging persons. NORC-SSPs have been found to contribute to reported improvements in participant social connections, community participation, service access, and perceived ability to age in place (Bedney et al. 2007). In addition, some NORC-SSPs, at least in New York, have been able to generate sufficient political capital to successfully advocate for ongoing funding from state and local governments for NORC-SSPs, as well as other age-friendly public policies. However, with the exception of New York, NORC-SSPs have proven relatively unstable, with many such collaborations fading following the termination of federal Older Americans Act demonstration funding.

17.4.3 Community Development Approaches

Community development efforts work to enhance community aging friendliness "from the inside out," focusing primarily on enhancing the supportive capacity of interest groups and their members, rather than large-scale system change. In so doing, such initiatives can support what has been called the "core economy," the various forms of social assistance typically provided by families, neighbors, and voluntary organizations, which have tremendous value to individuals and society, but for which money is not usually exchanged (Cahn 2006). John McKnight argues that society is like a four-legged stool, with associations being the fourth support (McKnight 2013). Unlike not-for-profit organizations, which usually are formal and hierarchical, associations tend to be informal and horizontal, relying primarily on the experience and knowledge of member citizens, and in so doing enhancing citizens' social and political capital. The great benefit of grassroots initiatives is that they can focus on ideas most appropriate to their own communities. Government and other sectors need the experience and knowledge of member citizens, who ultimately will implement, and benefit from, AFC efforts. In a fundamental way, this represents the basic idea behind civil society and "urban citizenship" (Buffel et al. 2012).

The Calgary Elder-Friendly Communities Project (*CEFCP*), for example, was a participatory, neighborhood-based, community development and research project designed to enhance capacity, foster meaningful participation, and promote aging in place for seniors and their caregivers in Calgary, Alberta. Developed by the University of Calgary Faculty of Social Work in collaboration with the Calgary Regional Health Authority, the City of Calgary, Calgary Family Services, Jewish Family Services, and the Calgary Catholic Immigration Society, the project targeted four neighborhoods that were home to high percentages of seniors and varied with regard to economic status and culture. With the help of social work students and faculty, the project engaged community members of various ages and ethnicities in community development projects such as intergenerational programs and infrastructure improvements (Austin et al. 2005). Neighborhood groups led by elderly community residents were responsible for community organizing activities, training workshops, social activities, and environmental improvements.

Community Without Walls (*CWW*), in Princeton, NJ, is a grassroots peer support organization started in 1992 by a middle-aged couple and two friends with personal experience caring for aging parents. Members assist one another with social support, information, education and advocacy, while also taking full responsibility for planning programs, recruiting new members, and submitting required government reports. CWW's 460 members are organized into six chapters, called Houses, each with 50 to 100 members who help one another as needed and attend a variety of small group social and educational activities. Through its advocacy efforts, CWW also has fostered improvements in local aging services, as well as senior housing, transportation, and health services.

The Village model is a grassroots approach that engages older community residents in developing membership associations that provide supportive services and social activities. In return for annual dues, Village members receive access to core services (e.g., weekly grocery shopping trips), referrals and discounts to outside services (e.g., home repair), social and educational activities, and opportunities to participate in organizational governance and peer support roles. Consumer engagement is a key feature of these organizations, including developing the initiative, providing oversight and governance, and offering support and services to other members. More than 85 % of Villages provide assistance with at least six of the eight WHO age-friendly domains, and 10 % have implemented features of all eight domains (Scharlach et al. 2014). Preliminary evidence from an ongoing study of California villages suggests a number of benefits of Village membership, including reported improvements with regard to social engagement, service access, and health and well-being, as well as increased ability to age in place (Graham et al. 2014).

Village members tend to be younger, more economically secure, and more likely to reside in higher SES communities than the general US population ages 65 and older (Scharlach et al. 2014). Consequently, concerns have been raised that Villages help older adults who already are relatively advantaged economically and socially to obtain a more personalized array of services and supports than is available to the general population. While Villages by their nature focus on select seniors within specific geographic areas, there appears to be substantial potential for replication in other, less-advantaged neighborhoods. Moreover, Villages have potential for using their available social and political capital to promote age friendliness in their larger communities. More than one-third of Villages are engaged in direct or indirect efforts to improve community physical or social infrastructures or improve community attitudes or policies toward older persons (Scharlach et al. 2014).

TimeBanking is another approach to creating intentional informal helping networks by and for community members. Unlike Villages, most TimeBanks include persons of all age groups. While many TimeBanks begin with particular groups defined by geography, neighborhood identification, or interest group, they generally aim to grow to be available community-wide. Based on informal helping networks that traditionally have existed within extended families, immigrant populations, religious institutions, and other interest groups, TimeBanks provide opportunities for members to get assistance when needed, provide assistance that contributes to other members' well-being, and reinforce a sense of community engagement and identity. Some of the impacts reported by TimeBank members include increased social connections, increased ability to age in place, and improved quality of life (Visiting Nurse Service of New York 2009). Related to this is an increased sense of belonging to a community, both as a recipient of support and as a contributor to the well-being of others and to the community as a whole. Of particular note, in one study, 93 % of TimeBank members said that they had interacted with or made friends with individuals of different ages, backgrounds, or cultures than their own.

Some TimeBanks, such as the Community Connections TimeBank of the Visiting Nurse Service of New York, include referrals to other community services beyond what is provided by TimeBank members. So 93 % of community connections members see the TimeBank as a place to obtain information about services in the community, and two-thirds of them report increased ability to access those services. TimeBanking has been shown to be especially beneficial for Spanish-speaking older adults and those with the lowest incomes (Visiting Nurse Service of New York 2009).

A major strength of community development models is their ability to activate and even expand underutilized community human and social capital. However, this strength may be their greatest limitation, as well, in that they may lack the expertise and financial resources needed to develop a successful organizational structure. While serious concerns have been raised regarding the sustainability of community development efforts, many of these initiatives have demonstrated surprising staying power. For example, nearly all Villages that have been developed are still in operation, although the jury is still out given that the majority of operating villages are less than three years old. It appears that community-based organizations are able to provide through volunteer labor what more formal not-for-profit organizations usually must pay for. In addition, many of them have established formal or informal collaborations with existing community organizations, in order to obtain needed expertise and administrative supports.

17.5 Unresolved Issues

17.5.1 Why Would a City Want to Become Age-Friendly?

This is perhaps the most fundamental question of all. Aging exists in social, cultural, economic, and political contexts, as do efforts to create AFC (Menec et al. 2011). From a purely rational economic perspective, and consistent with neo-liberal perspectives that assign value based on economic costs and benefits, communities will take steps to implement age-friendly principles when they perceive that it is in their best interest to do so, i.e., when the benefits of change, or the cost of not changing, exceed the perceived cost of those changes. This suggests the need to better understand the value proposition of having an age-friendly community, including the economic and/or social benefits of helping older adults to age in place.

Phoenix Mayor Greg Stanton, for example, has launched an initiative to make Phoenix the most desirable place in America for older adults to live, not primarily because of any particular desire to promote the well-being of individuals as they age, but rather because of the potential economic benefits of an influx of economically well-off individuals. For the same reason, other cities are trying to attract young professionals and the high-tech businesses that employ them. I dare say that few cities are going out of their way to promote an influx of economically disadvantaged, disabled older adults.

It also is important to note that even the individuals who are most likely to be affected by age-friendly improvements often are reluctant to take the necessary steps to implement those improvements in their own lives. Only a small minority of disabled older adults have made home modifications that would improve their ability to live safely and with fewer limitations in their current residences. Furthermore, houses with ramps, grab bars, and other obvious age-friendly design features actually sell for less than similar houses without those features. Until individuals are willing and able to adopt age-friendly improvements in their personal lives, it seems unlikely that their communities will do so. Individuals and communities both seem to be hindered by a desire to avoid facing the realities of aging, whether personal or societal.

The lack of strong consumer demand for age-friendly infrastructure improvements suggests the need for consumer engagement and empowerment (i.e., "consciousness raising"), as well as greater attention from media and governmental entities not only regarding existing options but also alternative possibilities, including age-friendly innovations that are being implemented throughout the world.

Aging advocates may have a role to play in raising public awareness not only of the limitations that come with age, but also of the potential value of investing individual and communal resources in enhancing the well-being and community participation of older community members. Seniors themselves need to be empowered to advocate on their own behalf and on behalf of forthcoming cohorts of older adults.

Governments have an important role to play in this public awareness process, and in creating models, standards, and fiscal incentives for individuals and communities to undertake aging-friendly improvements. This simply is too important to leave to the predilections and disparate interests of diverse local communities and their members.

Business also has an important role to play, not least of all in terms of the economic health of communities and their residents. Economic prosperity provides more financial resources for communities to employ in implementing age-friendly improvements in the built environment and social systems. Arguably, having a greater number of community members with sufficient time and resources to devote to helping older community members and advocating for community change can make a difference as well.

Finally, evaluation research is needed regarding the effectiveness of particular initiatives at achieving their disparate goals, as well as their secondary costs and benefits for a variety of stakeholders. As we have seen, some interventions designed to improve person–environment fit for older adults can actually have negative implications for younger adults, or even exacerbate accessibility barriers for some younger people with disabilities (and vice versa).

17.5.2 Is Aging-in-Place Inherently Beneficial?

Older adults' ability to age in place has been promoted as the primary objective of age-friendly initiatives, but it is not entirely clear that aging in place is always in the best interests of older adults or their communities. On the one hand, studies repeatedly find that most older adults want to continue to live in their current communities as they age, including 87 % of elderly respondents to AARP's 2014 livable communities survey (Harrell et al. 2014). However, if living in age-integrated communities is so desirable, why do people pay so much to live in senior communities?

Are age-segregated communities a sign of the failure to provide true age friendliness in age-integrated communities, or do some elderly "birds of a feather" truly want to "flock together," reflecting social stratification based on age- and cohort-related similarities?

Rather than sweeping generalizations regarding elders' preferences, further research is needed regarding the relative advantages and disadvantages of age-integrated environments (e.g., aging in place) and age-segregated settings. Clearly, at least some older adults believe that age-segregated settings afford opportunities for better living conditions and enhanced well-being, while others do not. Furthermore, some older adults have the financial and social resources to avail themselves of alternative residential options, while those with fewer resources may find themselves "stuck in place," unable to leave undesirable or unsupportive residences or neighborhoods.

17.5.3 Who Should Be Included in Age-Friendly Initiatives?

Age-friendly efforts need to draw upon all four legs of society's "four-legged stool" (McKnight 2013). Central and regional governments need to provide planning, leadership, and incentives. Local governments need to refine and implement those plans in accordance with local conditions. The private sector can offer emerging technologies and other business innovations and market place solutions. Not-for-profit organizations can help to provide and coordinate services, especially for persons who cannot compete in the marketplace or are not adequately served by public programs. Older adults, other community members, and their associations are best suited to implement solutions on the ground, while providing critical feedback regarding their ultimate effectiveness and sustainability.

Gonyea and Hudson (2012) have developed a classification scheme for illustrating the types and extent of inclusion in community-based age-friendly programs, using three dimensions: population (advantaged/privileged vs. disadvantaged/marginalized); environment (physical vs. social); and sector (public vs. private).

The role of older adults in creating AFC is especially important. Here, one is reminded of the disability rights mantra, "nothing about us without us." Older adults have an important role to play, not only as beneficiaries of age-friendly city efforts, but also as partners in their development and implementation. Yet, older adults typically are included primarily as informants, rather than as full participants in change processes, as initiators, actors, and overseers. At a minimum, are representative groups of older adults involved in evaluating what is and is not done? Simply put, is the process accountable to those whose interests it is designed to serve? Does it result in greater empowerment, more of a voice, for older community members (Buffel et al. 2012)?

17.5.4 What Skills Are Needed?

Age-friendly initiatives involve challenges that cross traditional disciplinary boundaries and address a variety of interconnected issues, affecting the health and well-being of individuals, families, organizations, and communities. Initiatives to promote aging-friendly community change inherently involve interventions at multiple levels, including the following: individual capacity building (e.g., enhanced competence, physical and psychosocial functioning, coping strategies, and empowerment); social networks (e.g., social engagement, social support, social capital development); person–environment fit (e.g., social inclusion, aging-in-place); organizational and system capacity building (e.g., service availability, accessibility, affordability, and appropriateness); community development (e.g., social capital development, local planning); and societal change (e.g., social and economic policies, culture change) (Scharlach 2009). Such efforts require expertise with regard to roles such as community organizing, resource brokerage, advocacy, group facilitation, consumer education, training, volunteer development, and management and administration (McDonough and Davitt 2011). The availability of necessary human and economic resources can be a challenge for communities that already are under resourced.

17.5.5 What Should Be the Primary Goal of Efforts to Create Age-Friendly Cities?

Age friendliness is typically conceptualized in terms of person–environment fit, suggesting the importance of reducing environmental barriers and improving environmental supports, in order to enable older adults to meet their needs more easily. Yet, is it sufficient simply to improve environmental fit by reducing barriers to participation in existing societal functions and programs, or does true age friendliness require transforming age-graded societies in order to overcome the forces that promote social exclusion of older adults? Do change processes result in systems that are more responsive to the needs and interests older adults? Do they result in greater empowerment, more of a voice, for older community members? Ultimately, do age-friendly initiatives result in fundamental changes in societal structures and values, designed to enhance the well-being of everyone, whether elderly currently or at some future time (Buffel et al. 2012)?

Ultimately, the goal of age-friendly initiatives should be not only the well-being of older adults, but also the well-being of all individuals and communities—a society for all ages. As former UN Secretary General Kofi Annan said in introducing the United Nations Year of Older Persons on October, 1998:

> A society for all ages is multigenerational. It is not fragmented, with youths, adults and older persons going their separate ways. Rather, it is age-inclusive, with different generations recognizing – and acting upon – their commonality of interest.

Age-friendly community initiatives have an important role to play in helping to achieve this vision.

References

AARP. (2013). *The policy book: 2013–2014*. Washington, DC: AARP.

Austin, C. D., Des Camp, E., Flux, D., McClelland, R. W., & Sieppert, J. (2005). Community development with older adults in their neighborhoods: The elder friendly communities program. *Families in Society: The Journal of Contemporary Social Services, 86*(3), 401–409.

Alley, D., Liebig, P., Pynoos, J., Banerjee, T., & Choi, I. (2007). Creating elder-friendly communities: Preparations for an aging population. *Journal of Gerontological Social Work, 49*, 1–18.

Bedney, B., Schimmel, D., & Goldberg, R. (2007). *Rethinking aging in place: Exploring the impact of NORC supportive service programs on older adult participants*. Paper presented at the Joint Conference of the American Society on Aging and the National Council on Aging, March 7–10, Chicago.

Buffel, T., Phillipson, C., & Scharf, T. (2012). Ageing in urban environments: Developing 'age-friendly' cities. *Critical Social Policy, 32*(4), 597–617.

Burau, V., Theobald, H., & Blank, R. H. (2007). *Governing home care: A cross-national comparison*. Northampton, MA: Edward Elger Publishing Inc.

Cahn, E. (2006). *Priceless Money: Banking Time for Changing Times*. Cheltenham: In2print. [Online]. Available: http://www.timebanking.org/files-list-publications.asp. Accessed June 23, 2014.

Chaskin, R. J. (1997). Perspectives on neighborhood and community: A review of the literature. *Social Service Review, 71*(4), 521–547.

Chavis, D. M., Hogge, J. H., McMillan, D. W., & Wandersman, A. (1986). Sense of community through Brunswik's lens: A first look. *Journal of Community Psychology, 14*(1), 24–40.

Farber, N., & Shinkle, D. (2011). *Aging in place: A state survey of livability policies and practices*. Washington, DC. AARP Public Policy Institute [Online]. Available: http://www.ncsl.org/documents/transportation/Aging-in-Place-2011.pdf. Accessed June 23, 2014.

Gonyea, J. G., & Hudson, R. (2012). *Emerging age-friendly communities: A conceptual framework for understanding inclusion*. Paper presented at the Annual Scientific Meeting of the Gerontological Society of America, November, San Diego.

Graham, C. L., Scharlach, A. E., & Wolf, J. P. (2014). The impact of the "village" model on health, well-being, service access, and social engagement of older adults. *Health Education and Behavior, 41*(1 suppl), 91S–97S.

Harrell, R., Lynott, J., Guzman, S., & Lampkin, C. (2014). *What is livable? Community preferences of older adults*. Washington, DC: AARP Public Policy Institute [Online]. Available: http://www.aarp.org/research/ppi/liv-com2/policy/Other/articles/what-is-livable-AARP-ppi-liv-com/. Accessed May 8, 2014

Healthy Aging Research Network Writing Group. (2006). The prevention research center's healthy aging research network. *Preventing Chronic Disease* [Online]. Available: http://www.cdc.gov/pcd/issues/2006/jan/05_0054.htm. Accessed May 8, 2014.

Keating, N., Eales, J., & Phillips, J. E. (2013). Age-friendly rural communities: Conceptualizing 'best-fit'. *Canadian Journal on Aging/La Revue canadienne du vieillissement, 32*(4), 319–332.

Kivnick, H. Q., & Wells, C. K. (2014). Untapped richness in Erik H. Erikson's rootstock. *The Gerontologist, 54*(1), 40–50.

Lawton, M. P. (1983). Environment and other determinants of well-being in older people. *The Gerontologist, 23*, 349–357.

Lui, C.-W., Everingham, J.-A., Warburton, J., Cuthill, M., & Bartlett, H. (2009). What makes a community age-friendly: A review of international literature. *Australasian Journal on Ageing, 28*(3), 116–121.

McDonough, K. E., & Davitt, J. K. (2011). It takes a village: Community practice, social work, and aging-in-place. *Journal of Gerontological Social Work, 54*, 528–541.

McKnight, J. (2013). *Four-legged stool*. Dayton, OH: Kettering Foundation.

Menec, V. H., Means, R., Keating, N., Parkhurst, G., & Eales, J. (2011). Conceptualizing age-friendly communities. *Canadian Journal on Aging/La Revue canadienne du vieillissement, 30*(3), 479–493.

Nutbeam, D. (1998). Health promotion glossary. *Health Promotion International, 13*, 349–364.

Partners for Livable Communities. (2013). *City resilient*. Washington, DC: Partners for Livable Communities [Online]. Available: http://livable.org/storage/documents/reports/city_resilient/city-resilient-july2013.pdf. Accessed May 8, 2014.

Partnership for Sustainable Communities. (n.d.) *Livability principles*. Washington, DC: Partnership for Sustainable Communities [Online]. Available: http://www.sustainablecommunities.gov/. Accessed May 8, 2014.

Putnam, R. D. (2004). Commentary: 'Health by association': Some comments. *International Journal of Epidemiology, 33*(4), 667–671.

Rizzo, V. M., & Rowe, J. M. (2006). Studies of the cost-effectiveness of social work services in aging: A review of the literature. *Research on Social Work Practice, 16*(1), 67–73.

Satariano, W. A., Guralnik, J. M., Jackson, R. J., Marottoli, R. A., Phelan, E. A., & Prohaska, T. R. (2012). Mobility and aging: New directions for public health action. *American Journal of Public Health, 102*(8), 1508–1515.

Scharlach, A. (2009). Frameworks for fostering aging-friendly community change. *Generations, 33*(2), 71–73.

Scharlach, A. E., Davitt, J. K., Lehning, A. J., Greenfield, E. A., & Graham, C. L. (2014). Does the village model help to foster age-friendly communities? *Journal of Aging and Social Policy, 26* (1–2), 181–196.

Smith, R. J., Lehning, A. J., & Dunkle, R. E. (2013). Conceptualizing age-friendly community characteristics in a sample of urban elders: An exploratory factor analysis. *Journal of Gerontological Social Work, 56*(2), 90–111.

Visiting Nurse Service of New York. (2009). *Impact of the TimeBank on its membership: Research study results* [Online]. Available: http://www.vnsny.org/community/in-the-community/community-connections-timebank/timebank-results/. Accessed May 8, 2014.

WHO. (2007). *Global age-friendly cities: A guide*. Geneva: WHO.

Weil, M., & Gamble, D. N. (1995). Community practice models. In R. L. Edwards (Ed.), *Encyclopedia of social work* (Vol. 1, pp. 483–494). Washington, DC: National Association of Social Workers Press.

[illegible]

Index

© Springer International Publishing Switzerland 2016

T. Moulaert and S. Garon (eds.), *Age-Friendly Cities and Communities in International Comparison*, International Perspectives on Aging 14,
DOI 10.1007/978-3-319-24031-2

MIX
Papier aus verantwortungsvollen Quellen
Paper from responsible sources
FSC® C105338

If you have any concerns about our products,
you can contact us on
ProductSafety@springernature.com

In case Publisher is established outside the EU,
the EU authorized representative is:
Springer Nature Customer Service Center GmbH
Europaplatz 3, 69115 Heidelberg, Germany

Printed by Libri Plureos GmbH
in Hamburg, Germany